Springer Optimization and Its Applications

VOLUME 128

Aims and Scope
Optimization has been expanding in all directions at an astonishing rate during the last few decades. New algorithmic and theoretical techniques have been developed, the diffusion into other disciplines has proceeded at a rapid pace, and our knowledge of all aspects of the field has grown even more profound. At the same time, one of the most striking trends in optimization is the constantly increasing emphasis on the interdisciplinary nature of the field. Optimization has been a basic tool in all areas of applied mathematics, engineering, medicine, economics and other sciences.
The series *Springer Optimization and Its Applications* publishes undergraduate and graduate textbooks, monographs and state-of-the-art expository works that focus on algorithms for solving optimization problems and also study applications involving such problems. Some of the topics covered include nonlinear optimization (convex and nonconvex), network flow problems, stochastic optimization, optimal control, discrete optimization, multi-objective programming, description of software packages, approximation techniques and heuristic approaches.

More information about this series at http://www.springer.com/series/7393

Athanasia Karakitsiou • Athanasios Migdalas
Stamatina Th. Rassia • Panos M. Pardalos
Editors

City Networks

Collaboration and Planning for Health and Sustainability

Springer

Editors
Athanasia Karakitsiou
Department of Business Administration
Technological Educational Institute
of Central Macedonia
Serres, Greece

Stamatina Th. Rassia
INSEAD
France

Athanasios Migdalas
Industrial Logistics
Department of Business Administration
Technology and Social Sciences
Luleå University of Technology
Luleå, Sweden

Panos M. Pardalos
Deparment of Industrial and Systems
Engineering
University of Florida
Gainesville, FL, USA

ISSN 1931-6828 ISSN 1931-6836 (electronic)
Springer Optimization and Its Applications
ISBN 978-3-319-65336-5 ISBN 978-3-319-65338-9 (eBook)
DOI 10.1007/978-3-319-65338-9

Library of Congress Control Number: 2017957980

Mathematics Subject Classification (2010): 37M10, 6207, 68R10, 90B06, 90C29, 90C30, 90C35, 90C90, 91CXX, 91D30, 94Axx, 94C15

Printed on acid-free paper

This Springer imprint is published by Springer Nature
The registered company is Springer International Publishing AG
The registered company address is: Gewerbestrasse 11, 6330 Cham, Switzerland

Preface

City Networks – Collaboration and Planning for Health and Sustainability is composed by research results, analyses, and new ideas, which focus on a diversity of interconnected factors targeting the development of urban network planning for enhanced human-centric sustainable city networks for healthy people and healthy future cities.

The topic of this book is premised on current knowledge that about 90% of our lives in the EU and USA are spent within environments which rely predominantly on newly developed infrastructure, technology, and networks. These networks, which can be conceptual in nature or physical systems, can centralize or decentralize urban functions as well as services and can impact not only transportation, communication, energy, and other resource distribution but also the overall citizens' experience of the urban metropolis and their lives within them. City functions and their respective network "smartness" correspond to their overarching city network planning, which can be highly complex and multifactorial. Due to vast urbanization and the need for a faster pace of living, theories and practices have been devised to actually develop networks according to specific interconnected elements of living, which cannot only impact on the urban functionality but also its livability.

Cities are both "living organisms" and "machines for living in"; thus, in this book, cities are examined as multifunctional unified technological advancements, computational models, natural and social environments, platforms, and more.

The book is comprised of 14 chapters which cover five major topics in the subject:

– Smart City and Its Networking
– Sustainable Transportation and Logistics
– Urban Health, Social Services, and Social Capital
– Architectural Design

The first issue is covered by the first three contributions which provide a general introduction and a broad overview of the subject area. Thus, the paper "Smart cities – Enabling Technologies for Future Living" by Peter Wlodarczak clarifies the concepts that have arose around the notion of smart city and describes the state-of-the-art technologies in information and communications that form the base of future living in smart cities.

In the second paper, "City Networking in Urban Strategic Planning," Federico Fontana contributes to the definition and classification of city networking and analyzes its contribution to the creation of public value.

The third paper, "Networks in Smart Cities from a Graph Theoretic Point of View" by Mohammed Haddad, reviews techniques based on graph theory that are used in optimizing the management of both network communication and route traffic in order to sustain smart city governance.

The next four contributions are concerned with specific issues relating to the transportation and road network.

Thus, in the fourth paper "An Evaluation of Measures for the Reduction of Carbon Dioxide Emissions from Automobile Traffic in Central Tokyo," Kayoko Yamamoto and Kuangtiao Shen evaluate measures for the reduction of carbon dioxide emissions from automobile traffic in Central Tokyo. Their approach is based on methods of spatial analysis, geographic information systems (GIS), and statistical analysis. Nine measures were extracted for the reduction of carbon dioxide and classified into three groups. The carbon dioxide emission levels before and after the implementation of the nine measures show that the selected nine measures were effective.

The fifth paper, "Using Social Media Data to Infer Urban Attitudes About Bicycling" by Justin B. Hollander and Yaqi Shen, explores the use of microblog data and sentiment and statistical analysis in order to determine the relationship between the "bikeability" of a place and people's chatting on microblog Twitter. The existence of relationship between people's attitudes, bicycling facilities, and physical environment factors is confirmed and leads to suggestions of strategies for planners and policy makers in order to further develop biking.

The sixth paper, "Sustainable Mobility" by Fotini Kehagia, is concerned with the challenge to make mobility sustainable. While transportation is a key driver in sustainable development, an integrated approach to tackle the environmental impacts of its function is needed. New automobile technology, improved public transportation, sustainable land-use planning, as well as information and communications technology are all important factors in achieving sustainable mobility. However, the transformation of the individual attitudes and values in people's travel choices constitutes a major element toward such a goal.

Beyond people, the road network is extensively used in the transportation of goods and commodities whose use can have far more consequences for the environment than the emissions produced by freight trucks. Thus, the seventh paper, "Sustainable Operations of Closed-Loop Logistics Chain from an Economic and Environmental Performance Perspective" by Biyu Liu and Haidong Yang, proposes a theoretical framework for closed-loop logistics and analyze the effect of such logistics on manufacturing, services, and people's lives with respect to recycling and remanufacturing. Closed-loop logistics as a strategic choice by enterprises results in reduction of waste and emissions and promotes environmental protection laws and regulations but also reduces production costs, improves economic efficiency, and enhances competitiveness.

Sustainable urban living in healthy cities that take into consideration the increasing number of older residents, the civic engagement, and the social capital poses challenges to urban architects, to city planners and decision makers, and to the infrastructures for health. Hence, the eighth paper, "Bridging Borders: Integrating Data Analytics, Modeling, Simulation, and Gaming for Interdisciplinary Assessment of Health Aspects in City Networks" by Jayanth Raghothama, Elhabib Moustaid, Vinutha Magal Shreenath, and Sebastiaan Meijer, is concerned with the design and planning of large-scale urban systems based on an interdisciplinary methodology that explores modeling, gaming, simulation, and data analysis.

In the ninth paper, "Aging in the Contemporary Urban Context: The Case of Older Residence in Genoa" by Mauro Palumbo and Stefano Poli, the authors assess the association between the living conditions of the older residents and the urban environments where these experience the aging process. A model is proposed for a secondary analysis of datasets provided by city registry offices aiming at the realization of a decision support system focused on a detailed geo-mapping representation of the city of Genoa in Italy.

In the tenth paper, "A Multicriteria Ranking of Thessaloniki's Public Hospitals Based on Their Infrastructure Adequacy" by Georgios Chatzipoulidis, Georgios Aretoulis, and Glykeria Kalfakakou, the authors develop a multicriteria methodology based on PROMETHEE in order to rank health-care units on the basis of the urgency for improvement actions with respect to the age of the building, the building area per bed, the field area per bed, the coverage of beds, and the population being served. They apply their methodology to data concerning the hospitals of Thessaloniki and the health infrastructure of the broader geographical regions Macedonia and Thrace in Northern Greece. They identify several hospitals of Thessaloniki that require immediate attention.

In the eleventh paper, "Sustainable Urban Living and Social Capital: Some Evidence from Crisis-Hit Greece" by Irene Daskalopoulou, the strengthening of the social capital levels as means to sustain institutional quality is considered in a spatial setting. The study analyzes the civic engagement pattern of urban residents in Greece and tests for the effect of soft social capital constructs such as social trust, social altruism, equality, tolerance, and humanitarianism. The results show that although the economic crisis has negatively affected the social capital scores, urban residents show higher levels of civic engagement compared to nonurban ones.

There can be no smart or healthy cities without the electrical networks evolving into smart grids that integrate clean and renewable resources or without availability of sustainable water resources and their optimized management. Thus, the following two papers are considering these issues.

In the twelfth chapter, "Power Distribution Networks Planning in Smart Cities" by V. Dumbrava, Th. Miclescu, and G. C. Lazaroiu, the meeting of the demands imposed by the transition from current state of networks to smart grids is considered in the planning process of power distribution networks that are required to operate under the new conditions. The authors propose a method of optimizing urban power distribution networks using the concept of idealized networks. In particular, location and capacity of transformer substations and stations as well as location of supply points and their connections are optimized within this context.

The thirteenth paper, "Sustainable Urban Water Management" by Antigoni Zafirakou, is concerned with the quality and distribution of water with respect to the availability and management of this utmost valuable resource. Water distribution should not depend only on its availability but also on its efficient use. Water quality plays a significant role in this context since if different water uses are taken into consideration, more sustainable water distribution could be the result. The author establishes that water quantity and quality should be jointly studied and managed to acquire sustainable water environment for urban areas.

Housing in the form of sustainable building design is a crucial element of the sustainable smart city. In most studies, building sustainability is considered with respect to optimized building energy consumption, to structural design with -based direct search methods. Based on empirical results, they conclude the superiority of model-based global direct methods over metaheuristics and multi-objective, Pareto-based optimization algorithms.

Clearly the present book can by no means claim to completely cover the subject. However, we do hope that the material presented here will sufficiently intrigue researchers and planners alike and motivate further cooperation and research in such an exciting multidisciplinary subject.

The special characteristic of this book is that it acts as an open platform of expression for all its authors to present their views, understanding, and results on this complex yet fascinating topic. As the research on the topic of the book is rapidly growing worldwide, in many different research directions and disciplines, the present book offers a framework where international experts present eminent work and highlight special approaches to this topic. In acknowledgment of all chapter contributions in this book, we would like to express our special thanks to all authors who participated in this collective effort.

Last but not least, we wish to acknowledge the superb assistance that the staff of Springer has provided during the preparation of this book.

Serres, Greece — Athanasia Karakitsiou
Luleå, Sweden — Athanasios Migdalas
INSEAD, France — Stamatina Th. Rassia
Gainesville, FL, USA — Panos M. Pardalos

Contents

Contributors

Georgios Aretoulis Department of Civil Engineering, Aristotle University of Thessaloniki, Thessaloniki, Greece

Georgios Chatzipoulidis Department of Civil Engineering, Aristotle University of Thessaloniki, Thessaloniki, Greece

Irene Daskalopoulou Department of Economics, University of Peloponnese, Tripoli, Greece

Virgil Dumbrava University POLITEHNICA of Bucharest, Bucharest, Romania

Federico Fontana Department of Economics and Business Studies, University of Genoa, Genoa, Italy

Mohammed Haddad Université de Lyon, Université Claude Bernard Lyon 1, Villeurbanne Cedex, France

Justin B. Hollander Tufts University, Department of Urban and Environmental Planning & Policy, Medford, MA, USA

Glykeria Kalfakakou Department of Civil Engineering, Aristotle University of Thessaloniki, Thessaloniki, Greece

Fotini Kehagia Aristotle University of Thessaloniki, School of Engineering, Department of Civil Engineering, Thessaloniki, Greece

George Cristian Lazaroiu University POLITEHNICA of Bucharest, Bucharest, Romania

B.Y. Liu School of Economics and Management, Fuzhou University, Fuzhou, China

Sebastiaan Meijer School of Technology and Health, KTH Royal Institute of Technology, Huddinge, Sweden

Theodor Miclescu University POLITEHNICA of Bucharest, Bucharest, Romania

Elhabib Moustaid School of Technology and Health, KTH Royal Institute of Technology, Huddinge, Sweden

Giacomo Nannicini IBM T.J. Watson Research Center, NY, USA

Mauro Palumbo Department of Education, University of Genoa, Genoa, Italy

Stefano Poli Department of Education, University of Genoa, Genoa, Italy

Jayanth Raghothama School of Technology and Health, KTH Royal Institute of Technology, Huddinge, Sweden

Kuangtiao Shen Graduate School of Information Systems, University of Electro-Communications, Tokyo, Japan

Yaqi Shen Tufts University, Department of Urban and Environmental Planning & Policy, Medford, MA, USA

Vinutha Magal Shreenath School of Technology and Health, KTH Royal Institute of Technology, Huddinge, Sweden

Peter Wlodarczak University of Southern Queensland, Toowoomba, QLD, Australia

Thomas Wortmann Singapore University of Technology and Design, Singapore, Singapore

Kayoko Yamamoto Graduate School of Informatics and Engineering, University of Electro-Communications, Tokyo, Japan

H.D. Yang School of Economics and Management, Fuzhou University, Fuzhou, China

Antigoni Zafirakou Civil Engineering Department, Faculty of Engineering, Aristotle University of Thessaloniki, Thessaloniki, Greece

Chapter 1
Smart Cities – Enabling Technologies for Future Living

Peter Wlodarczak

Abstract The significant increase in urbanization has led to cities of unprecedented sizes and densities. The trend towards mega-cities is likely to continue in the next decades and will lead to even higher concentrations of populations in urban areas. They pose challenges from an economic, social, and environmental point of view. To face these challenges, smarter solutions are needed. Specifically, new approaches in information and communication technology are required. New paradigms such as the Internet of Things, advances in wireless technologies, and smart devices have fostered the emergence of what is called the Smart City. Smart Cities aim to improve supply, waste disposal, transport, security, and the overall quality of life in the towns of the future. This chapter describes the state-of-the-art technologies that form the base of future living.

Keywords Smart City • IoT • Cloud computing • Fog computing • Edge computing • Predictive analytics • Machine learning

Introduction

Urbanization has significantly increased over the past years. Contemporary cities are the scenes of sudden and numerous changes from social, economical, and environmental points of view [10]. Towns and cities are accommodating half of the world's population, thereby creating tremendous pressure on every aspect of urban living [13]. These cities are responsible for three-fourths of the world's energy consumption and green house gas emissions [8]. The highly dense populations within increasingly big cities pose challenges for urban governance such as traffic and resource management, the environment, public health, and security. Cities are required to deliver thriving economies, great quality of life, political stability, business friendliness, and a reduced environmental impact in order to be competitive,

---

P. Wlodarczak (✉)
University of Southern Queensland, West Street, Toowoomba, QLD 4350, Australia
e-mail: W0095194@umail.usq.edu.au

A. Karakitsiou et al. (eds.), *City Networks*, Springer Optimization and Its Applications 128, DOI 10.1007/978-3-319-65338-9_1

not only on a regional or national scale, but globally [14]. These challenges require novel approaches to meet the requirements of future living. New applications have been appearing through the interconnectedness of smart devices such as sensors, portables, or wearables to form what is called the Internet of Things (IoT). These applications can optimize traffic and waste management, improve public health, education and security, and increase the overall quality of life in future cities. Advances in wireless technologies, smart devices such as sensors, smartphones or tablets, and cloud computing enabled the creation of services that start to penetrate every aspect of urban life. IoT, smart devices, and the cloud form the foundation of what is called a Smart City. A Smart City can effectively process networked information to improve outcomes on any aspect of city operations [3]. In Smart Cities, Information and Communication Technology (ICT) is seamlessly integrated into the city infrastructure and "invisible" for the citizens. IoT, ubiquitous cell coverage, and smart devices allow services that can be accessed by all and reduces the digital divide for citizens and Smart Cities. Research in the area of IoT and Smart Cities has become a very active field. It is inherently interdisciplinary and spans areas such as life sciences, computer science, environmental sciences, and economics. This chapter describes the enabling technologies of Smart Cities and shows some future trends.

Smart Cities

There is no agreed upon definition for Smart City in literature. There is, however, a general consensus that Smart Cities are those that are trying to solve their long-term challenges such as population growth, transport constraints, and budget pressures [14]. The definition of Smart Cities usually focuses on technology in literature. However, in times of slow growth, Smart Cities also offer new business opportunities and promise the creation of wealth. The need to balance social development and economic growth in a context of high urbanization is the main driver of the worldwide interest in Smart Cities [2]. Here we use a broad definition of a Smart City, which we define as a city in which ICT is merged with traditional infrastructures, coordinated and integrated using new digital technologies [1]. Smart City refers to new socioeconomic environments in which citizens, enterprises, and governments can more efficiently access services and resources [2]. Technology and governance are the instruments in a Smart City to prevent pollution, congestion, inequality, and to foster economic development and inclusion. Several properties characterize a Smart City:

1. Citywide real-time sensing
2. Data aggregation of various data sources and types
3. Seamless data processing
4. Knowledge discovery through the analysis of all aggregated data

Examples of ICT usage in a Smart City are:

- **Smart mobility:** ICT is used not only to collect, process, and spread information about the traffic in a city, but also to implement Intelligent transport systems;
- **Smart energy:** ICT is used to manage and automate the production, consumption, and delivery of energy, optimizing the use and reducing waste and pollution;
- **Smart buildings:** several ICT instruments can be used for optimizing both public buildings and private dwelling especially to reduce energy waste and consumption [5]

It is important to notice that ICT plays a pivotal role in Smart City initiatives, but it is not exclusive. The human factor has to be accounted for since people need to possess the skills to use the technology in a Smart City. Also, ICT technology may be implemented specifically for Smart City initiatives, but they also use existing technologies. For instance, wireless networks such as 3G, LTE, and 5G are heavily employed in Smart Cities, but they were not originally installed to supply Smart City initiatives. Smart Cities encompass the following areas:

- Smart people
- Smart government
- Smart economy
- Smart living
- Smart traffic
- Smart environment

Figure 1.1 on page 4 summarizes the Smart City areas and their applications.

Smart People

The reliance on technology in Smart Cities requires people to possess primary skills first in order to use this technology but also to follow on the advances in day-to-day activities [8]. People in Smart Cities have to be "smart" in the sense that they have the qualification, learning capacity and diversity to benefit form their Smart City as well as to improve it. Smart people also need the openness to embrace new paradigms such as the sharing economy so Smart Cities can evolve. However, some people such as older, less educated, or unemployed people might not have the skills or resources to benefit from smart cities.

Smart Government

A smart government needs the infrastructure for effective communication with its citizens. It encourages its citizens to take over traditional government tasks such as public transport through innovative new business models such as care sharing or social lending.

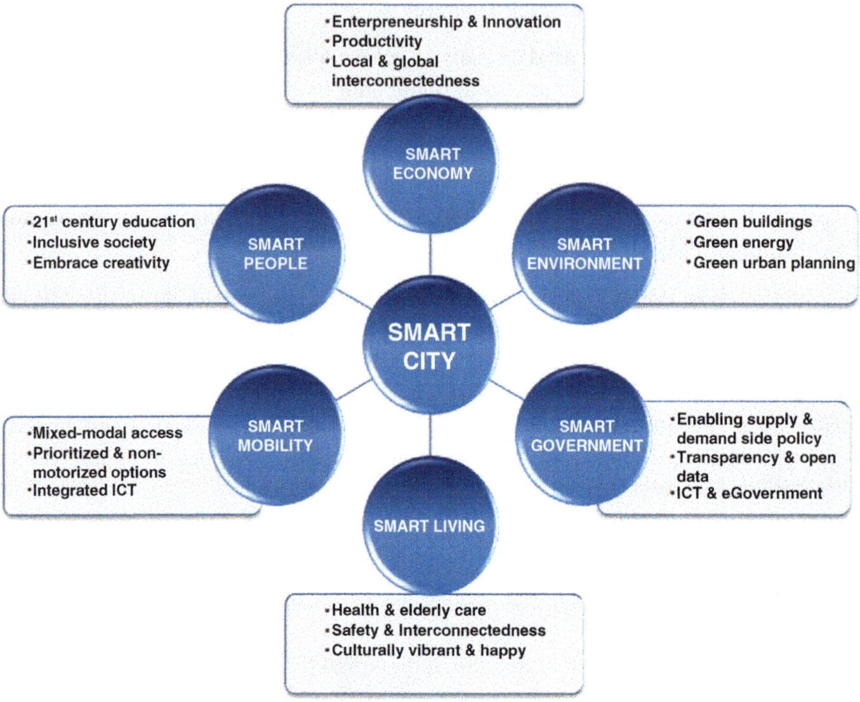

Fig. 1.1 The Smart City model [5]

Smart Economy

A smart economy is efficient and relies on renewable energy sources. It recycles waste to produce energy and is based on sustainable growth. It also makes heavy use of the sharing economy, whether it is car sharing or pooling, peer-to-peer accommodation or crowdsourcing.

Smart Living

Smart living focuses on providing a desirable place to live, work, and spend time for its citizens in a Smart City. Cities are more and more in competition with each other and quality of life is crucial to a city's success. Smart living includes health, safety, housing, and education.

Smart Traffic

Smart mobility is an essential aspect of a Smart City. Its aim is to avoid congestion, optimize transportation routes, reduce time to find a parking space, and improve traffic and pedestrian safety. It can reduce the commuting hours and thus reduce fuel consumption. For instance, it can help an ambulance or fire truck reduce time to the emergency room or fire by switching the traffic lights to green when it approaches and improve garbage collection or postal delivery efficiency by optimizing the routes. Smart traffic promotes sustainable mobility through the use of advanced smart public transport means and employment of advanced smart public transportation systems, such as smart applications for ticketing, and walking and cycling [9].

Smart Environment

A Smart environment lets humans interact with objects and smart devices in a seamless, nonintrusive way to improve usability in a Smart City. For instance, a movement detector to open a door or start a faucet should be almost unnoticed by the user. A Smart environment is typically adaptive, acts autonomously, and interacts with humans in an easy way.

The Internet of Things

A key enabling technology of Smart Cities is the Internet of Things. The Internet of Things (IoT) is a network of interconnected physical and virtual "things." These things include sensors, actuators, and smart devices such as smartphones or tablets that have communication capabilities. Since humans often carry a smartphone or smart watch around, they also become "things" in the IoT world. The IoT vision is to have a global ecosystem of highly interconnected objects that communicate with each other and deliver a wide range of services to the user. New devices can be seamlessly deployed into the IoT infrastructure to provide additional data. IoT objects can be home appliances such as refrigerators and microwaves, security systems or machines. In a Smart City, the IoT can be considered the nervous system since it contains the sensing capabilities and transmits the signals to an IoT platform for analysis that can be considered the brain.

IoT is not only used in Smart Cities, but it is also one of the key enabling technologies in Smart Homes, in Industry 4.0 and the Industrial Internet of Things (IIoT) and will ultimately lead to the Smart Planet.

IoT relies on devices, networking, middleware, and applications. Devices encompass intelligent appliances, cell phones, and setup boxes, among others [8]. In IoT, a device can be part of several applications. For instance, a thermometer might be used to control the heating in a building, but might also be part of a fire detection system.

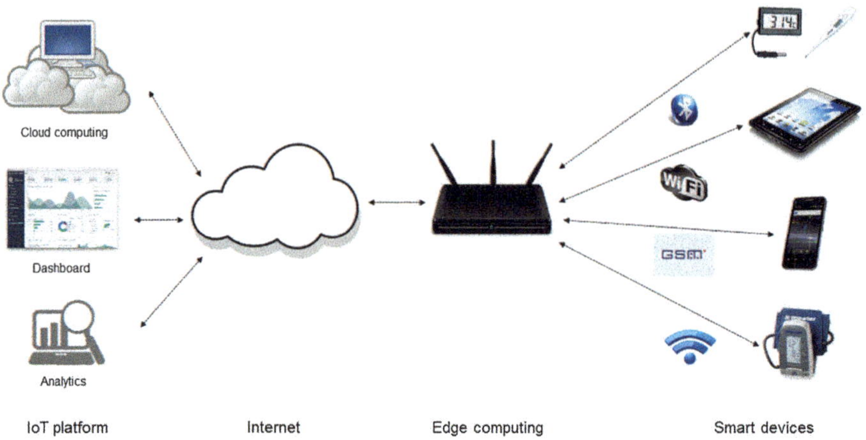

Fig. 1.2 The Internet of Things

Devices collect data and typically sent them to a cloud system for analysis, reporting, alerting, and archiving. The data doesn't necessarily have to be processed by a cloud system, on-premises solutions or hybrid, partly in the cloud and partly on-premises, exist. The communication usually passes through a middleware. Message Queuing Telemetry Transport (MQTT) is a lightweight messaging protocol used in IoT. It has a small footprint and is suitable for devices with limited resources such as sensors and actuators and where limited network bandwidth is available. IoT uses public Wi-Fi or the now ubiquitous cellular coverage, UMTS or LTE and 5G networks. Figure 1.2 on page 6 shows a typical IoT architecture.

An IoT platform can visualize data in a dashboard, has reporting functionality, and performs real-time analytics and alerting and event management. It also has a database for data persistence and historization. An IoT platform usually also has remote device maintenance functionality.

Often there is a distinction in literature between the Internet of Things (IoT), and the Internet of Everything (IoE). IoT consists of physical things connected to the Internet. IoE includes virtual "things" and services, for instance, a cloud file storage service or digital albums, users connected to the Internet or processes.

Fog Computing

Fog computing is a vital architectural paradigm for the IoT and IoE. The aim of fog computing is to minimize latency and to conserve bandwidth. Fog computing supports the IoE concept, where most devices are connected to each other such as smartphones, wearables, augmented reality devices or cars. In literature, there is no clear distinction between fog and edge computing. Sometimes it is used interchangeably. Fog computing is a recent research field that has substantial overlap with edge-centric computing [7]. In fog computing, a significant amount of storage, management, measurement, and control is not done in the cloud but on end-user or near-user edge devices. For instance, data collect in a car is directly stored and analyzed in a device deployed in a car for predictive maintenance. Only data needed for visualization or historization is transmitted to the IoT cloud platform. This reduces the data volumes transmitted, and when the moving vehicle is out of network coverage, it can still perform data storage and analytic tasks.

Edge computing is a topological paradigm that pushes applications such as analytics and knowledge discovery to the boundaries of the network [7]. It has several advantages:

- Decrease data volumes
- Improved security since there are clearly defined ingress points
- Doesn't need constant network coverage

This makes edge computing ideal for moving devices such as cars or airplanes since network coverage might get lost while moving and bandwidth might be limited. Fog computing is not only used in Smart Cities. Fog computing applications are smart buildings, smart grid and vehicle networks and software defined networks. Edge computers are sometimes called cloudlets. A cloudlet is a small-scale, mobility enabled cloud data center at the edge of the Internet. Cloudlets have also been called "micro data centers" and "fog" in the literature to refer to these small, edge-located data centers. Cloudlets have four characteristics:

1. **Only soft states:** A cloudlet does not have a hard state, but it might contain a cached state from the cloud
2. **Powerful, well-connected and safe:** A cloudlet has enough computing resources and has excellent connectivity; its integrity as a computing platform is assumed
3. **Close at hand:** A cloudlet is at the proximity of the mobile devices, meaning there is low latency and high bandwidth
4. **Builds on standard cloud technology:** A cloudlet encapsulates offload code from mobile devices in virtual machines

Mobile devices send the service requests to the closest cloudlets before the requests reach the cloud servers [6]. Cloudlets have enabled a new class of applications called wearable cognitive assistance. They can be used in Smart Homes and support its users in their ability to interact with their surrounding.

Architectures

There are many proposed reference architectures for IoT. At its simplest manifestation, an IoT architecture consists of physical devices connected to an IoT platform through the Internet. However, due to the constrained nature of many devices and the reduced bandwidth in many areas, an IoT architecture needs to address these

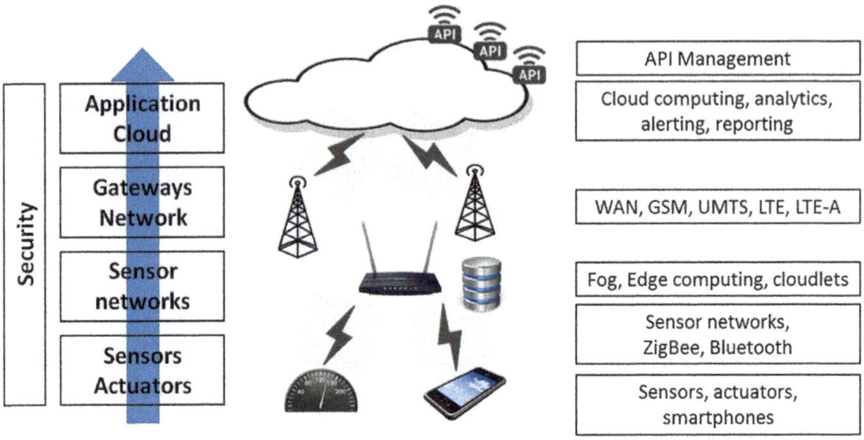

Fig. 1.3 IoT layers

limitations for availability reasons. Edge computing mitigates these shortcomings by providing an additional layer close to the devices that can filter sensor data, can convert data formats to reduce heterogeneity, and can provide scalability by allowing new devices to connect. An Edge gateway provides:

- Connectivity
- Routing
- Data filtering and aggregation
- Data transformation
- Data analytics
- Software management and provisioning
- Event management and alerting
- Interoperability between devices

An IoT platform, cloud or on premises, provides Application Programming Interfaces (API) so it can be integrated into an existing portal solution and for a mashup. An IoT platform can thus be operated standalone, or integrated into an other application through its APIs. Figure 1.3 shows a standard IoT architecture for a Smart City.

Typically a Service Oriented Architecture (SOA) is used. A SOA architecture has many advantages in the IoT context. SOA typically adopts widely used Internet standards such as REST or Web Services. This enables interoperability between devices that are otherwise incompatible. It saves resources since in a SOA architecture the devices connect on-demand to avoid bandwidth overuse. SOA can be used on the device side, on the Edge gateway or the IoT platform side. SOA has the advantage that it can include legacy systems by exposing its functionality using REST or Web Services. If the device itself doesn't have the capability for creating Web interfaces, it can be exposed on the gateway or the IoT platform itself.

Data Analysis

The collected sensor data has to be analyzed to find actionable patterns. Data analysis, or data mining, is one of the core tasks of an IoT platform. It is used, for instance, to make predictions about traffic conditions or predictions that engine failure is imminent and parts or the whole machine has to be replaced. Data analysis goes through several steps. They are divided into a data conditioning or data preprocessing phase and a data analysis phase [16]. Data conditioning starts with the selection of the data sources to determine, which ones are relevant for a specific problem. For instance, sensors in an engine might measure the fuel consumption to minimize wastage, other sensors measure material fatigue for predictive and preventive maintenance. Real world data is seldom in a form suitable for analysis. A problem that plagues practical data mining is poor quality of the data [15]. It is often noisy, has duplicates or outliers or even contains fake data. Data cleansing is a crucial step to getting a good predictive performance. The data pre-processing phase is at least as important as the predictive analysis phase. Data pre-processing is domain specific and depends heavily on the type of data. The data can be discrete sensor measurements, it can be streaming video data from a surveillance camera, or natural language from a microphone. Data pre-processing can include data deduplication, entity resolution and record linkage, outlier removal or sentiment analysis.

During the predictive analysis phase, the data mining algorithms have to be selected. If the rules are simple and limited in number, a rule engine with rules written by a developer are sufficient. If the number of rules grows or the rules become too complex to be developed by a data scientist, machine learning techniques are adopted. A typical data analysis task is classification. For instance, sensor data of an alarm system is classified into normal and suspicious data. If suspicious data is detected, a security guard can be alarmed. The conventional models used for classification are decision trees, neural network, statistical and clustering techniques [12]. Machine learners have to be trained first. Usually, historic data is used for training. The data is divided into training data and test data. Only the training data is used in learning, and the test data serves as ground truth for testing [18]. The test data is then used to evaluate the trained learner. If the predictive performance is not satisfactory, more data pre-processing or more learning iterations might be needed. Poor predictive performance might be a high number of false positives of an alarm system. The data analysis steps are summarized in Figure 1.4.

Classification is usually done using supervised learning. In supervised learning, the class label is known. For instance, data might be classified into "normal traffic" and "traffic jam." The label here is "normal traffic" or "traffic jam". If the class label is unknown, unsupervised methods apply. Unsupervised learning is the unsupervised division of instances into groups of similar objects [18]. Many data mining algorithms exist. They are well documented in literature and beyond the scope of this chapter.

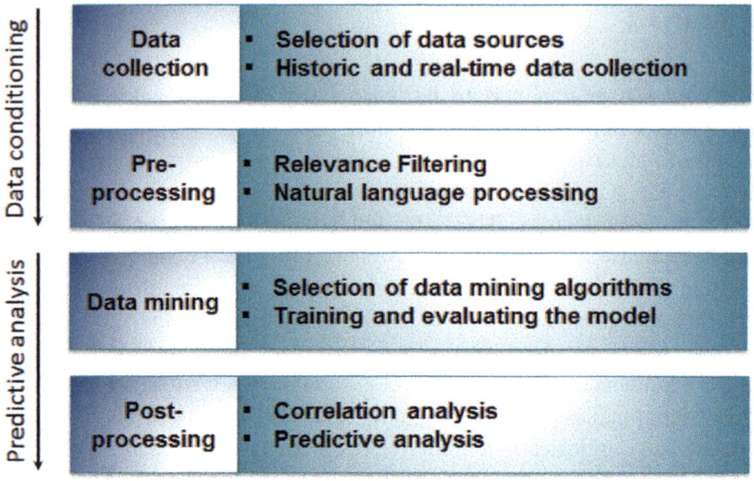

Fig. 1.4 Data mining steps [16]

Standards

To be able to connect to and communicate through the Internet, IoT devices have to support Internet protocols. However, many Internet standards are not optimized for devices with limited resources. Also, many devices have legacy or proprietary interfaces that are not compatible with Internet protocols. Incompatibility remains one of the main impediments of IoT. Since interconnectedness is a key requirement for IoT, a lot of effort has been put into standardizing the communication interfaces. However, there are other areas where standardization efforts have been made. Following list summarizes some of the most prevalent standards grouped into their functional affiliation:

1. **Device:** Identification (IPv6, ucode), Device discovery (mDNS, DNS-SD), Device management (OMA-DM)
2. **Communication:** 6LowPAN, Wifi, LPWAN, ZigBee, NB-IOT
3. **Semantic:** JSON-LD
4. **Data protocols:** MQTT, CoAP, AMQP

Figure 1.5 on page 11 shows a layered representation of IoT and their corresponding protocols and functions.

Device

IPv6 is the most recent Internet Protocol (IP). IPv6 overcomes the addressing limits of IPv4 and is suitable for IoT since it is capable of addressing the billions of nodes,

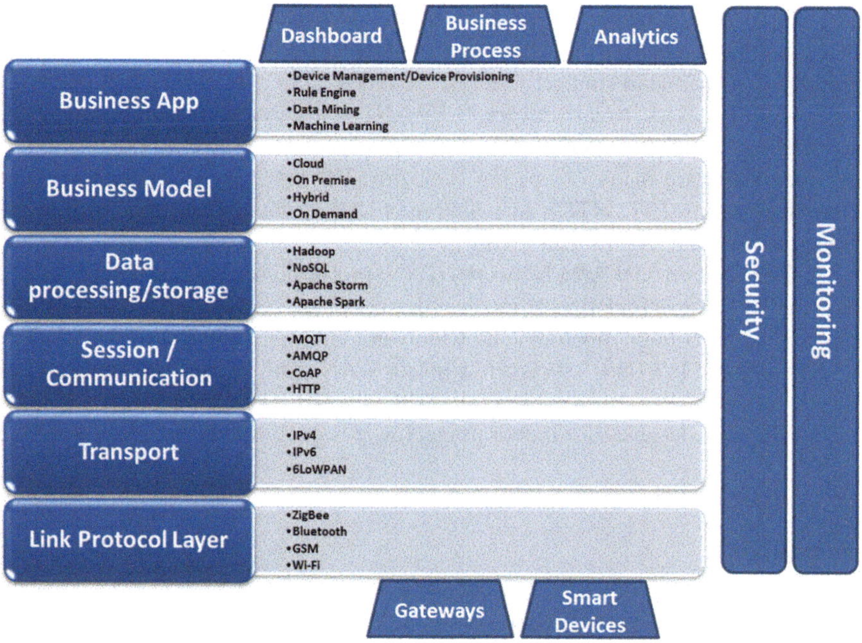

Fig. 1.5 IoT protocols

each of which shall be (in principle) uniquely addressable [19]. With billions of devices already connected to the Internet and billions more being connected in the near future, IPv6 with approximately 3.4×10^{38} addresses is capable of addressing all these connected devices. IPv6 also provides security capabilities such as encrypted communication and identification. ucode is an identification number system that is used to uniquely identify physical devices in IoT. Other standards such as OMA Device Management (OMA-DM) are used to configure devices and update its firmware. It has been specified by the Open Mobile Alliance (OMA).

Communication

6LowPAN stands for IPv6 over Low power Wireless Personal Area Networks and is an adaption layer for IPv6. It has been designed so the Internet Protocol can be used by the smallest devices. It defines encapsulation and header compression mechanisms that allow IPv6 packets to be sent over low bandwidth networks.

Wi-Fi is a wireless Local Area Network (LAN) technology widely used in computers, laptops, digital cameras or home electronics such as set-top boxes and video-game consoles. Wi-Fi enabled devices can connect to the Internet through Wireless LAN (WLAN) networks and hotspots. Hotspots can cover rooms, homes and in open areas a range of several kilometers. Wi-Fi Direct is a Wi-Fi technology that simplifies the pairing of devices and requires only one device to be Wi-Fi Direct compliant to be able to connect.

Low-Power Wide-Area-Network (LPWAN) is a wireless communication network designed for devices on battery such as sensors to allow long range at a low bit rate. There are many competing standards. For instance, LTE Machine Type Communications (LTE-MTC), an evolution of LTE for the IoT specified by The 3rd Generation Partnership Project (3GPP).

ZigBee is a Personal Area Networks (PAN) protocol used for low power devices as found in home automation or medical devices. It is a simpler and less expensive protocol than Wi-Fi or Bluetooth and widely used in Smart Homes.

NarrowBand IoT (NB-IoT) is a low power WAN radio technology, standardized by the 3rd Generation Partnership Project (3GPP). NB-IOT is a Mobile IoT (MIoT) technology and designed for indoor usage, low power consumption, and low cost.

Semantic

JSON for Linking Data (JSON-LD) is based on JavaScript Object Notation (JSON), a text-based format to transmit data objects consisting of attribute-value pairs. JSON-LD is based on JSON but adds the concept of context. JSON can easily be transformed into JSON-LD. It is used for exchanging information, for instance, sensor descriptors or sensor data.

Data Protocols

Message Queue Telemetry Transport (MQTT) is a lightweight message oriented middleware using a publish/subscribe model. It is designed to use a small code footprint on devices with limited resources and bandwidth. MQTT is an ISO standard and has been widely used in new IoT platforms. MQTT is a message oriented middleware (MOM). A variant of MQTT is MQTT-SN (MQTT For Sensor Networks). It is also a lightweight MOM designed for machine-to-machine communication and mobile applications.

Like MQTT, Advanced Message Queuing Protocol (AMQP) is a MOM that supports publish/subscribe, but also point-to-point communication and has reliability and security as an integral part of the standard.

Constrained Application Protocol (CoAP) is also designed for constrained devices. CoAP, an IETF standard (RFC 7252), is a radically simplified UDP-based analog to HTTP. It is an Internet Application Protocol and can be easily translated into HTML for inclusion in web sites.

Many more protocols and standards exist in the IoT ecosystem. For instance, Radio-frequency identification (RFID), barcodes or Quick Response Code (QR code) and Near-field communication (NFC) have been widely adopted by smart devices manufacturers. Some protocols such as Bluetooth and ZigBee have a short range and are suitable for Smart Homes, whereas Wi-Fi or LTE can be used at a global level in Smart Cities or for the Smart Planet. It is beyond the scope of this chapter to cover all standards.

Challenges

Future cities will need to adapt to, or in some cases work to mitigate against many challenges. Climate change is a major concern today. When we argue whether the anthropogenic impact is the main reason for this change of environment and ecology or not, the data of climate and power consumption are collected by a large number of sensors located almost everywhere at the same time [4]. Population growth has an impact on the transport system, the living conditions and resource consumption. A larger population does not only consume more resources, it also requires a larger spatial and temporal coverage of energy and consumer goods supply as well as services such as health care or waste disposal.

The aging of population has caused big challenges to the health care systems all over the world [11]. It has an even bigger impact on countries such as India and China because in these countries the welfare and health care systems are not well established.

There are also technical challenges inherent to IoT. IoT is a distributed environment where device management, time synchronization, security, and reliability is a major concern. Smart Cities have a high dependency on technology and its proper functioning. Insecurity about energy, water or food supply might leave many users with a feeling of unease and inhibit them from adopting these technologies.

Smart Cities have to address these challenges and solutions need to be implemented. However, Smart Cities will need huge investments to develop appropriate applications, and economic considerations might be a limiting factor. Mega-cities such as Lagos or New Delhi would profit from a Smart City infrastructure. However, these cities might not have the capital to invest in IoT or to educate its citizens to be able to use it. This economic gap can also increase inequality between cities and countries and lead to social tensions. IoT initiatives need the support of management and politicians and a long-term strategy has to be developed. Despite growing interest, the management and marketing literature practically ignores the process of strategizing Smart Cities [2].

One of the main challenges of Smart Cities and IoT is device incompatibility which leads to interoperability problems. Many standards have been proposed to allow interconnectivity and interoperability of devices. Some have found broader acceptance and started to be widely deployed such as MQTT, AMQP, and CoAP. However, it is currently difficult to predict which ones will establish itself and be adopted by device manufacturers and which ones will disappear. Many devices are legacy devices that have no or limited communication capabilities and their interfaces might be out of life.

Many devices have limited resources, limited bandwidth, and operate on batteries. They are designed for low power usage and battery lifetime is a critical factor. Also, network coverage varies in different places. Especially moving devices might get in and out of wireless range. This can be particularly problematic, for instance, for security devices such as wearables that raise an alarm in case of a medical emergency.

Security remains a challenging problem for IoT. Many devices remain vulnerable to cyber attacks and due to their mobile character, securing them poses many problems. Especially with limited resources, it is difficult to install effective endpoint protection solutions. They need to be kept up to date with, for instance, the newest antivirus definitions and security patches, and due to their constrained nature, firewall or intrusion prevention systems cannot be deployed.

The privacy of the citizens has to be a central concern for any IoT infrastructure in a Smart City. Through IoT people can be easily tracked and identified through video cameras. User data has to be anonymized where possible and a clear data retention policy has to be defined. Also access to the data has to be clearly regulated and it has to be transparent for the citizens who has access to what data and how long it will be stored.

There are also regulatory challenges. If an autonomously driving car is involved in an accident or if a robot causes a damage, liability needs to be regulated. Otherwise, manufacturers face legal uncertainty.

Future Development

Smart Cities will become smarter, more efficient, greener, more economical, and will self-regulate the energy consumption, adjust traffic patterns to respond to congestion, improve security and more. However, there will be an ever increasing amount of data that cannot be processed using traditional ICT systems. The data needs to be correlated to increase accuracy and to improve the predictive capabilities of an IoT platform. For instance, better traffic security can be achieved by correlating traffic infos with weather and street conditions. IoT in Smart Buildings, Smart Cities and Smart Factories have implications on how a building, a city or factory is designed. It has to be adaptive and scalable through software and added functionality.

Technology becomes an integral part of our environment and the volumes and heterogeneity of the data that need to be processed will increase. The ramifications are that the correlation and predictive rules cannot be implemented by developers anymore since they are becoming too complicated and we will see more and more IoT platforms that are AI enabled. Machine Learning techniques will learn from historic data to improve the efficiency of a Smart City. The recent successes of Deep Learners will also influence the future development of IoT platforms. Deep Learners are in essence artificial neural networks with many layers. They have been particularly successful in analyzing multimedia data and natural language processing [17]. IoT platforms will be enhanced with object and face recognition capabilities in videos and will be able to interpret voice commands. For instance, a security system only gives access to premises using face or voice recognition. Deep learners need a large volume of training data for fitting; the inner workings of a fitted deep learner cannot be easily understood anymore. The ramifications are a loss of control and interpretability.

Security is paramount in Smart City applications. Many devices are vulnerable to cyber attacks and need to be secured. Since constantly new vulnerabilities are discovered and exploited, IoT solutions need adaptive security systems that can fend off not only known cyber threats but also new, unseen attacks. Future security solutions will be Artificial Intelligence enabled using Machine Learner to detect suspicious patterns in IoT traffic and can detect so far unknown attack patterns such as zero-day exploits.

Smart Cities require massive investments in infrastructure and development. Since predicting the future is impossible, IoT solutions will have to be flexible enough to adapt and scalable to process ever increasing data volumes. Edge computing reduces the hardware complexity substantially and makes IoT architectures simpler. The same principle must be applied to the software platforms that operate a Smart City. It has to focus on innovation and flexibility, not on the underlying hardware.

References

1. Batty, M., Axhausen, K.W., Giannotti, F., Pozdnoukhov, A., Bazzani, A., Wachowicz, M., Ouzounis, G., Portugali, Y.: Smart cities of the future. Eur. Phys. J. Spec. Top. **214**(1), 481–518 (2012)
2. Ben Letaifa, S.: How to strategize smart cities: revealing the SMART model. J. Bus. Res. **68**(7), 1414–1419 (2015)
3. Celino, I., Kotoulas, S.: Smart cities. IEEE Internet Comput. **17**(6), 8–11 (2013)
4. Chun-Wei, T., Chin-Feng, L., Ming-Chao, C., Yang, L.T.: Data mining for Internet of Things: a survey. IEEE Commun. Surv. Tutorials **16**(1), 77–97 (2014)
5. Dameri, R.P.: Using ICT in Smart City: Smart City Implementation: Creating Economic and Public Value in Innovative Urban Systems, pp. 45–65. Springer, Berlin (2017)
6. Gai, K., Qiu, M., Zhao, H., Tao, L., Zong, Z.: Dynamic energy-aware cloudlet-based mobile cloud computing model for green computing. J. Netw. Comput. Appl. **59**, 46–54 (2016)
7. Garcia Lopez, P., Montresor, A., Epema, D., Datta, A., Higashino, T., Iamnitchi, A., Barcellos, M., Felber, P., Riviere, E.: Edge-centric computing: vision and challenges. ACM SIGCOMM Comput. Commun. Rev. **45**(5), 37–42 (2015)
8. Guelzim, T., Obaidat, M.S., Sadoun, B.: Chapter 1 – Introduction and overview of key enabling technologies for smart cities and homes: Smart Cities and Homes, pp. 1–16. Morgan Kaufmann, San Francisco (2016)
9. Kylili, A., Fokaides, P.A.: European smart cities: the role of zero energy buildings. Sustain. Cities Soc. **15**, 86–95 (2015)
10. Mattoni, B., Gugliermetti, F., Bisegna, F.: A multilevel method to assess and design the renovation and integration of Smart Cities. Sustain. Cities Soc. **15**, 105–119 (2015)

11. Pang, Z., Zheng, L., Tian, J., Kao-Walter, S., Dubrova, E., Chen, Q.: Design of a terminal solution for integration of in-home health care devices and services towards the Internet-of-Things. Enterp. Inf. Syst. **9**(1), 86–116 (2015)
12. Pardeep, K., Kumar, N., Sehgal, V.K., Chauhan, D.S.: A benchmark to select data mining based classification algorithms for business intelligence and decision support systems. Int. J. Data Min. Knowl. Manag. Process **2**(5), 25–42 (2012)
13. Perera, C., Zaslavsky, A., Christen, P., Georgakopoulos, D.: Sensing as a service model for smart cities supported by internet of things. Trans. Emerg. Telecommun. Technol. **25**(1), 81–93 (2014)
14. Vaquero-García, A., Ãlvarez-García, J., Peris-Ortiz, M.: Urban Models of Sustainable Development from the Economic Perspective: Smart Cities. In: Peris-Ortiz, M., Bennett, D.R., Pérez-Bustamante Yábar, D. (eds.) Sustainable Smart Cities: Creating Spaces for Technological, Social and Business Development, pp. 15–29. Springer, Berlin (2017)
15. Witten, I.H., Frank, E., Hall, M.A.: Data Mining, 3rd edn. Elsevier, Amsterdam (2011)
16. Wlodarczak, P., Ally, M., Soar, J.: Data Process and Analysis Technologies of Big Data: Networking for Big Data, pp. 103–119. Chapman and Hall/CRC, London (2015)
17. Wlodarczak, P., Ally, M., Soar, J.: Multimedia data mining using deep learning: 2015 Fifth International Conference on Digital Information Processing and Communications (ICDIPC), Sierre, pp. 190–196 (2015)
18. Zafarani, R., Abbasi, M.A., Liu, H.: Social Media Mining. Cambridge University Press, Cambridge (2014)
19. Zanella, A., Bui, N., Castellani, A., Vangelista L., Zorzi, M.: Internet of Things for Smart Cities. IEEE Internet Things J. **1**(1), 22–32 (2014)

Chapter 2
City Networking in Urban Strategic Planning

Federico Fontana

Abstract City networks are systems of relations through which cities can cooperate in order to face the challenges of economic development, social protection, and environmental sustainability. Consequently, city networking assists local authorities in carrying out their most distinctive function, i.e. the creation of public value in a financially sustainable way. This function has become increasingly complex, leading to a wide and growing interest in urban strategic planning, which, in fact, may offer a useful contribution to local government, as long as it is set and implemented in an authentic and substantial way. This paper aims to contribute to the definition and classification of city networking, analyse its involvement in the creation of public value, and underline its importance in urban strategic planning, identifying some critical points and providing some useful suggestions for the future.

Keywords City networks • Creation of local public value • Urban strategic planning

Introduction

The process of concentration in urban areas of population and activities – and hence social and economic flows – is expected to continue to grow.

Cities are the primary sites of production and provision of goods and services, as well as the principal centres of social and cultural progress. Yet, cities are faced with a great pressure to solve serious problems: financial crisis, economic conversion, unemployment, poverty, traffic congestion, and pollution, just to name a few. In other terms, they are key players in economic development, social protection, and environmental sustainability, which are increasingly interdependent worldwide, due to globalisation, and for this reason even more relevant on a local basis. This is true for all cities, small and large, although the relative strengths and weaknesses are of course different and, to some extent, symmetrical.

F. Fontana (✉)
Department of Economics and Business Studies, University of Genoa, Genoa, Italy
e-mail: fontana@economia.unige.it

A. Karakitsiou et al. (eds.), *City Networks*, Springer Optimization and Its Applications 128, DOI 10.1007/978-3-319-65338-9_2

To protect and improve their competitive advantages and effectively face the challenges inherent to their role, cities may cooperate among each other, through networks of relations that can be both local and global, increasing in this way their chances of success.

For this reason, city networking, although broadly classifiable, plays a central role in the distinctive function of local authorities, i.e. the creation of public value in a financially sustainable way, which integrates the effective satisfaction of public needs with the efficient use of public resources. The fulfilment of this function is very complex, due to both the growing proliferation of public needs and the progressive lack of available resources and also because the roles played by and the relations established among all the social actors, both within and beyond urban boundaries, significantly affect it. These are the main reasons for the wide and growing interest in urban strategic planning, which, in fact, may offer a useful contribution to local government.

The aim of this paper is twofold. On a theoretical level, it aims (1) to contribute to the definition and classification of city networking, (2) analyse its role in the creation of local public value, and (3) underline its importance in urban strategic planning. On a practical level, the paper assesses the inclusion of city networks in urban strategic plans by looking at a significant sample of major Italian cities. In essence, the paper aims to draw a critical and empirically informed analysis of the actual complexity and potential benefit of city networking planning.

A Typology of City Networks

City networks are systems of relations that spread among urban centres. The latter can be connected by links of different nature and may exchange various kinds of socio-economic flows.

This is a deliberately broad and adaptable definition of city networking, which, although dated, encompasses alternative explanation proposed by scholars of urban studies and may be helpful due to the multidisciplinary nature of the matter.

Networking is widely considered an effective response to the challenges posed by globalisation and its antithesis localisation to both metropolises/big cities and medium/small ones. It is seen as a way for cities to cooperate by sharing or exchanging information, objectives, visions, policies, projects, best practices, resources, and knowledge, increasing their chances of success.

Among the main types of city networks, it is worth emphasising the following classifications, based on the kind of relations, members, and themes involved.

A first distinction concerns the spatial scale of relations, according to which city networks can be divided into intra-urban, interurban, and interregional [4, 14, 29]. The intra-urban scale is typical of metropolitan areas and is the result of processes either of decentralisation, with the birth of sub-centres, or of incorporation/fusion of previously existing urban entities. The interurban network characterises polycentric urban regions, and it mainly occurred with the functional integration of historically separate urban areas. The networks that belong to the highest spatial scale, the interregional ones, are typologically similar to the interurban ones, except for their size and the fact that they often develop at an international and possibly global level. This last phenomenon, when referring to major cities, has been the subject of particular attention by scholars [43], who coined the expressions 'world city network', mainly used by geographers and planners (from the contributions of Hall and Friedmann [22, 24]), and 'global city network', especially recurrent in economic and sociological studies (from the contributions of Sassen and Castells [10, 40]).

Another classification, partially linked to the previous one, considers the nature of relations between cities, intended as network nodes [17]. In this perspective, city networks can be characterised by hierarchical relationships or not. The former, typical of the past, but still possible at the local level, can be traced back to the classical model of the central place [12] that activates a spatially contiguous, asymmetric, and gravity-type network. The non-hierarchical type, gradually widespread because of globalisation and possibly operating at any level (local and global), responds to a polycentric model, based on mutual cooperative flows. These two kinds of relations can also variously combine activating complex and multicentric networks that are characterised by both horizontal and vertical links [6].

Obviously, the operation of these networks needs a multidimensional infrastructure of connectivity: multimodal transport on air, water, and land, information technologies and telecommunication channels, advanced information systems, and a wide range of support services [27].

Also based on the type of urban relations is the distinction of complementarity, synergy, and innovation city networks [5]. A complementarity network links specialised centres, each of which focuses on different aspects and specific ways of tackling common problems, so as not to compete directly and increase their chances of success. A synergy network connects like-minded centres, which intend to integrate their initiatives, based on shared strategies, methods, and procedures, and take advantage of economies of scale as well as of the support provided by the network. In an innovation network, cooperation between cities covers critical topics or specific projects that benefit from the sharing of resources and skills and through which it is possible to find original solutions and design effective policies.

Another key factor in the typology of city networks is their membership. In this regard, networks can be classified as those in which cities are the main actors and cooperation is between local governments and those, more informal, in which cities are strategic places for interconnected and often global players, whether they are corporates, non-governmental organisations, universities, or other entities [3]. Among the former category, networks can be classified according to whether their members are only cities (e.g. Metropolis, World Association of the Major Metropolises, gathering 138 global cities) or they also include other types of actors (e.g. INTA – International Urban Development Association – which has five membership groups: national authorities, local authorities, public and private corporations, academic and research institutions, and individuals).

The cities' representatives participating to the network can be politicians – elected officials and members of governing bodies – or members of technical staff, i.e. experts having managerial responsibilities. In cases of greater involvement, both subjects can obviously be engaged.

Finally, city networks can be distinguished according to whether they arise voluntary and unprescribed or are promoted and supported by intergovernmental organisations (such as UN, OECD, or EU) or other entities. The former can be informal or variously formalised, contractualised, and directed networks. The latter typology, and in general the formalised networks, can also be classified according to whether or not they require membership fees and provide membership benefits, like information services, organisation of meeting, consulting, funding, and advocacy [25].

No matter their typology, city networks, unlike other networks, do not simply link entities. Rather, they connect urban spaces, whose identity is defined by the coexistence and interaction of multiple – social, political, economic, and technical – networks. In other words, they are networks of networks [36].

A third important criterion to classify city networks is their thematism [28]. Urban cooperation can refer to a range of topics variously related to economic development, social protection, and environmental sustainability and also to urban planning and management. This cooperation supposes that single successful ideas and projects can incite and produce virtuous practices in other places or may be shared and integrated in a common process of growth.

The so-called club networks or web networks arise [31]: in the former, cities focus on mutual interests and share common objectives or services, while in the latter actors develop different complementary activities with the aim to obtain a vertical synergy (this distinction recalls the previous classification of city networks based on relation type).

The themes on which the networks focus may or may not be in areas where the cities have the necessary competence and power to act directly. On this basis, a distinction is possible between 'functional city networks' and 'political city networks' [11]: in the former, cities can cooperate to technically solve or improve their local situations (e.g. utilities management); in the latter, cities politically promote certain causes that often can only be resolved at a higher level of government (e.g. reduction of armament). However, there are cases in which the two options coexist (e.g. environmental sustainability) and wherein both political and technical commitments are possible (this distinction is to some extent linked to the previous one based on member type).

The above classification, while not intending to be exhaustive, highlights the complexity of city networking by revealing not only the variety of alternatives in the field but also the multitude of possible combinations between them.

Anyway, city networking shows that cooperation is a more efficient and effective way to operate for cities, than isolation or direct competition: mutual benefits – deriving from shared information, common strategies, joint actions, and complementary or synergic solutions – increase cities' chances of success both locally and globally.

Finally, city networks can play an active role in the implementation of increasingly widespread urban models, such as the knowledge city and the smart city. In particular, a knowledge city is designed to encourage and nurture the collective knowledge, i.e. the intellectual capital of the community, seen as a determinant factor for socio-economic development [9, 19]. In turn, a smart city is actively engaged in improving the quality of life of its citizens and in pursuing sustainable socio-economic development, thanks to the wide and innovative use of information and communication technologies [8, 13]. In both cases, the cooperation resulting from the development of city networks can contribute to speeding up the introduction, enhancing the dissemination, and supporting the improvement of the targeted innovations.

Challenges in Creating Public Value

City networking plays a central role in the distinctive function of local authorities, i.e. the creation of public value in a financially sustainable way [16, 33]. This function integrates the effective satisfaction of citizens' public needs, thus contributing to the socio-economic development, with the efficient use of increasingly scarce public resources.

This function is both specific and complex [20], starting from the possible convergence, but also divergence between the creation of public value and the financial performance of local authorities.

In fact, the formation of a positive difference between the benefits produced for and the sacrifices required to citizens is not in itself guarantee of financial sustainability. This is true not only because the nature of benefits and sacrifices is both economic and noneconomic [41], but also because these effects correspond to accounting records of opposite sign. On the other hand, financial results do not necessarily measure public value outcomes, because of the various forms of remuneration of local authorities, which only partly require users to pay profitable prices. Most often, local authorities' remuneration comes from political prices and direct or transferred taxation [34]. Nevertheless, local authorities must jointly pursue public value and financial performances by so fulfilling both their own reason to exist and the possibility to operate [37].

To continue, this function is affected by the high diversification of the activities of local authorities [2]. Specifically, the plurality of services they manage is very significant in terms of areas of needs and target groups, composite nature of problems to solve, modalities of intervention, and multidisciplinary/multisectoral skills required.

The relationships of opposite sign – cooperative and competitive – that characterise urban administration are even more important in terms of operational complexity [44]. First, the governing bodies of local authorities are the expression of the ideas, values, and claims of only one section of the community. Second, in many cases, taxpayers do not coincide with users of public services. Finally, even among users, there are often contending interests, which are functionally antagonistic or competitive in the allocation of scarce resources. Therefore, the systemic process of bringing together divergent expectations is a critical condition for the creation of local public value.

Besides, the socio-economic, scientific-technological, and political-cultural changes imply that public needs and public policies significantly evolve [1]. This means that the identity of the city is often questioned, if not completely doubted, because of phenomena that give it uncertainty and discontinuity.

Furthermore, the creation of public value only partially depends on local authorities. The roles played by other social actors – citizens, businesses, other public administrations, and not-for-profit organisations – and their contributions in terms of resources, experiences, ideas, and actions are equally determinant [26]. These players constitute a rather fragmented framework, but their attitudes and behaviours nevertheless affect the cities' socio-economic development. Therefore, local authorities have to adopt a public governance approach [35, 39], attracting, involving, and promoting the activities and interconnections of other actors, both within and beyond urban boundaries. In this way, local authorities can leverage the very nature of cities, as nodes of internal and external networks, promoting their opportunities for socio-economic development.

Finally, even the targeted geography is variable and almost never corresponds to the administrative boundaries. Classical examples concern the fields of transports and logistics, public utilities such as water and gas, electricity distribution and also waste management, and tourism promotion. The same is true, more generally, for economic development, social protection, and environmental sustainability. Local authorities, therefore, need to develop a multi-level city networking in order to pursue cooperation and share goals, projects, resources, and skills.

For the foregoing reasons of complexity, city networking can offer a complementary or synergic and often innovative contribution to the creation of public value, by providing information, knowledge, methods, and practices useful to devise and implement solutions at the same time efficient and effective for socio-economic development. Among the possible advantages are included: achieving critical mass and economies of scale in many areas of activity, horizontal and vertical subsidiarity that reduces overlap and unproductive competition, private and public sector innovation, attracting greater investments, increasing employment, administrative simplification, reducing costs and improving quality of public services, spread of local skills, and adaptation to international standards. Together, these results enhance the attractiveness of the territory, stimulating economic growth, favouring social inclusion, and promoting environmental safeguard.

Urban Strategic Planning Beyond Boundaries

The above-mentioned features of specificity and complexity in the creation of local public value are the main reasons for the growing interest in urban strategic planning [30]. This process may offer a useful contribution to local government, as long as it is set and carried out in an authentic and substantial way. It needs to define a clear model of development guided by a long-term and widespread vision and be able to explain the meaning of its foundational choices, identifying possible courses of action, prior projects, and related solutions.

The effectiveness of the urban strategic plan requires the adoption of some proper conditions concerning both the objects and the subjects of planning [20, 21], which are even more important from a city networking point of view.

The contents of the plan must be selective and integrated, both far-reaching and perspective, sustainable and flexible [42].

A selective approach, limited to a few themes, objectives, and projects that are relevant to an area's socio-economic development, is essential to focus – rather than disperse – attention, resources, energies, and actions on crucial and decisive issues, such as those of greater impact on future scenarios. Besides, an integrated approach, aware of interdependence and codetermination of the various policy interventions and partner contributions, is essential to make them consistent and coordinated, to generate useful complementarities or synergies, and to create systemic value.

From a space-time viewpoint, a far-reaching and long-term horizon is necessary to take account of complementarities between cities/regions and various levels of government, to achieve important goals, to coagulate significant resources, and to enable innovative processes. Nevertheless, the best implementation of the plan requires the setting of short-term and medium-term objectives, as well as long-term ones, fostering rapid results and early successes that keep all the players active and committed [32].

The sustainability and flexibility of the planning contents are further key features to take into account. The former, which is the result of the winning correlation between goals and resources, makes the plan rational and realistic, at once ambitious and achievable, avoiding idealistic temptations as well as the propensity to abandon. The latter, which matches the dynamism of the context, makes the plan adaptive and constantly updated both in terms of operational and geographical issues.

In essence, all these features of the urban strategic plan allow to identify (1) the areas in which the city, on the basis of its identity, vocation, and resources, can excel autonomously; (2) those in which, in order to be successful, it must weave cooperative and partnership relations with other cities or larger territories; and (3) those where it does not have or get the conditions to play a significant role and would be reasonable to give up.

Excluding the last case, the activation of city networks is essential in the second one, to enable the city to acquire the appropriate operating conditions and achieve suitable socio-economic performances, through cooperative relations of either or both complementary, synergic, and innovative nature. Nevertheless, even in the first case, the city networking can play a significant role, allowing the city to make available its winning resources and skills to other urban contexts, improving its image, and strengthening its competitive position, locally, regionally, and/or globally.

The criteria to assess urban strategic plans with reference to the actors involved in them are openness, partnership, and leadership [18, 38].

Preparing and subsequently carrying out an urban strategic plan require an open and transparent approach, which is at the same time relational and communicative, engaging and participatory, in order to promote the fruitful interaction between the plurality of public and private, local and global key players. In this way, it is possible to balance all powers involved, to positively deal with conflicts of interest, and to promote mutual trust among the different actors, thus encouraging proactivity and collaboration.

The development of partnership is very important to make strategic public-private and city-to-city alliances. These need to be the result of both voluntary/unprescribed and promoted/sponsored agreements, guided by clear rules and by negotiation skills that allow for a fair distribution of responsibilities, tasks, risks, and benefits among all players.

The leadership role of local governments remains fundamental, presupposing their authority and their legitimacy and resulting in the construction of truly shared and consensual scenarios.

On this basis, the plan can be an actual tool of public governance and strategic management, dynamically able to integrate the needs for socio-economic development with the technical means necessary to achieve the shared selected and priority goals, on which it is possible to gravitate interests, generate resources, and promote responsibilities, both within and beyond urban boundaries.

Therefore, the plan can be a potentially high-impact instrument, both ambitious and challenging, which requires the acquisition, development, and dissemination of appropriate knowledge and skills among all actors involved and which can follow different methodological approaches according to the specificities of the context [7, 23].

Nevertheless, a risk often found in practice is that the approach to urban strategic planning remains formal and rhetorical, unrelated to the actual processes of local government and missing the opportunity for making a real change [15]. In this situation, the plan is likely to appear indifferent to decision-making process and its contents.

Consequently, the effectiveness of the strategic planning is neither simple nor obvious. It is indeed rather ambitious and challenging, though potentially rewarding for the positive results in terms of creation of local public value.

An Empirical Study About City Networking and Urban Strategic Planning

In view of the contribution that city networking can offer to the creation of local public value, it may be interesting to assess if and how this concept is taken into account in urban strategic plans and if consideration is given to its various typology. The following analysis looks at the inclusion of city networking in the urban strategic plans of the Italian regional capitals, as reflected in the documents published on their institutional websites (or other dedicated specific ones).

The focus on regional capitals lets us assess a relatively limited but significant sample of institutions that, although having some common features, differ in several aspects, notably their sizes, geographical characteristics, and socio-economic conditions. They constitute an adequately representative sample of the institutional variety that characterises the system of local authorities in Italy, since they cover all the significant areas of the country.

The documents analysed include all urban strategic planning tools – regardless of their denominations and methodological approaches – available on the web as of December 31, 2016.

In general, all the analysed administrations fully commit to urban strategic planning (the data shown represent an update to those reported in [21]). All municipalities have now started a strategic process (from the initial experience of Turin, whose first plan was approved in 2000, to the last one of Trieste, whose plan is dated 2014). There is also a high level of disclosure, since 18 out of 21 municipalities (86%, with the sole exceptions of Campobasso, Catanzaro, and Potenza) publish online their urban strategic plans (Table 2.1).

The same data can be read with reference to the demographic classes defined by the Ministry of the Interior and to the geographical areas identified by the Italian Institute of Statistics (ISTAT) (Table 2.2).

From a demographical point of view, all the regional capitals with more than 100,000 inhabitants establish and publish on the web their strategic planning documents. Relatively smaller, however, is the online disclosure of the municipalities of smaller size (up to 100,000 inhabitants), in which only 40% of the plans are published (the others being anyway available from other online sources).

Geographically, the cities in the North and Centre plus the Islands show full strategic commitment and transparency, with the publication of urban plans by all regional capitals. The southern area is characterised by a lower level of disclosure (50%), even if all regional capitals have begun the process of strategic planning.

After this general picture regarding the preparation and publication of urban strategic plans, it is interesting to analyse some other qualifying features.

First, even if all the documents show a strategic breath and a medium to long-term perspective (usually 10 years long), 3 out of 21 plans (those of Genoa, Milan, and Trieste, representing 14% of the total) focus on urban-regulation aspects, even if they are the results of participative decision-making processes.

Typically, local authorities take care directly of urban strategic planning. Nevertheless, there are cases – like those of Turin and Florence – where the process is promoted by a separate organisation (viz. ‘Strategic Turin Foundation’ and ‘Future Florence Association’), which gathers both public and private actors, but has no management power. In these situations, the plan may contain highly sophisticated analyses and proposals and be perceived as the privileged site for the meeting and engagement of all key players and for the establishment of an effective communication strategy through which developing the social capital of the community. However, it is doubtful that the plan can be an effective tool of government, if the local authority does not formally endorse it.

Table 2.1 Urban strategic plans of Italian regional capitals: overall framework

Cities	Demographic classes and geographical areas	Titles of the urban strategic plans	Year
Turin	From 500,000 to 1,000,000 inh.North-west	1) City strategic plan – international Turin 2) 2nd strategic plan of Turin metropolis 3) 3rd strategic plan of Turin metropolis	2000 2006 2015
Genoa	From 500,000 to 1,000,000 inh.North-west	1) Plan of the city of Genoa 2) The city changes 3) Pact for the city of Genoa	2002 2009 2016
Florence	From 250,000 to 500,000 inh.Centre	1) Strategic plan Florence 2010 2) There is more than one Florence	2002 2009
Trento	From 100,000 to 250,000 inh.North-east	1) Strategic plan 2010 2) Strategic agenda 'Trento 2020'	2003 2007
Bolzano	From 100,000 to 250,000 inh.North-east	Ideas for 2015: thinking the city	2004
Perugia	From 100,000 to 250,000 inh.Centre	Perugia – Europe from 2003 to 2013	2004
Venice	From 250,000 to 500,000 inh.North-east	Venice metropolitan area	2004
Potenza	Up to 100,000 inh.South	1) Strategic project of Potenza's hinterland 2) Strategic plan of urban development	2005 2015
Naples	From 500,000 to 1,000,000 inh.South	Strategic plan	2006
Bari	From 250,000 to 500,000 inh.South	BA2015 – metropolitan area of Bari	2008
Cagliari	From 100,000 to 250,000 inh.Islands	1) Strategic plan 2) Interurban strategic plan	2008 2012

Campobasso	Up to 100,000 inh.South	Territorial strategic plan	2008
Catanzaro	Up to 100,000 inh.South	Strategic plan	2008
L'Aquila	Up to 100,000 inh.South	1) L'Aquila 2020 2) Strategic plan	2008 2012
Ancona	From 100,000 to 250,000 inh.Centre	1) A plan for Ancona: the changing city 2) Strategic Ancona 2025	2009 2015
Rome	Over 1,000,000 inh.Centre	Strategic plan for the development of Rome Italian capital	2009
Aosta	Up to 100,000 inh.North-west	Future of Aosta: Strategic plan of Aosta and of La Plaine	2010
Palermo	From 500,000 to 1,000,000 inh.Islands	Palermo, capital of the Euro-Mediterranean area	2010
Milan	Over 1,000,000 inh.North-west	City plan	2011
Bologna	From 250,000 to 500,000 inh.North-east	1) Metropolitan strategic plan 2) Metropolitan strategic plan – PSM 2.0	2013 2016
Trieste	From 100,000 to 250,000 inh.North-east	City plan	2014

Author's elaboration

Table 2.2 Urban strategic plans of Italian regional capitals: data by demographic classes and geographical areas

	No. of cities	Percentage of population	No. (%) of strategic plans			
			Approved		Online	
Total	21	100%	21	(100%)	18	(86%)
Demographic classes						
Up to 100,000 inh.	5	3%	5	(100%)	2	(40%)
From 100,000 to 250,000 inh.	6	9%	6	(100%)	6	(100%)
From 250,000 to 500,000 inh.	4	14%	4	(100%)	4	(100%)
From 500,000 to 1,000,000 inh.	4	32%	4	(100%)	4	(100%)
Over 1,000,000 inh.	2	42%	2	(100%)	2	(100%)
Geographical areas						
North-west	4	30%	4	(100%)	4	(100%)
North-east	5	11%	5	(100%)	5	(100%)
Centre	4	35%	4	(100%)	4	(100%)
South	6	16%	6	(100%)	3	(50%)
Islands	2	8%	2	(100%)	2	(100%)

Author's elaboration

Most of the urban strategic plans are recent (the oldest one was approved in 2000, and only 12 out of 21 plans, representing 57% of the total, are more than 5 years old). It would therefore be premature to assess their socio-economic effect.

However, among the older experiences, nine (43% of the total) are significant, having already moved to the second or even third generation of urban strategic plans. In many cases, the second- or third-generation plan stems from a critical analysis of the structure, content, status of implementation, and impact of the previous one, showing the actual managerial nature of the planning process. Nevertheless, in a few cases (those of Genoa and Potenza), the different generation documents are not sequential and rather highlight a discontinuity of both strategic and administrative nature.

Also, in 8 out of 21 cases (38% of the total), equally distributed between municipalities belonging to different demographic classes and geographical areas, the urban strategic plans, after their approval, are not followed by subsequent administrative acts and appear to have been set aside, marking a possible loss of strategic tension in the local government.

Within this framework and considering the nevertheless high levels of strategic planning and disclosure recorded by the generality of Italian regional capitals, it is interesting to analyse if, how, and what of the city networking concept is reported in their urban strategic plans.

First, 17 out of 21, i.e. a large majority (81%) of the urban strategic plans formulated by the Italian regional capitals, explicitly consider city networking.

Demographically, all the regional capitals with more than 250,000 inhabitants refer to city networks in their plans. Furthermore, also in the smaller demographic classes, city networking is qualified as an issue of strategic importance for most of the municipalities, increasing with their size (60% in the class up to 100,000 inhabitants and 67% in the class from 100,000 to 250,000 inhabitants).

On a geographical basis, all the cities in the North-West and the Islands insert city networks in their strategic planning documents. In the other geographical areas, the strategic importance given to city networking is still significant, increasing with latitude (67% by the southern cities, 75% by those of the Centre, and 80% by the north-eastern ones).

The high level of reference to city networks in these strategic documents is not surprising, given the central role, sometimes of metropolitan nature, of most of the Italian regional capitals. This feature is indeed contained in the title itself of many urban strategic plans, as in 'Strategic Plan of Aosta and of La Plaine', 'Strategic Plan of Turin Metropolis', 'Venice Metropolitan Area', 'Metropolitan Strategic Plan' of Bologna, 'Territorial Strategic Plan' of Campobasso, 'Metropolitan Area of Bari', 'Strategic Project of Potenza's Hinterland', and 'Interurban Strategic Plan' of Cagliari. Also the international commitment of these cities recurs in some titles, as in 'International Turin', 'Perugia–Europe', and 'Palermo, Capital of the Euro-Mediterranean Area'.

From this overall picture, taking into account the contents of urban strategic plans and also other information available on institutional websites – including twin cities lists, network projects promoted by the EU, and other cities' international relations – the following types of city networks are found (Tables 2.3 and 2.4).

Classification of city networks according to the types of relationships they have with each other:

- On a spatial scale, intra-urban networks prevail, both overall (76%) and in the various demographic classes (from 60% to 100%) and geographical areas (from 50% to 100%), being typical of metropolitan areas and other urban concentrations. Interregional networks still exist (recording an overall 38%), especially in the case of the major north-central cities, more interested also in international cooperation, for both dimensional and geographical reasons. Interurban networks showed a smaller presence in the analysed documents (29% in total) while recurring quite frequently both in the intermediate demographic classes and in certain geographical areas (especially in the Islands).
- These relations are mainly non-hierarchical, both in general (76%) and in all demographic classes (from 40% to 100%) and geographical areas (from 60% to 100%), consistent with the polycentric nature of the cities involved. There are however also hierarchical relationships (52% overall), as a consequence of the central role played by some cities on a territorial basis.
- As to the type of possible cooperation, the relationships are mostly synergic (76%) and complementary (71%); slightly less frequent are the innovative ones (43%). This distribution is due to the previous classifications and considers the possibility of major Italian cities to cooperate with both similar and different urban contexts, on either a national and international basis.

Table 2.3 The city network typology in the urban strategic plans of Italian regional capitals

				Demographic classes					Geographical areas				
Key factors	Aspects	Types of city networks	Overall situation	Up to 100,000	100,000–250,000	250,000–500,000	500,000–1000,000	Over 1000.000	North-West	North-East	Centre	South	Islands
Relationship	Spatial scale	Intra-urban	16 (76%)	3 (60%)	4 (67%)	3 (75%)	4 (100%)	2 (100%)	4 (100%)	4 (80%)	2 (50%)	4 (67%)	2 (100%)
		Interurban	6 (29%)	–	3 (50%)	1 (25%)	2 (50%)	–	–	3 (60%)	–	1 (17%)	2 (100%)
		Interregional	8 (38%)	–	–	3 (75%)	3 (75%)	2 (100%)	3 (75%)	2 (40%)	2 (50%)	1 (17%)	–
	Nature	Hierarchical	11 (52%)	1 (20%)	3 (50%)	1 (25%)	4 (100%)	2 (100%)	4 (100%)	3 (60%)	2 (50%)	1 (17%)	1 (50%)
		Non-hierarchical	16 (76%)	2 (40%)	4 (67%)	4 (100%)	4 (100%)	2 (100%)	4 (100%)	3 (60%)	3 (75%)	4 (67%)	2 (100%)
	Cooperation	Complementary	15 (71%)	2 (40%)	4 (67%)	3 (75%)	4 (100%)	2 (100%)	3 (75%)	4 (80%)	2 (50%)	4 (67%)	2 (100%)
		Synergic	16 (76%)	3 (60%)	4 (67%)	4 (100%)	3 (75%)	2 (100%)	3 (75%)	4 (80%)	3 (75%)	4 (67%)	2 (100%)
		Innovative	9 (43%)	–	2 (33%)	3 (75%)	2 (50%)	2 (100%)	3 (75%)	4 (80%)	1 (25%)	1 (17%)	–

Membership	Role of the city	Actor	17 (81%)	3 (60%)	4 (67%)	4 (100%)	4 (100%)	2 (100%)	4 (100%)	4 (80%)	3 (75%)	4 (67%)	2 (100%)
		Strategic place	14 (67%)	–	4 (67%)	4 (100%)	4 (100%)	2 (100%)	3 (75%)	4 (80%)	3 (75%)	2 (33%)	2 (100%)
	Representatives	Politicians	17 (81%)	3 (60%)	4 (67%)	4 (100%)	4 (100%)	2 (100%)	4 (100%)	4 (80%)	3 (75%)	4 (67%)	2 (100%)
		Members of technical staff	17 (81%)	3 (60%)	4 (67%)	4 (100%)	4 (100%)	2 (100%)	4 (100%)	4 (80%)	3 (75%)	4 (67%)	2 (100%)
	Origin	Voluntary/unprescribed	14 (67%)	3 (60%)	4 (67%)	2 (50%)	3 (75%)	2 (100%)	3 (75%)	4 (80%)	2 (50%)	3 (50%)	2 (100%)
		Promoted/sponsored	3 (14%)	–	–	2 (50%)	1 (25%)	–	1 (25%)	–	1 (25%)	1 (17%)	–
Thematism	Nature of topics	Socio-economic development	17 (81%)	3 (60%)	4 (67%)	4 (100%)	4 (100%)	2 (100%)	4 (100%)	4 (80%)	3 (75%)	4 (67%)	2 (100%)
		Urban planning/management	6 (29%)	2 (40%)	1 (17%)	2 (50%)	1 (25%)	–	2 (50%)	1 (20%)	–	2 (33%)	1 (50%)
	Contents	Club networks	15 (71%)	2 (40%)	4 (67%)	3 (75%)	4 (100%)	2 (100%)	3 (75%)	4 (80%)	2 (50%)	4 (67%)	2 (100%)
		Web networks	17 (81%)	3 (60%)	4 (67%)	4 (100%)	4 (100%)	2 (100%)	4 (100%)	4 (80%)	3 (75%)	4 (67%)	2 (100%)
	Competences	Functional	17 (81%)	3 (60%)	4 (67%)	4 (100%)	4 (100%)	2 (100%)	4 (100%)	4 (80%)	3 (75%)	4 (67%)	2 (100%)
		Political	2 (10%)	–	–	1 (25%)	1 (25%)	–	1 (25%)	–	–	1 (17%)	–

Author's elaboration

Table 2.4 The city networking orientation in the urban strategic plans of Italian regional capitals: a few examples

Urban strategic plans (year) pp.	Orientation to city networking (selected excepts)
3rd strategic plan of Turin metropolis (2015) 68, 192	Adopting a metropolitan approach means ensuring a farther development, reducing costs, making systems and services more efficient, improving the private and public skills, and being more competitive worldwide. Having an international outline, being open and connected to the world, grasping the talents, promoting global investments, acquiring a knowledge of tongues lead nowadays to a concrete competitive advantage. Turin has spent a lot and gained great results in internationalisation. The strategic plan makes a prior commitment continuing along this path, enhancing the resources and the skills of the city and making the internationalisation a development key lever for the future
Metropolitan strategic plan of Bologna – PSM 2.0 (2016) 6	According to the areas of intervention, guidelines, and cross factors we adopted, our strategic plan aims to be a tool of choice, of distinction, of sharing, and of interconnection. Following this approach, the PSM 2.0 is meant to be able to deal with variables and geographical boundaries, operating both in limited areas (. . .) and along broader lines consistent with the nature of different actions and sectors. It should also make it possible to cross the borders of our metropolitan area whenever a better result can be achieved by cooperating with the surroundings
There is more than one Florence (2009) 10	The constant search for consistency among the various levels of government is the only meaning to get a systematic administrative action, able to produce virtuous skills and start the overall development of the territory (. . .). This programme is not intended to interfere with local identities but on the contrary to allow the governance of the entire metropolitan area by the improvement of and the synergy among the very own growth factors of each area
Strategic plan of Naples (2006) 11	The vision 'Naples focus of the Mediterranean area' means that Naples and its metropolitan area are a growth engine for the entire region, especially being a territory where to find many potentialities required to promote regional strategies for socio-economic development. These strategies are aimed to a future 'reconquest' of the Mediterranean area (both as a natural basin of influence and trade and, by the Suez-Gibraltar axis, as a bridge towards China, India, and the Far East), to a connection with the Balkans, Turkey, and the Middle East, and to a strong linkup with the very heart of Europe
Palermo, capital of the Euro-Mediterranean area (2010) 12	By launching the operational phase of the strategic plan, the municipality intends to adopt a governance approach. Indeed, the plan is not only an administrative act, it is a working programme with a metropolitan and subregional breadth, to which all powers – Institutional, economic, social, and cultural ones – Are required to cooperate in defining shared guidelines for the development of the local community as the node of a wider network

Author's selection and translation

Classification of city networks according to their membership:

- The predominant role of cities is of network actors, both in general (81%), demographically (from 60% to 100%), and geographically (from 67% to 100%); this is consistent with the strategic nature of the analysed documents, where the governing bodies define objectives to be achieved primarily by their institutions. However, the role of cities as strategic places emerges significantly (67% overall), showing an awareness of the nature of cities as nodes of several socio-economic networks.
- Cities' representatives in networks are simultaneously politicians and members of technical staff (81% overall; from 60% to 100% in the different demographic classes; from 67% to 100% in the various geographical areas); this is due to the nature of strategic management tool of the analysed documents.
- The city networks emerging from urban strategic plans are mainly voluntary/unprescribed (67% overall; from 50% to 100% in both demographic and geographic clusters). However, in some cases (14% overall), the city networks, and possibly the same planning processes, are the results of projects promoted/sponsored by the EU. In hindsight, almost all the regional capitals are engaged in this kind of projects, which here count only if explicitly mentioned in the urban strategic plans.

Classification of networks according to their theme/focus:

- The city networks we analysed mainly deal with topics of socio-economic development, both in general (81%) and in all demographic classes (from 60% to 100%) and geographical areas (from 67% to 100%). They deal with a wide range of urban policies, including housing, mobility, social security, health care, hospitality, social inclusion, entrepreneurship, employment, pollution reduction, utilities provision, education, culture, sports, entertainment, urban quality, and tourism promotion. More marginal is the role of those networks that are engaged in urban planning and management (accounting for 29% overall), which come into play where the strategic plans are developed jointly by several municipalities belonging to the same area and where the planning process is itself a result of city networking.
- Either the development of complementary activities or the sharing of common objectives/services is widely represented, activating, respectively, web networks (81% overall) and club networks (71% overall); after all, this distribution is consistent with the previous one related to the type of city-to-city cooperation.
- Finally, the city networks reported in the urban strategic plans are almost all functional, concerning matters falling within the competence and power to act of local authorities (81% overall; from 60% to 100% in the different demographic classes; from 67% to 100% in the various geographical areas). However, in some cases (10% overall), these documents represent an opportunity to address policy issues, in particular to the national government, such as that of the level of autonomy granted to municipalities.

Summarising the results of the analysis, the planning processes of Italian regional capitals seem to confirm the strategic nature of city networks. However, a few critical considerations can be made.

First, the fact that 8 out of 21 urban strategic plans have been abandoned and two second-generation plans are not following up their previous versions suggests that the strategic orientation of about half of the analysed cities is mostly formal and rhetorical and a similar attitude is also likely to characterise the approach to city networking. In addition, the presence of four other documents that neglect the issue further weakens the approach to city networks, which seems to be of a strategic nature in only one third of the considered cities.

Furthermore, the highest concentration in the smallest cities, mainly located in the Centre-South, of the failure to take account of city networks in urban strategic plans represents a strong element of strategic short-sightedness. In fact, just such cities, because of their size and of belonging to the less developed areas of the country, could significantly benefit from city-to-city cooperation. This is true at both intra-urban, interurban, and interregional level, achieving greater economies of scale and reaching the necessary critical mass in many areas of activity, combining complementarity, taking advantage of synergies, and experiencing innovations. In essence, actively managing the city networking, these cities, even more than others, could enlarge their attractiveness, improve their competitive advantages, and increase their chances of success in socio-economic development.

Finally, the analysed urban strategic plans are characterised by the general lack of concrete performance indicators, also, but obviously not only, with reference to city networking. These indicators would be needed to define, measure, and monitor the planned investments, activities, and goals. In reality, the few indicators included in these plans are mainly of financial and timeline nature, covering, at best, only the dedicated resources and the scheduled processes. It is quite rare the use of output, outcome, and impact indicators, which would be necessary to explain ex ante which targets are expected, to check on an ongoing basis whether the plan is on the right track, and to verify ex post which results are obtained. The lack of these elements contributes to weaken the strategic meaning and the managerial nature of the analysed urban plans and their contents, city networking included.

Conclusion

The contribution of city networking to the creation of local public value and the utility of its inclusion in urban strategic planning are evident from the analysis of the literature and from the empirical data collected on this subject.

In fact, city networks are systems of relations through which cities can cooperate, in different ways and at various levels, in order to face the challenges of economic development, social protection, and environmental sustainability.

By promoting and building these networks, local authorities can exchange or share information, objectives, visions, policies, projects, best practices, resources, and knowledge. This would enable them to better contribute to creating local public value and increasing their chance to succeed in the effective satisfaction of public needs and the efficient use of public resources.

Consequently, city networking, as a fundamental enabling condition of the distinctive function of local authorities, should be included in their systems of governance and, especially, in their urban strategic plans.

However, city networks, to be effectively set up and carried out, require the application of appropriate knowledge and skills and the adoption of true strategic vision and approach. They cannot be managed in an improvised or episodic way.

To this end, city networking has to be clearly stated in the urban strategic plans and, in an integrated and convergent way, in the operational programmes and budgets of local authorities. This condition is, in fact, essential to make meaningful, relevant, and functional the overall system of governance and to avoid the proliferation of a multitude of independent and distinct planning tools. These could perhaps be singularly well designed but risk to compose a too crowded set, which can be redundant and wasteful, inevitably rigid, and of little value, since its results are essentially alien to the effective processes of government and management.

The inclusion of city networking in urban strategic planning should avoid some additional weaknesses.

For what concerns the object of networking, at least two perils must be avoided: first, the excessive generality of goals, which is typical of relationships that tend to be all inclusive, and second, the opaque definition of the agreement, because of non-rational or nontransparent choices. In both cases, the result is a weak version of the plan, which would end up being considered unrealistic or too generic. Therefore, it would not be convincing or compelling nor able to produce results.

With reference to decision-making, the main limits are the oftentimes only apparent openness of the process and the purely fictitious involvement of some partners (e.g. the smallest cities in the metropolitan area). When this occurs, a weak version of the plan arises, unable to inspire confidence and willingness to cooperate, because of poor communication and negotiation skills, which do not adequately support the institution's reputation. The urban strategic plan runs the risk of having no contact with the reality of the context and no real impact on behaviour.

Another risk is the lack of coherence between the city networking orientation pursued in the strategic plan of a local authority and (1) that of contiguous/similar territorial contexts and (2) that of higher levels of government. This aspect is particularly important for the urban realities of smaller size, which are increasingly, and per se commendably, testing strategic planning. If the need for an integrated approach is not taken into account, these initiatives might be characterised more by their audacity than by their probability of success.

In addition, it is important to counter the risk of lessening over time the motivation behind the planning effort, which is highly likely especially in the absence of short-term and medium-term goals, in addition to long-term ones. This could be due to the difference between the time horizon of the urban strategic plan and the length of the administrative mandate and also to the difficulty of keeping all partners engaged. In essence, this difficulty refers to the lack of awareness by key players of the evolving nature of the strategic planning process. The latter develops over successive phases of diagnosis, planning, implementation, and evaluation and never sees an end due to the constant and dynamic redesigning of the desired future scenario.

Finally, a very frequent weakness is the lack of measures to assess city networking and other contents of urban strategic plans. Without the ability to measure the expected results, it is not possible to assess the ex ante quality of the decisions taken. In addition, without the measurement of performances, it is not possible to assess the ongoing progress of planned actions and outputs, nor the ex post effectiveness of outcomes and impacts. This relationship, at any time but especially ex ante (where – as previously seen – it is mostly missing), is particularly relevant in the context of urban strategic planning, given the lack of revocability of the decisions taken, which makes it too expensive and risky to proceed by trial and error.

In all such cases, urban strategic plans are only formally 'for governance'. In reality, they are dominated by rhetoric, fashion, or fiction; they can be self-referential, short-sighted, emulative, unrealistic, and, in any case, unable to effectively activate city networks and contribute to creating local public value.

On the contrary, in order to be useful to city networking and, more generally, to socio-economic development, urban strategic planning must build a clear, strong, distinctive, and long-term vision and formulate specifically selected, yet at once flexible and adaptive, goals and projects. For this, it requires the prior definition of appropriate rules concerning openness, transparency of information and communication flows, solutions for the involvement and cooperation of partners, relationship arrangements, and the exercise of leadership. These are essential rules to try to reduce and overcome – with the awareness of never succeeding completely – the above-mentioned risks.

The result will be an agenda for local government that is significant enough to make a difference and streamlined enough to be effectively implemented and shared among relevant partners. This will allow the local authority to mobilise interests, build consensus, attract resources, and produce positive results, actually contributing to creating local public value.

References

1. Anselmi, L. (ed.): Principi e metodologie economico-aziendali per gli enti locali. L'azienda comune. Giuffrè, Milano (2005)
2. Borgonovi, E.: Principi e sistemi aziendali per le amministrazioni pubbliche. Egea, Milano (2004)
3. Bouteligier, S.: Cities, Networks, and Global Environmental Governance. Spaces of Innovation, Places of Leadership. Routledge, New York (2012)
4. Brezzi, M., Veneri, P.: Assessing polycentric urban systems in the OECD: country, regional and metropolitan perspectives. Eur. Plan. Stud. **23**(6), 1128–1145 (2015)

5. Camagni, R.: From city hierarchy to city network: reflections about an emerging paradigm. In: Lakshmanan, T.R., Nijkamp, P. (eds.) Structure and Change in the Space Economy, pp. 66–87. Springer, Berlin (1993)
6. Capello, R.: The city network paradigm: measuring urban network externalities. Urban Stud. **37**(11), 1925–1945 (2000)
7. Capello, R., Nijkamp, P. (eds.): Urban Dynamics and Growth. Elsevier, Amsterdam (2004)
8. Caragliu, A., Del Bo, C., Nijkamp, P.: Smart cities in Europe. J. Urban Technol. **18**(2), 65–82 (2011)
9. Carrillo, F.J. (ed.): Knowledge Cities. Butterworth-Heinemann, Boston (2005)
10. Castells, M.: The Rise of Network Society. Blackwell, Oxford (1996)
11. Chiu, N.: Networked cities' responses to global problems: a typology. J. Comp. Polit. **4**(1), 111–125 (2011)
12. Christaller, W.: Die Zentralen Orte in Süddeustschland. Fischer, Jena (1933)
13. Dameri, R.P.: Smart City Implementation. Springer, Heidelberg (2017)
14. Davoudi, S.: Polycentricity in European spatial planning: from an analytical tool to a normative agenda. Eur. Plan. Stud. **11**(8), 979–999 (2003)
15. Davoudi, S.: Evidence-based planning. Rhetoric and reality. diSP Plan. Rev. **165**(2), 14–24 (2006)
16. Deidda Gagliardo, E.: La creazione del valore nell'ente locale. Giuffrè, Milano (2002)
17. Dematteis, G.: Sistemi locali nucleari e sistemi a rete. In: Bertuglia, C.S., La Bella, A. (eds.) I Sistemi Urbani. Franco Angeli, Milano (1991)
18. Donna, G.: La pianificazione strategica nell'ambito pubblico. Impresa Progetto Electron. J. Manag. http://www.impresaprogetto.it (2010)
19. Edvinsson, L.: Aspects on the city as a knowledge tool. J. Knowl. Manag. **10**(5), 6–13 (2006)
20. Fontana, F.: The intellectual capital in urban strategic planning. J. U.S. China Public Adm. **9**(7), 791–811 (2012)
21. Fontana, F.: Smart city and creation of local public value. In: Dameri, R.P., Rosenthal-Sabroux, C. (eds.) Smart City. How to Create Public and Economic Value with High Technology in Urban Space. Springer-Verlog, Heidelberg (2014)
22. Friedmann, J.: The world city hypothesis. Dev. Chang. **17**, 69–83 (1986)
23. Gibelli, M.C., Curti, F. (eds.): Pianificazione strategica e gestione dello sviluppo urbano. Alinea, Firenze (1996)
24. Hall, P.: The World Cities. Heinemann, London (1966)
25. Keiner, M., Kim, A.: Transnational city networks for sustainability. Eur. Plan. Stud. **15**(10), 1369–1395 (2007)
26. Kickert, W.J.M.: Public governance in the Netherlands: an alternative to Anglo-American 'managerialism'. Public Adm. **75**(Winter), 731–752 (1997)
27. Kiyoshi, K., Lakshmanan, T.R., Anderson, W.P.: Structural Changes in Transportation and Communication in the Knowledge Society. Edward Elgar, Northampton (2006)
28. Labaeye, A., Sauer, T.: City networks and the socio-ecological transition. A European inventory. WWWforEurope, Working Paper no. 27 (2013)
29. Masip-Tresserra, J.: Polycentricity, performance and planning. Concepts, evidence and policy in Barcelona, Catalonia. Archit. Built Environ. **6**(7), 1–348 (2016)
30. Mazzara, L.: Il piano strategico nell'ente locale. Ipsoa, Milano (2009)
31. Meijers, E.: Polycentric urban regions and the quest for synergy: is a network of cities more than the sum of the parts. Urban Stud. **42**(4), 765–781 (2005)
32. Mohlund, O., Forsman, A.: Citywide Strategic Planning: a Guideline. United Nations Human Settlements Programme (UN-HABITAT), Nairobi (2010)
33. Moore, M.H.: Creating Public Value. Harvard University Press, Cambridge (1995)
34. Passaponti, B.: I prezzi politici nei servizi di pubblica utilità. Servizio Editoriale Universitario, Pisa (1986)
35. Pendenza, M.: Cooperazione, fiducia e capitale sociale. Liguori Editore, Napoli (2000)
36. Pflieger, G., Rozenblat, C.: Introduction. Urban networks and network theory: the city as the connector of multiple networks. Urban Stud. **47**(13), 2723–2735 (2010)

37. Pollitt, C., Bouckaert, G.: Public Management Reform. University Press, Oxford (2000)
38. Pugliese, T., Spaziante, A. (eds.): Pianificazione strategica per le città. Franco Angeli, Milano (2003)
39. Putnam, R.: The Evolution of Social Capital. University Press, Oxford (2005)
40. Sassen, S.: The Global City. Princeton University Press, Princeton (1991)
41. Sòstero, U.: L'economicità delle aziende. Giuffrè, Milano (2003)
42. Tanese, A., Di Filippo, E., Rennie, R. (eds.): La pianificazione strategica per lo sviluppo dei territori. Rubbettino, Soveria Mannelli (2006)
43. Taylor, P.J.: Specification of the world city network. Geogr. Anal. **33**, 181–194 (2001)
44. Volpatto, O.: Amministrare oggi l'ente locale. Giuffrè, Milano (1987)

Chapter 3
Networks in Smart Cities from a Graph Theoretic Point of View

Mohammed Haddad

Abstract The advances in wireless communication technologies and the proliferation of mobile devices have enabled the emergence of intelligent environments for citizen to interact with each other, receive information in real time, and access a large variety of services, thus making the cities they live in smarter. Internet of things (IoT) materialized with the appearance of connected objects that are becoming smarter and more powerful. IoT dramatically modified our daily lives since it allowed progress in health monitoring, transportation management, pollution surveillance, etc. Communication is a key point in such environments. Another important problem in governance of a smart city is traffic management. In this chapter, we present a review of techniques based on graph theoretical concepts to optimize flows management in both network communication and route traffic.

Keywords Smart city • Graph theory • Flow management • Network communication • Route traffic

Introduction

A smart city tries to conciliate social, cultural, and environmental objectives through systemic approaches that combine participatory governance and informed management of natural resources in order to meet the needs of institutions, businesses, and citizens. Internet of Things (IoT) is a global infrastructure for the information society that provides advanced services by interconnecting objects (physical or virtual) with existing or evolving inter-operable information and communication technologies. Thanks to its ability to collect data, IoT now offers governance the opportunity to analyze citizen's needs in real time and develop a new range of targeted and personalized actions (traffic control, population health monitoring, air quality, emergency management, etc.). With this huge amount of knowledge, governance can redefine services offered by the city regularly according to population needs and demands.

M. Haddad (✉)
Université de Lyon, Université Claude Bernard Lyon 1, F-69622, Villeurbanne Cedex, France
e-mail: mohammed.haddad@univ-lyon1.fr

A. Karakitsiou et al. (eds.), *City Networks*, Springer Optimization and Its Applications 128, DOI 10.1007/978-3-319-65338-9_3

Optimizing flows is a crucial task in order to sustain smart city management. Flows may vary from data flows as gathering and collecting information from sensors and connected objects to vehicular traffic and congestion management within the city. Recent years have witnessed significant advances in wireless communication and connected objects which emerge as one of the most promising technologies for the twenty-first century [1]. In fact, they present huge potential in several domains ranging from health care applications to military applications. Distributed in irregular patterns across remote environments, objects or sensor nodes will autonomously aggregate into collaborative and asynchronous communication mode. Indeed, the asynchronous mode presents the major advantages of allowing more flexible communication schemes. They are less sensitive to the communication delays and to their variations. Moreover, they also present some tolerance to the loss of data messages since that losses do not prevent the progression of the collection process on both the sender and destination nodes. In general, the primary objective of connected objects or wireless sensors is to collect data from the monitored area and to transmit it to a base station or a sink node for processing. During this phase, resource failures are more likely to occur and can have an adverse effect on the application. Hence, they must be robust and survivable despite individual node and link failures [29, 32, 39]. The advent of wireless sensor networks and its conception constraints, has posed a number of research challenges to the networking and distributed computation communities. Several studies were proposed to construct and maintain some properties of the communication topology. Graph based network topologies aim to give better organization to the network. The presence of such structures ought to decrease the impact of connectivity changes, optimize broadcasts, diffusion and data gathering, take into account load balancing mechanisms, reduce response times, and improve scalability, which is a direct consequence of the statements above. A good network organization simplifies the design in terms of scale considerations.

In this chapter, we present a review of techniques based on graph theoretical concepts to optimize flows management in both network communication and route traffic applications. Section gives a short presentation of problems that involve flow management thus making them very suitable for graph based techniques. Section provides the reader with all necessary definitions and notations for good understanding of the present chapter. Then, graph based techniques are presented according to three classes: Dominating sets based techniques (Section), Spanning subgraphs based techniques (Section), and Partitioning based techniques (Section). Each technique is illustrated with a practical application. Section finishes the chapter.

Some Flow Management Problems

Distributed data gathering in networks emerged from connected objects is an important issue in distributed agreement and synchronization problems [20] and is also a central topic for load balancing (with divisible tasks) in parallel computing [30].

More recently, it has also found applications in distributed coordination of mobile autonomous agents and distributed data fusion in sensor networks [24, 27].

In the literature, this problem has been formulated and studied in various ways. The first approaches were based on flooding. For instance, in [18], each sensor node broadcasts all its stored and received data to its neighbors. After sometime, each node will hold all the data of the network and acts as a fusion center to compute the estimate of the unknown parameter. Some approaches like [26, 28] compute the average of the sensor measurements combined with local Kalman filtering and/or mobile agents. Other works [18, 35, 38] consist of distributed linear iterations, where each connected node updates its current state by a weighted aggregate of its current neighbors' states (which are distorted when they reach it) and these weights decrease to zero in an appropriate way, as time progresses. To perform communication at a larger scale, nodes must cooperate to form routes between sources and their destination. Due to external environment variations, network connectivity may frequently change. Many communication protocols like routing, resources sharing, service discovery or location, prediction, etc. use flooding mechanism to diffuse or gather information. Because of this, recent works try to optimize data flows and exchanges. Another major problem in Smart City management is to make road traffic more efficient. Here also graph theoretical approaches give several solutions to a large variety of problems. Sometimes city governance seeks to make every street in a traffic network one-way. In graph-theoretic terms, assuming all streets are two-way to begin with, we seek to find an orientation of an undirected graph, and in particular we seek a strongly connected orientation [31]. Other works like [22] address the problem of designing street directions and lane additions in urban road networks. The solution is based on the concept of reserve capacity which is defined as the largest multiplier applied to a given existing demand matrix that can be allocated to a network without violating the capacities of arcs. Having a two-way streets base network, modeled by an undirected graph, and the allowable street lane additions, the problem is to find the optimum configuration of street directions and two-way street lane allocations, and the optimum selection of street lane addition projects, in a way that the reserve capacity of the network is maximized.

Definitions and Notations

Graph theory is the study of the properties of graphs. We begin by introducing the concept of graph as well as its common terminology. Most of this terminology is standard and can be found in any classical book on graph theory [5, 6, 37]. In this section, we give a short overview of standard graph terminology used throughout this chapter.

Graph and Directed Graph A graph G is a pair of sets $(V(G), E(G))$, where $V(G)$ is the set of vertices (*a.k.a.* nodes) and $E(G) \subseteq V(G) \times V(G)$ is the set of edges, formed by pairs of vertices. If e is an edge that connects u and v. Vertices u and

v are then called extremities of the edge e. The cardinality of the vertex set $V(G)$, denoted by $|V(G)| = n$, is called the *order* of G. The cardinality of the edge set $E(G) = m$, we call the *size* of G and we denoted by $|E(G)|$. A graph G is connected if there exists a path between any two distinct vertices of G. Otherwise, the graph G is disconnected. A directed graph (*a.k.a.* digraph) is a graph, denoted by G^*, where the edges have a direction associated with them. In other words, its set of edges is represented by a set of ordered pairs of vertices, called arrows, directed edges or arcs.

Degree Degree is the number of edges incident to the vertex. It's also called the *local degree* or *valency*. The degree of a vertex v in the graph G is denoted by $\deg(v)$. If $\deg(v) = 0$, a vertex v is called an *isolated vertex*. A vertex of degree one is called a *leaf* or *pendant vertex*. The maximum degree of a graph G, denoted by $\Delta(G) = max\{deg(v) : v \in V(G)\}$, and the minimum degree of a graph, denoted by $\delta(G) = min\{deg(v) : v \in V(G)\}$.

Subgraph A graph H is a subgraph of a graph G, denoted $H \subseteq G$, if the vertex set $V(H)$ of H is contained in the vertex set $V(G)$ of G and all edges of H are edges in G, i.e., $V(H) \subseteq V(G)$ and $E(H) \subseteq E(G)$. For any vertex subset $S \subseteq V(G)$, the *subgraph induced* by S denoted by $G[S] = (S, E_s)$ contains all the edges of $E(G)$ whose extremities belong to S. As particular subgraphs, we have clique and independent set defined below.

Distance and Diameter The distance between two vertices v and u in a graph G, denoted by $dist(u, v)$, is the number of edges of a shortest path connecting them. The diameter d of G is the maximum distance between any two vertices of G.

Neighborhood (Vertex and Edge) The (open) neighborhood of a vertex v in a graph G is the set of all vertices adjacent to v, denoted by $N(v) = \{u : uv \in E(G)\}$. The number of neighbors of v corresponds to the degree of v in G, then $|N(v)| = \deg(v)$. The closed neighborhood of v is $N[v] = N(v) \cup \{v\}$. For a set $S \subseteq V$, $N(S) = \bigcup_{v \in S} N(V)$ and $N[S] = \bigcup_{v \in S} N[V]$. The neighborhood of an edge e in a graph G is the set of all edges having at least a common extremity with the edge e, denoted by $N(e)$. The k-neighborhood (also, k-hop neighborhood) of the vertex u to be the set of all vertices which are at most at distance k from u.

Cycle Graph A cycle C_n with $n \geq 2$ is a connected graph having n vertices (n is also called the length of the cycle). It consists of a sequence of vertices $(v_1, v_2, \ldots, v_n)$ starting and ending at the same vertex, with each two consecutive vertices in the sequence adjacent to each other in the graph. In other words, for every $i \in \{1, \ldots, n-1\}$, the edge $v_i v_{i+1}$ exists in $E(C_n)$ and $v_n v_1$ also. A simple cycle is a cycle with no repetitions of vertices and edges. A simple cycle containing all the vertices of a graph G is called a *Hamiltonian cycle* of G. A cycle (not necessarily simple) containing all the edges of a graph G is called an *Eulerian cycle* of G.

Tree Graph A tree T_n is a connected graph with no cycles and having n vertices. Recall that a vertex with degree one is called a *leaf* and a vertex of degree at least two is called an *internal vertex*. A tree is called a *rooted tree* if one of its vertices has been designated the root, in which case the edges have a natural parent–child orientation, towards the root. A vertex v in a rooted tree is a descendant of a vertex u if u lies on the unique path from the root to v. The parent of a vertex v is the last vertex before v in a path from the root to v. The depth of a vertex v in a rooted tree is the length of the path from the root to v.

Let $G = (V, E)$ be a graph. In our review of techniques based on graph theoretical concepts we were brought to classify them into three main classes:

- *Dominating sets*: This class is based on subsets of the vertex set V that verifies some properties (like covering property, dominating property, etc.). It regroups Multipoint relays (MPR sets), Dominating Sets (DS), Independent Sets (IS), etc.
- *Spanning subgraphs*: This class is based on subgraphs of G, generally, $G' = (V, E')$ where E' is a subset of E. It regroups Neighborhood Graphs (NG), Spanning Trees (ST), Rings, etc.
- *Partitionings*: Approaches of this class try to partition the vertex set V into subsets of vertices sharing some common properties. It regroups Clusterings, Communities, Colorings, etc.

Dominating Sets Based Techniques

Algorithms and protocols of this class usually choose a subset of nodes which will be considered as network leaders or to form a backbone of the topology. In other words, the chosen nodes will have some privileges and responsibilities according to the problem. The most known techniques which are based on electing special nodes are:

Dominating and Independent Sets

A Dominating Set (DS) of the graph G is a subset S of V such that every vertex of V is either in S or adjacent to at least one vertex of S (i.e., for each $v \in (V \backslash S)$: $N(v) \cap S \neq \emptyset$). A Minimal Dominating Set (MDS) is a dominating set such that no subset of it satisfies the dominating property. (see Figure 3.1).

An independent set (*a.k.a. stable set*) is a set of vertices in a graph such that no two of which are adjacent. In other words, an Independent Set (IS) in the graph G

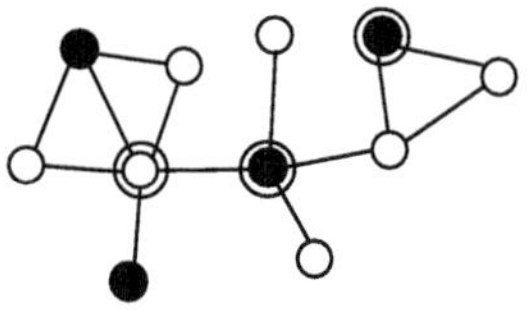

Fig. 3.1 Dominating set (surrounded)/maximal independent set (*black*)

is a subset S of V such that S does not admit any pair of adjacent vertices (i.e., for each $v \in S : N(v) \cap S = \emptyset$). The independent set S is said Maximal Independent Set (MIS) if there is no independent set S' such that $S \subset S'$. For any maximal independent set, any pair of complementary subsets of a MIS are separated by either two or three hops. By consequence, a MIS is also a DS.

A graph $G = (V, E)$ is k-vertex connected if it is connected, and removing any $1, 2 \ldots k - 1$ vertices from G will not cause partition in G. The concept of k-connectivity is usually used in the context of fault tolerance support. A subset S of V is a k-Dominating Set (k-DS) of G if every vertex not in S has at least k neighbors in S. A d-Dominating Set (d-DS) of the graph G is a subset S of V such that each vertex of V is either in the d-DS S or in the d-neighborhood of at least one vertex of S. Let $G = (V, E)$ be a connected graph, the d-closure of G is the graph $G_d = (V, E_d)$ such that $E_d = E \cup D$, where D is the set of all virtual edges connecting each node with all others in its d-neighborhood.

In a recent work [33], the authors propose a novel virtual coordinate construction technique using graph-theoretic concept of dominating sets. They tackled performance issues of existing virtual coordinate assignment techniques that are constrained by topological situations such as low density or having voids/holes, and where greedy forwarding suffers due to local minima when no neighbor is found closer to the destination or low-quality routes comprised of long distance hops.

Connected Dominating Sets

A Connected Dominating Set (CDS, see Figure 3.2) of the graph G is a subset S of V such that:

- Vertices of S form a dominating set of G.
- The induced subgraph by the set S is connected.

Constructing a connected dominating set generally aims to get a better topology control strategy by reducing the network communication overhead. In [25], an energy efficient distributed connected dominating set algorithm based on a new degree-based greedy approximation algorithm named as Connected Pseudo Dominating Set Using 2-Hop Information. This is done to prolong the network lifetime and balance energy consumption.

A k-DS is a k-Connected k-Dominating Set (k-CDS) if the subgraph induced by the k-DS is k-vertex connected. For a graph G and its d-closure G_d, a set S is a d-DS

Fig. 3.2 A connected dominating set

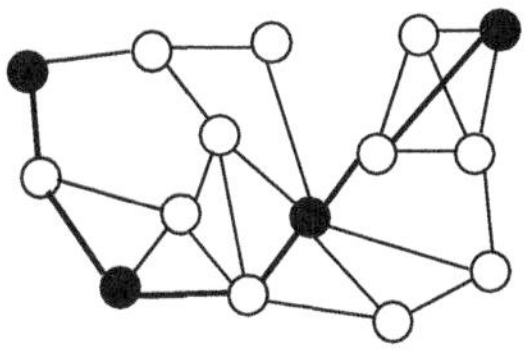

Fig. 3.3 A weakly connected dominating set

in G, if it is a DS in G_d, and a set S is d-hop connected in G, if is connected in G_d. If also S is a d-DS, then we say that S is a d-CDS. The notion of d-CDS is also used to induce a clustering of the graph G. In [40], the authors deal with large-scale wireless network. They showed there is a need for effective approaches to construct d-hop connected dominating sets (d-CDS), so that the size of the dominating sets can be reduced greatly and improve the scalability of previous work.

Weakly Connected Dominating Sets

A Weakly Connected Dominating Set (WCDS, see Figure 3.3) of the graph G is a subset S of V such that vertices of S form a dominating set of G and by connecting every two vertices of S that are at distance one or two the induced subgraph is connected. [19] uses the clustering concept of WCDS to propose an improved ant-based on-demand clustering routing protocol for wireless ad-hoc networks. Network information is obtained from the ant agents, and is only broadcast by the head of every cluster, thus decreasing the communication overheads. To increase the efficiency, a pseudo-random-proportional-selection strategy is used to evaluate the best path from the source node to the destination node.

Multipoint Relays

The set of multipoint relays (MPR) of a node is a minimal size subset of its direct neighbors that connects it to all nodes that are at distance two from it. This technique restricts the number of retransmitters to a subset of neighbors instead of all of them (like in pure flooding). Each node selects, using its local information, a subset of its neighborhood which retransmits its data. Figure 3.4 shows an example of an MPR set for the node a. When a wants to broadcast a message, only m_1, m_2, and m_3 forward it. An example of work relying on concept of MPR is the one proposed by Cervera et al. [8]. They proposed a Multipath routing scheme to increase resilience against network failures or improve security in mobile wireless networks. As authors state, the Optimized Link State Routing (OLSR) protocol has been adopted by several multipath routing strategies. They implement MPR nodes as a flooding mechanism for distributing control information. The construction

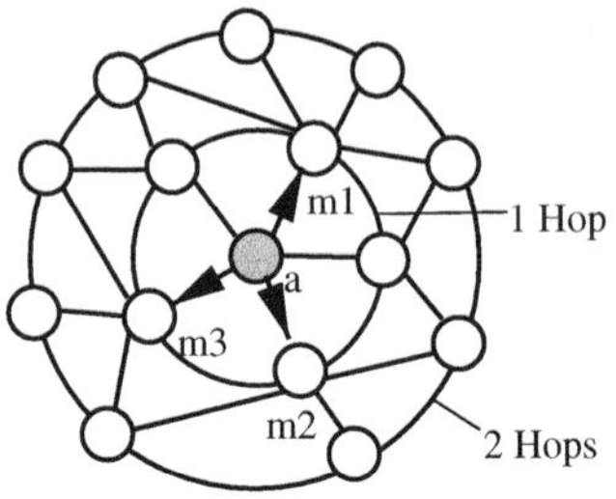

Fig. 3.4 Multipoint relay set example

of multiple disjoint paths helps to increase resilience against network failures or malicious attacks. They present a strategy to compute multiple strictly disjoint paths between any two nodes in OLSR-based networks and provide mechanisms to improve the view of the network topology by the nodes, as well as handling potential flooding disruption attacks to the multipath construction mechanism in OLSR-based networks.

Spanning Subgraphs Based Techniques

This class is based on constructing subgraphs of the network graph. Contrary to the first one, this class aims to minimize the number of used edges or links, thus minimizing the communication overheads or to force flows to more often choose specific routes over others. The most known link based topologies are:

Spanning Trees

A tree is an undirected graph $T = (V, E)$ which is connected and has no cycle. If V is finite and $|V| = n$, then $|E| = n - 1$. Let $G = (V, E)$ be a connected graph. A spanning tree of G is any subgraph $T = (V, E')$ which is a tree. Figure 3.5 shows a spanning tree example. Spanning trees are very useful in broadcasting and multicasting applications. Indeed, the tree structure of the topology guarantees the broadcast reachability of all vertices with a minimum number of messages. Several variants of spanning trees were developed in the literature: Shortest Path trees, Steiner trees, Minimum Diameter Spanning trees, Maximum Leaf Spanning trees, etc. When the tree is rooted at the collecting node, a shortest path tree will try to reduce the global latency between connected nodes and the root. A Steiner tree in a graph is a minimum weight spanning tree but covering only a subset S of V. The maximum leaf spanning tree problem has been proved to be equivalent to the minimum connected dominating set problem. In fact minimizing the number of nodes of a connected dominating subgraph is the same as maximizing the number of nodes that are outside of the subgraph, thus maximizing the number of leaves.

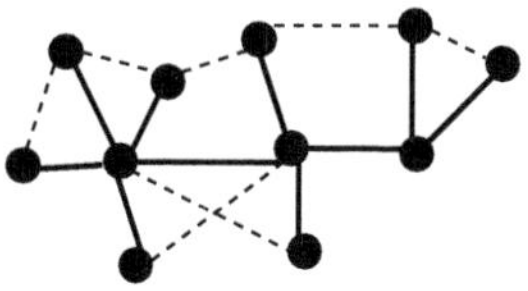

Fig. 3.5 A spanning tree

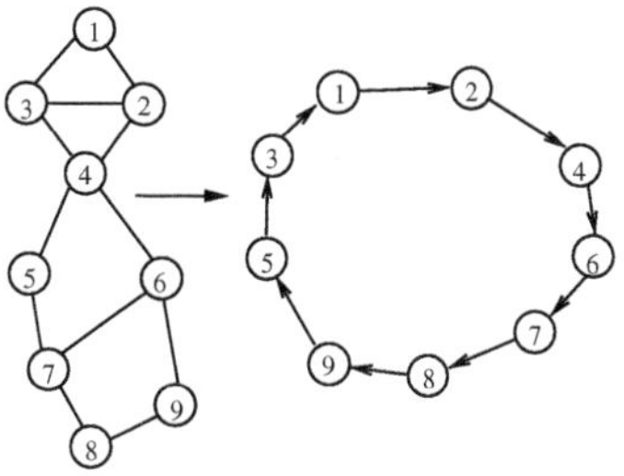

Fig. 3.6 Virtual ring construction

In [36], the authors propose the monitoring of the proportion of air pollution with a wireless inspection and notification system through the concept of Internet of Things (IoT). A Radio frequency identification (RFID) technology which is a low-cost and mature wireless communication method is adopted to collect and transmit emissions information of vehicles. The RFID devices need to be installed on the traffic lights. To take into consideration the real environment, an efficient and innovative maximum spanning tree algorithm is presented to select suitable traffic lights and aims to reduce the number of RFID devices.

Rings

A network is a ring topology if every node has exactly two branches connected to it, i.e. there is only one cycle between a node and itself. Further, this cycle contains all nodes in the network.

Contrary to other topologies, which are based on physical network connectivity, this topology is more virtual. In fact, the construction of this kind of topologies relies on the routing layer (see Figure 3.6). Rings are in general used to distribute circularly a privilege or to provide an order on communication, resource access, etc. Although it is a classical approach, many modern token and agent oriented solutions still use ring topologies or virtual rings.

Jiang et al. [13] provides a novel dynamic ring-based routing scheme for correlation data aggregation. In their model, the authors first route nodal data to rings which have abundant energy in minimum distance and then all data aggregation is processed along the ring before aggregated data is routed to the sink with shortest path. They also show their scheme to have higher network lifetime than previous related work.

Fig. 3.7 A core of the tree

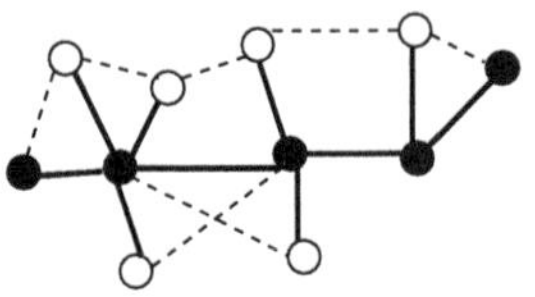

k-Tree Core

Given a tree T, a "k-Tree core" of T is defined as the subtree T' with exactly k leaves and which minimizes the distances sum to all other nodes in T. If $k = 2$, then T' is simply a core path of T (see Figure 3.7).

This topology offers a nontrivial d-domination property for the induced subgraph by the k-tree core.

Dvir and Segal [10] presents new efficient strategy for constructing a wireless tree network satisfying QoS requirements. They use the centdian function which is a convex combination of bandwidth and delay (QoS). Given a tree network T, the authors define a coredian path as a path in T that minimizes the centdian function, a k-coredian tree is a subtree of T with k leaves that minimizes the centdian function, and a (k, l)-coredian tree is a subtree of T with k leaves and diameter l at most that minimizes the centdian function. The authors give efficient constructions of coredian path cores and (k, l)-coredian tree cores.

Neighborhood Graphs

Given the graph G, neighborhood graphs construct a subgraph of G which keeps its connectivity but try to minimize the graph degree (essentially by eliminating transitive edges). Here, we give two examples of these graphs. Relative Neighborhood Graphs and Gabriel Graphs. A Relative Neighborhood Graph (RNG) contains an edge uv if the intersection of the two disks respectively centered at u and v with radius equals to dist(u,v), is empty of other vertices. A Gabriel Graph (GG) contains an edge uv if the disk with uv as diameter is empty of other vertices (see Figure 3.8).

Cartigny et al. [7] considers both topology control and broadcast oriented protocols. It describes localized protocols where nodes require only local information about their neighborhood (distances or geographic positions). The proposed solutions are based on the use of neighbor elimination scheme applied on the relative neighborhood graph (RNG) and local minimum spanning tree (LMST). In [15], the authors propose a framework for WSN which combines clustering, routing, and topology control. In their framework, they implement an energy efficient zone-based topology and routing protocol and propose a new set of graphs referred to as the Mini Gabriel (MG) graphs. The proposed framework demonstrates the best performance in terms of the network energy consumption. Moreover, the MG demonstrates that it achieves the connectivity property.

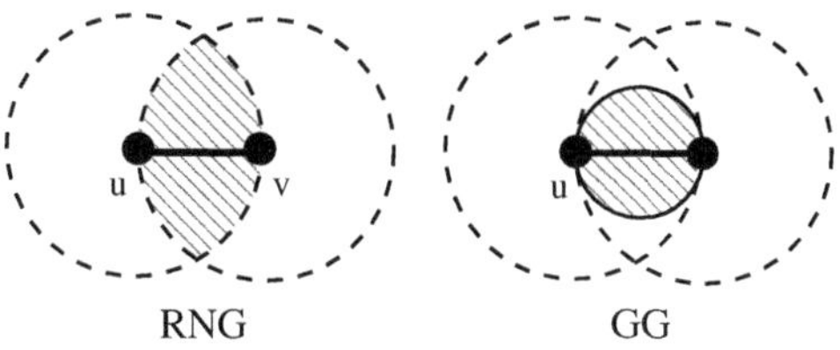

Fig. 3.8 RNG and GG

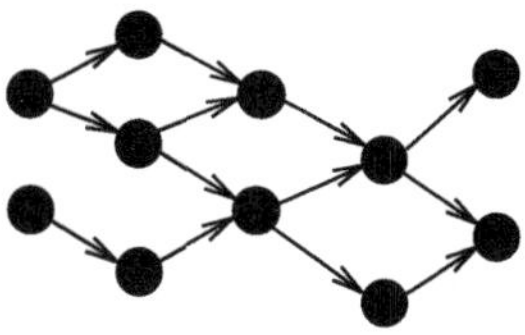

Fig. 3.9 A directed acyclic graph

Directed Acyclic Graphs

In this kind of topologies, we consider a directed graph $G = (V, A)$ where A is a set of arcs (*directed edges*). If G does not have any directed cycle, then G is a Directed Acyclic Graph (DAG, see Figure 3.9). These topologies are much used in Link Reversal (LR) approaches. An example is the Link Reversal Routing (LRR). LRR algorithms construct a DAG Rooted at the destination. This involves that there is no loop between a source and its destination. Also, only the destination may have not outgoing links. Gaddour et al. [11] addresses the issue of mobility support in the Routing Protocol for Low power and lossy networks (RPL), the recently adopted IETF routing protocol standard for low power wireless sensor networks. Since RPL was originally designed for static networks with no support for mobility, the authors address this gap and propose Co-RPL as an extension to RPL based on the Corona mechanism to support mobility and study the impact of node speed, packet transmission rate, and number of DAG roots on network performance.

Another interesting study deals with graph orientation/reorientation problems. The authors of [2] study the influence of graph orientation has on the network flow over time. This problem is also known as the contraflow or lane reversal problem. They introduce and analyze the price of orientation, i.e. how much flow is lost in any orientation of the network if the time horizon remains fixed?

Partitionings Based Techniques

Algorithms and protocols of this class try to regroup nodes into classes or families sharing some common properties or interest. The goal is to give a certain hierarchy to the network or extract some information related to interactions within the network to improve the global behavior as well as performance. The most known partition based techniques are:

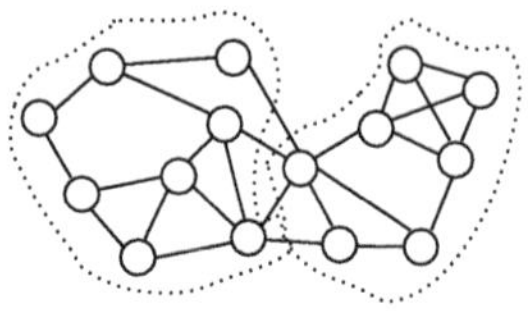

Fig. 3.10 A 4-clustering of the graph

Clusters and k-Clusters

A cluster is a subset of nodes of the underlying network that satisfies or shares a certain property. The precise definition of this property varies within different contexts. Most node-centric clustering schemes insist on the existence of a central node adjacent to all the remaining nodes in the cluster. This central node is referred to be cluster-head. In this case, the corresponding property is known as the dominance property. In the presence of a central node, consensus is reached trivially: it is decided by the central node. However, cluster-heads should cooperate between them to perform larger scale operations (for example, locating a service, etc.). The k-Clustering technique consists in partitioning the network in a minimal number of clusters whose respective diameters do not exceed k hops (see Figure 3.10).

Layuan and Chunlin [17] provides QoS-sensitive routes in a scalable and flexible way in network environment with mobility. In the proposed scheme, each local node needs only to maintain local multicast routing information and/or summary information of other clusters (or domains) but does not require any global ad hoc network states to be maintained. Mehdi Afsar et al. [21] presents a comprehensive survey on clustering approaches. It defines the objectives and characteristics of clustering as well as a classification on the clustering algorithms in WSNs.

Communities

As defined by the IEEE world forum of Internet of Things, the Social Internet of Things emerges from the convergence of the Internet of Things and Social Networking paradigms for the creation of social networks in which things are nodes that establish social links. Community structures is an important property of social graphs. Semantically, we can define a community as a group of individuals sharing the same interest. A community structure is the partition of the graphs into groups of nodes having more links with each other than with the rest of the graph [9]. A community in a social graph is then a set of nodes sharing same interests and having more interactions with each other. Detecting these communities aims to get a better understanding of the studied networks. Moreover, since the partition of the network into communities carries information on how the social entities interact within the social network, it is important to study the community structure.

Atzori et al. [3] gives an interesting review of works that investigated the potentialities of integrating social networking concepts into Internet of Things

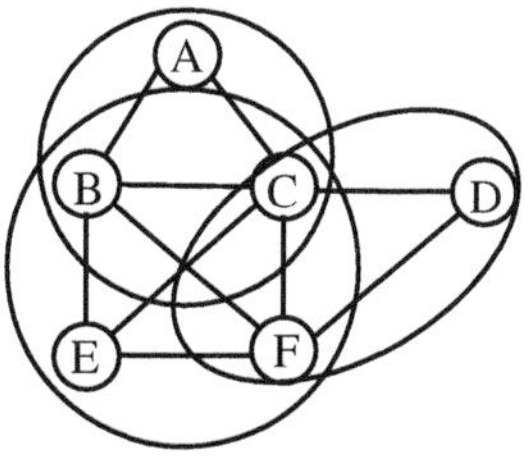

Fig. 3.11 Maximal cliques ABC, BCEF, CDF

solutions. The authors claim the resulting paradigm, they named Social Internet of Things (SIoT), to have the potential to support novel applications and networking services for the IoT in more effective and efficient ways. Misra et al. [23] presents a community detection method in an integrated Internet of Things and social network architecture using a graph mining approach in which the formation of communities is considered only if the occurrence of two nodes is connected and has at least two common neighbors. Their approach considers mutual friends as a metric for suggesting new friends.

Cliques

Let $G = (V, E)$ be a graph. A clique is a subset of vertices of a simple graph G such that its induced subgraph is complete, that is, every two distinct vertices in the clique are adjacent. In other words, if a subgraph G' of G is a complete graph (i.e., there is an edge between any two vertices of G'), then G' is called a Clique. A maximal clique of G is one that is not contained in any other clique (see Figure 3.11).

Gupta et al. [12] uses a conflict graph that models the interference relationship between links to determine if a set of flow rates can be accommodated. Using the cliques of the conflict graph, the authors derive constraints that are sufficient for a set of flow rates to be feasible and guaranteed to be within a constant bound of the optimal. They also extend the ad hoc network model to incorporate variations in the interference range, and obstructions in the network. Lakhlef [16] presents a new hierarchical partitioning method based on hierarchical scheme that takes into account the size of cliques and clusters, and minimizes the distance between the cluster head and its nodes. Author motivates his proposition to respond to the energy and memory constraints for WSNs.

Colorings

The node coloring consists in choosing different colors for any two neighboring nodes in an arbitrary graph. This problem issued from classical graph theory finds applications in distributed computing as coloring algorithms are mainly used for resource allocation. A coloring defines a partial order on nodes allowing them to execute critical tasks or to share time slots according to the colors. Another problem where graph coloring is useful is preventing potential interference when assigning frequencies to wireless devices where nodes that are sufficiently close must have different colors. As frequencies or colors are a valuable resource, it is also desirable to use as few colors as possible. In this variant of the problem, colors should be not different for direct neighbors but also for nodes at further distance according to practical constraints.

In [14], the authors propose a distributed graph coloring algorithm for time division multiple access scheduling. The goal is to consider real-time performance for a clustering-based sensor network in order to determine the smallest possible length of conflict-free assignment of time slots within cluster's transmissions. More recently in [34], the authors present multi-threaded algorithms for graph coloring suitable to the shared memory programming model. They first focus on shared memory implementations to the algorithms such as Jones Plassman graph coloring. Then, they propose new approaches to solve the problem of coloring using mutex locks while making sure that deadlocks do not occur. Other work extend applications to city related problems like electricity and water networks management or air traffic flow management using graph coloring algorithms [4].

Conclusion

Various researches on smart city networks gave much interest to use dynamic graph topologies to improve performance. In this chapter, we presented popular graph techniques used to provide more organization within the studied networks or to enforce some suitable properties according to applications. In addition, we described these techniques as well as their usefulness of city networks like wireless sensor networks or traffic networks. The main aim of dominating sets based techniques is to determine a subset of all nodes in the network that can be used as a support or a backbone that allows its nodes to have more privileges and/or responsibilities in the network. These approaches may suffer from lack of load balancing if not implemented in the solution. On the other hand, Spanning subgraphs techniques try to give a certain orientation to data flows by reducing number of used links in larger scale communication. This class of approaches may suffer from robustness issues if the chosen subgraph does not offer good connectivity or induce a distortion within the graph forcing flows to take longer paths than what they could have but this is generally done to prevent congestions. Partitioning based techniques regroup nodes into classes or groups in order to either allocate resources according to groups or to install a hierarchy based behavior. This is generally done to improve scalability.

References

1. Akyildiz, I., Su, W., Sankarasubramniam, Y., Cayirci, E.: A survey on sensor networks. IEEE Commun. Mag. 102–114 (2002)
2. Arulselvan, A., Groß, M., Skutella, M.: Graph orientation and flows over time. Networks **66**, 196–209 (2015)
3. Atzori, L., Iera, A., Morabito, G., Nitti, M.: The Social Internet of Things (SIoT) – when social networks meet the Internet of Things: concept, architecture and network characterization. Comput. Netw. **56**(16) (2012)
4. Barnier, N., Brisset, P.: Graph coloring for air traffic flow management. Ann. Oper. Res. **130**(1), 163–178 (2004)
5. Berge, C.: The Theory of Graphs. Courier Corporation (1962)
6. Bondy, J.A., Murty, U.S.R.: Graph Theory with Applications. Macmillan, London (1976)
7. Cartigny, J., Ingelrest, F., Simplot-Ryl, D., Stojmenović, I.: Localized LMST and RNG based minimum-energy broadcast protocols in ad hoc networks. Ad Hoc Netw. **3**(1), 1–16 (2005)
8. Cervera, G., Barbeau, M., Garcia-Alfaro, J., Kranakis, E.: A multipath routing strategy to prevent flooding disruption attacks in link state routing protocols for MANETs. J. Netw. Comput. Appl. **36**(2), 744–755 (2013)
9. Clauset, A., Newman, M.E.J., Moore, C.: Finding community structure in very large networks. Phys. Rev. E **70**(6) 066111 (2004)
10. Dvir, A., Segal, M.: The (k,l) Coredian Tree for Ad Hoc Networks. In: 2008 The 28th International Conference on Distributed Computing Systems Workshops, Beijing, pp. 267–272 (2008)
11. Gaddour, O., Koubâa, A., Rangarajan, R., Cheikhrouhou, O., Tovar, E., Abid. M.: Co-RPL: RPL routing for mobile low power wireless sensor networks using Corona mechanism. In: Proceedings of the 9th IEEE International Symposium on Industrial Embedded Systems (SIES 2014), Pisa, pp. 200–209 (2014)
12. Gupta, R., Musacchio, J., Walrand, J.: Sufficient rate constraints for QoS flows in ad-hoc networks. Ad Hoc Netw. **5**(4), 429–443 (2007)
13. Jiang, L., Liu, A., Hu, Y., Chen, Z.: Lifetime maximization through dynamic ring-based routing scheme for correlated data collecting in WSNs. Comput. Electr. Eng. **41**, 191–215 (2015)
14. Kang, H., Zhao, Y.-n., Mei, F.: A graph coloring based TDMA scheduling algorithm for wireless sensor networks. Wirel. Pers. Commun. **72**(2), 1005–1022 (2013)
15. Karim, L., El Salti, T., Nasser, N.: Routing on Mini-Gabriel graphs in Wireless Sensor Networks. In: 2011 IEEE 7th International Conference on Wireless and Mobile Computing, Networking and Communications (WiMob), Wuhan, pp. 105–110 (2011)
16. Lakhlef, H.: A multi-level clustering scheme based on cliques and clusters for wireless sensor networks. Comput. Electr. Eng. **48**, 436–450 (2015)
17. Layuan, L., Chunlin, L.: A QoS multicast routing protocol for clustering mobile ad hoc networks. Comput. Commun. **30**(7), 1641–1654 (2007)
18. Legg, J.A.: Tracking and sensor fusion issues in the tactical land environment. Technical Report TN.0605 (2005)
19. Li, K.-H., Leu, J.-S.: Weakly connected dominating set-assisted ant-based routing protocol for wireless ad-hoc networks. Comput. Electr. Eng. **48**, 62–76 (2015)
20. Lynch, N.: Distributed Algorithms. Morgan Kaufmann Publishers, San Francisco (1996)
21. Mehdi Afsar, M., Tayarani-N, M-H.: Clustering in sensor networks: a literature survey. J. Netw. Comput. Appl. **46**, 198–226 (2014)
22. Miandoabchi, E., Farahani, R.Z.: Optimizing reserve capacity of urban road networks in a discrete Network Design Problem. Adv. Eng. Softw. **42** (2011)
23. Misra, S., Barthwal, R., Obaidat, M.S.: Community detection in an integrated Internet of Things and Social Network Architecture. In: IEEE Global Communications Conference (GLOBECOM) (2012)

24. Moallemi, C.C., Roy, B.V.: Consensus propagation. IEEE Trans. Inf. Theory **52**(11), 4753–4766 (2006)
25. Mohanty, J.P., Mandal, C., Reade, C., Das, A.: Construction of minimum connected dominating set in wireless sensor networks using pseudo dominating set. Ad Hoc Netw. **42**, 61–73 (2016)
26. Olfati-Saber, R., Fax, J., Murray, R.: Consensus and cooperation in networked multi-agent systems. Proc. IEEE 215–233 (2007)
27. Olfati-Saber, R., Murray, R.M.: Consensus problems in networks of agents with switching topology and time-delays. IEEE Trans. Autom. Control **49**(9), 1520–1533 (2004)
28. Olfati-Saber, R., Shamma, J.S.: Consensus filters for sensor networks and distributed sensor fusion. In: 44th IEEE Conf. on Dec. and Cont. CDCECC (2005)
29. Paradis, L., Han, Q.: A survey of fault management in wireless sensor networks. JNSM **15**(2), 171–190 (2007)
30. Rabani, Y., Sinclair, A., Wanka, R.: Local divergence of Markov chains and the analysis of iterative load-balancing schemes. In: IEEE Symposium on Foundations of Computer Science, Palo Alto (1998)
31. Roberts, F.S., Xu, Y.: On the optimal strongly connected orientations of city street graphs IV: four east-west avenues or north-south streets. Discret. Appl. Math. **49**(1–3), 331–356 (1994)
32. Saleh, I., Eltoweissy, M., Agbaria, A., El-Sayed, H.: A fault tolerance management framework for wireless sensor networks. J. Commun. **2**(4), 38–48 (2007)
33. Shukla, S., Misra, R., Agarwal, A.: Virtual coordinate system using dominating set for GPS-free adhoc networks. Ann. Telecommun. (2017)
34. Singhal, N., Peri, S., Kalyanasundaram, S.: Practical multi-threaded graph coloring algorithms for shared memory architecture. In: Proceedings of the 18th International Conference on Distributed Computing and Networking (ICDCN '17). ACM, New York, NY, USA, Article 44 (2017)
35. Talebi, M.S., Kefayati, M., Khalaj, B.H., Rabiee, H.R.: Adaptive consensus averaging for information fusion over sensor networks. In: IEEE International Conference on Mobile Adhoc and Sensor Systems (MASS), pp. 562–565 (2006)
36. Vong, C.M., Wong, P.K., Ma, Z.Q., Wong, K.I.: Application of RFID technology and the maximum spanning tree algorithm for solving vehicle emissions in cities on Internet of Things. 2014 IEEE World Forum on Internet of Things (WF-IoT), Seoul, pp. 347–352 (2014)
37. West, D.B.: Introduction to Graph Theory. Prentice Hall, Upper Saddle River, NJ (2001)
38. Xiao, L., Boyd, S., Lall, S.: A space-time diffusion scheme for peer-to-peer least-squares estimation. In: Proceedings of Fifth International Conference on Information Processing in Sensor Networks (IPSN 2006), pp. 168–176 (2006)
39. Ye, F., Zhang, H., Lu, S., Zhang, L., Hou, J.C.: A randomized energy conservation protocol for resilient sensor networks. Wirel. Netw. **12**(5), 637–652 (2006)
40. Zheng, C., Yin, L., Sun, S.: Construction of d-Hop Connected Dominating Sets in Wireless Sensor Networks. Advanced in Control Engineering and Information Science. Procedia Eng. **15**, 3416–3420 (2011)

Chapter 4
An Evaluation of Measures for the Reduction of Carbon Dioxide Emissions from Automobile Traffic in Central Tokyo

Kayoko Yamamoto and Kuangtiao Shen

Abstract This study aims to evaluate measures for the reduction of carbon dioxide (CO_2) emissions from automobile traffic in the 23-ward district of Tokyo (central Tokyo), Japan, using the methods of spatial analysis by Geographic Information Systems (GIS) and statistical analysis. All of the measures to prevent global warming in Japan were first investigated to extract nine measures to reduce CO_2 emissions and then classified into the three groups. Secondly, using the GIS functions for database creation and information analysis, a digital map databases reflecting the real situation of automobile traffic in the study area was created. Using this database, the reduction of CO_2 emissions was calculated before and after the implementation of the above nine measures (2005 and 2011) in each road unit. Furthermore, in order to evaluate these measures, the situation of the reduction of CO_2 emissions was visualized in each road unit on the digital maps using GIS, and the CO_2 emissions before and after the implementation effects of the above nine measures were grasped by t-tests using R language. Based on the evaluation results, it was clear that CO_2 emissions are the highest on the main roads in the city center, while the roads with a high reduction of CO_2 emissions differed according to each measure. It was also found that all the above nine measures were effective, except for the promotion of eco-drive for small-size automobile. This is because, in the case of small-size automobiles, the diffusion rate of eco-drive did not increase greatly, the reduction effect of CO_2 emission was small, and the reduction of CO_2 emissions by fuel consumption was low.

Keywords Carbon dioxide reduction • Automobile traffic • Spatial analysis • Statistical analysis • Japan

---

K. Yamamoto (✉)
Graduate School of Informatics and Engineering, University of Electro-Communications, Tokyo, 1-5-1 Chofugaoka Chofu-shi, Tokyo, 182-8585, Japan
e-mail: k-yamamoto@is.uec.ac.jp

K. Shen
Graduate School of Information Systems, University of Electro-Communications, Tokyo, 1-5-1 Chofugaoka Chofu-shio, Tokyo, 182-8585, Japan
e-mail: yqb@si.is.uec.ac.jp

A. Karakitsiou et al. (eds.), *City Networks*, Springer Optimization and Its Applications 128, DOI 10.1007/978-3-319-65338-9_4

Introduction

According to the 5th Assessment Report [6] of the Intergovernmental Panel on Climate Change (IPCC), in the 200 years after the Industrial Revolution, the concentration of greenhouse gases has increased exponentially along with the increase of human activity. Greenhouse gases include carbon dioxide (CO_2), nitrous oxide (N_2O), methane (CH_4), ozone, and water vapor, which are gases that already exist in the atmosphere, in addition to hydrofluorocarbon (HFC), perfluorocarbon (PFC), and sulfur hexafluoride (SF_6), which are artificial gases. Although the greenhouse effects per a certain amount of methane, N_2O, and fluorocarbons are higher than that of CO_2, because of the high amount of CO_2 emissions, it is the highest contributing gas toward global warming and accounts for 60% of all greenhouse gases exhausted worldwide.

Japan exhausts 5% of the total amount of CO_2 emissions worldwide and ranks number 4 after the United States, China, and Russia. Therefore, it can be said that Japan has a major responsibility toward other countries in the prevention of global warming. In addition, according to the Ministry of Environment [22], the international temperature average has risen by approximately 0.7 °C over the past 100 years, and the cause of this is considered to be global warming. Though the average temperature in Japan has risen by approximately 1.1 °C in the past 100 years, it has also risen by 3 °C in Tokyo and 2 °C in large cities such as Sapporo, Nagoya, Osaka, and Fukuoka.

According to the Ministry of Land, Infrastructure, Transport and Tourism [18], of the CO_2 emissions (1.265 billion tons) of Japan in 2014, the shipping department accounted for 17.2% (217 million tons) which ranked second after the business departments which accounted for 20.6% (260 million tons). Of the total amount of CO_2 emissions for the shipping department, automobiles accounted for 86.0%, which is 14.7% of the total amount of CO_2 emissions for Japan. This is considered to be due to the increase in the number of cars owned by each household and the decrease in automobile occupancy of each car. In this way, because the amount of CO_2 exhausted from automobile traffic is significant, the implementation effects of measures and regulations are anticipated. Under such present situations, calculations of CO_2 emissions for the automobile traffic departments using detailed space scale units as well as considering the implementation effects of measures for the reduction of CO_2 emissions are essential.

From the background described above, regarding cities in Japan, especially central Tokyo, the reduction of CO_2 emissions is urgent both as an environmental issue and a social issue. This study aims to evaluate measures for the reduction of CO_2 emissions from automobile traffic in the 23-ward district of Tokyo (central Tokyo), Japan, using the methods of spatial analysis by Geographic Information Systems (GIS) and statistical analysis.

Related Work

Similar to this study, preceding studies concerning global warming can be categorized into three groups: (1) discernment of actual conditions and future predictions, (2) impact and issues, and (3) measures and effect. The significance of this study will be presented after the introduction of representative studies in recent years on the CO_2 emissions from automobile traffic of the aforementioned three groups.

Regarding studies concerning the discernment of actual conditions and future predictions, Oda [26] made estimation on the amount of CO_2 emissions from automobile traffic in Japan and considered signal control in order to reduce the amount of CO_2 emissions. Moon et al. [24] estimated the amount of air pollutant and CO_2 emissions from automobile traffic according to fuel type. Wang et al. [29] estimated the amount of air pollutant and CO_2 emissions from automobile traffic in China during the period of 2000–2015. Kang et al. [8] estimated the amount of CO_2 reduction concerning 30 provinces in China during 2005–2009. Matsumoto and Iwashita [14] predicted the amount of CO_2 emissions based on the population prediction as well as the characteristics of automobile utilization according to gender, age, and area in Japan.

Regarding studies on impact and issues, Lin [10] estimated the amount of CO_2 emissions that could be reduced by using public transportation in five national parks located in Taiwan. Ando et al. [2] proposed a system for safe driving which can reduce the amount of CO_2 emissions. Ueda et al. [28] grasped the impact of the shift toward online shopping instead of actually going out to shop can have on CO_2 emissions and the city retention time. Matsumoto et al. [13] analyzed the impact ecological driving can have on following automobiles. Morita et al. [25] analyzed the impact introducing low-carbon technology, while taking into consideration individual life schedules, can have on the amount of CO_2 emissions.

Regarding studies concerning measures and effects, Matsuhashi et al. [12] organized measures toward mid- and long-term drastic reduction of CO_2 emissions in traffic department. Hickman et al. [5] evaluated traffic measures for the reduction of CO_2 emissions in London. Ito et al. [7] considered traffic measures to achieve their goal of reducing the amount of CO_2 emissions from regional passenger traffic in Japan by 80% during the period of 2000–2050. Brink [3] considered measures for the reduction of CO_2 emissions from automobile traffic in EU countries. Masaka et al. [11] evaluated low-carbon traffic policies of provincial cities in Japan.

Concerning the many preceding studies listed above, it is evident that versatile surveys and analyses in various research fields were set as challenges. Additionally, among the three groups mentioned above, this study falls under the category of studies concerning measures and effects. However, preceding studies in this field have not evaluated measures for the reduction based on the results of the estimated CO_2 emissions from automobile traffic with detailed space scale units, using each road unit. Therefore, unlike such preceding studies, the significance of this study will be demonstrated through points such as the calculation of the amount of CO_2 emissions from automobile traffic according to each road unit using various open big data, as well as the evaluation of measures for the reduction of CO_2 emissions using GIS and statistical analysis.

Evaluation Method

Framework and Method of Evaluation

In this study, measures for the reduction of CO_2 emissions from automobile traffic will be systematically organized into three categories. Regarding the implementation effects of the measures for the reduction of CO_2 emissions from automobile traffic, initial functions and effect functions will be set in order for road traffic conditions to be taken into consideration when analyses are conducted. These functions use road traffic conditions data, and the coefficients of the functions use administrative data and experimental results as well as estimates. When each datum or research information are updated by means of future technical or research advancements, the previous datum will be replaced with the latest and more precise data. In addition, although evaluation areas are set for the 23-ward district of Tokyo concerning this study, if the same type of data or information is obtainable, evaluation by applying the evaluation method proposed in this study will be also made possible in other metropolitan areas. As stated above, the evaluation method can be expected to obtain temporal and spatial versatility and reproducibility.

Regarding the implementation of the measures for the reduction of CO_2 emissions from automobile traffic, the initial functions are set for the initial stage (2005), and the effect functions are set for the medium stage (2010). Then, the amount of CO_2 reduction by each road unit at these two points will be calculated using the functions, and it will be visualized on the digital maps using the GIS. In addition, an evaluation will be conducted by comparing the amount of CO_2 reduction at these two points and grasping the implementation effect of each measure using statistical analysis. These two points in time are when the road traffic census, which is the main data for this study, was issued, and the 2010 data is the latest information at present. Additionally, 2005 was the year the Kyoto Protocol, which contains information about reducing the average for the 5 years from 2008–2012 by 6% from the results in 1990, went into effect, and the "Kyoto Protocol Target Achievement Plan" [21] was approved by the cabinet office in Japan. Therefore, by comparing the amount of CO_2 reduction at the initial (2005) and medium (2010) implementation stages of the measures for the reduction of CO_2 emissions, the implementation effect of each measures can be seen.

Additionally, the calculation methods of the amount of CO_2 reduction from automobile traffic can be categorized into two types based on the calculation equation and necessary data: the method based on fuel consumption (top-down method) and the method based on traveled kilometers (bottom-up method). According to the Ministry of Economy, Trade and Industry/the Ministry of Land, Infrastructure and Transport [15], as it has become evident that the top-down method using measured values for the amount of fuel consumption has the highest analysis precision, this study will adopt this method.

Furthermore, as it will be mentioned in detail in the next section, using the digital maps that can display road in line forms as the base map, a database, added and integrated with necessary data for the evaluation methods in this study, will be created and used for analyses and evaluation. This will result in calculations of the amount of CO_2 reduction according to each road unit being more accurate than before. Therefore, careful consideration for the characteristics of road traffic conditions as well as the evaluation of measures for the reduction of CO_2 emissions from automobile traffic will be made possible.

Outlines of Evaluation Areas

The area to be evaluated in this study is the 23-ward district of Tokyo (central Tokyo). In Tokyo, as mentioned in section "Introduction", recorded temperatures have risen by approximately 3 °C in the past 100 years. Therefore, the Tokyo Metropolitan Government is aiming to reduce CO_2 emissions by 38% compared to those in 2013 by 2030. This target for Tokyo is considerably higher than the national target of 26%.

The CO_2 emission from automobile traffic is the highest in metropolitan areas where the absolute quantity of traffic demand is large. CO_2 emission per unit travel quantity also tends to be larger in metropolitan areas with heavy traffic congestion. Therefore, measures for the reduction of CO_2 emissions must be immediately implemented, focusing on metropolitan areas with large quantity of CO_2 emission. According to the Ministry of Land, Infrastructure and Transport [16], regarding metropolitan areas, it has become evident that roads especially in central Tokyo have a high amount of CO_2 emission per kilometer of one car. Therefore, in central Tokyo, in order to reduce CO_2 emission from automobile traffic, there is an urgent need for highly effective measures to be implemented.

As a measure for the reduction of CO_2 emissions from automobile traffic, the Automobile Environmental Management Plan System and the Cargo Transportation Evaluation System have been already implemented in Tokyo Metropolis. The purpose of these systems is to promote ecological driving and disseminate low-pollution/fuel-efficient automobiles. In addition, in Tokyo Metropolis, the basic policies for CO_2 emission reduction from automobile traffic are "lowering fuel consumption of each automobile," "controlling automobile traffic demands," and "expanding road traffic capacity," and the measures for the reduction of CO_2 emissions are implemented from these three aspects.

Systemization of Measures for the Reduction of CO_2 Emissions and the Setting of Functions

Systemization of Measures for the Reduction of CO_2 Emissions

An investigation of global warming measures listed in documents of the Ministry of the Environment; the Ministry of Land, Infrastructure, Transport and Tourism; and the Tokyo Metropolitan Government and the extraction of measures for the reduction of CO_2 emissions from automobile traffic which have been implemented or are considering implementation were conducted, and nine measures were found. These measures can be categorized into three groups; A. measures for automobiles only, B. measures for traffic facilitation, and C. measures for road facilities.

A. Measures for Automobiles Only

Because the reliable promotion of single automobile measures is necessary, the advancement of fuel-efficiency improvement through the reinforcement of automobile fuel economy standards is important. The development and diffusion of next-generation automobiles (hybrid, electric, plug-in hybrid, fuel-cell, CNG automobiles, etc.) should be promoted while ascertaining the environmental burden of the lifecycle. More specifically, "No.1: Updating to fuel-efficient automobiles" and "No.2: Promotion of eco-drive systems" are the two points that can be mentioned.

B. Measures for Traffic Facilitation

Because traffic congestion increases CO_2 emission per mileage, smoothing of traffic flows is necessary. As the main method of traffic facilitation measures is the adjustment of the automobile traffic demand, use of public transportation systems must be promoted. More specifically, "No.3: Reduction of automobile traffic," "No.4: Acceleration of utilization of public transportation," and "No.5: Commutation traffic management" can be mentioned.

C. Measures for Road Facilities

It is an effective measure concerning all automobile traffic, and although it is unrelated to automobile type, by means of traffic environment including road expansion and rail laying, the degree of implementation effects is considered to differ. Therefore, targeted roads are limited. More specifically, "No.6: Promotion of intelligent transport systems (ITS)," "No.7: Improvement of the environment for using bicycles," "No.8: Reduction of road construction," and "No.9: Conversion to LED lights" are four points that can be mentioned.

Setting of Functions

For initial functions and effect functions, regarding the "Kyoto Protocol Target Achievement Plan" [21], an equation to calculate the amount of CO_2 reduction will be constituted with three points including "diffusion ratio," "fuel-efficiency change," and "power consumption," which were set as subjects to the implementation effects of measures for the reduction of CO_2 emissions, taken into consideration. Additionally, emission factors of greenhouse gases resulting from fuel consumption, which have been publicly announced by the Ministry of the Environment based on the "Order for Enforcement of the Act on Promotion of Global Warming Countermeasures" (1999), will be used. More specifically, typical CO_2 emission factors per unit fuel consumption amount by fuel type based on the "calorific value per unit fuel consumption amount" and "CO_2 emission factors" refer to system of calculation, report, and publication of greenhouse gas emission (enacted by Ministry of the Environment in 2006, updated annually). Road traffic census (2005 and 2010) and data concerning signal lights by the National Police Agency are primarily used as road traffic condition data.

Configuration of the Equation for the Amount of CO_2 Reduction

The calculation guideline for greenhouse gases reduction amount [20] and the manual of the calculation and report of greenhouse gases reduction [23] are primarily referred to, and the equation for the amount of CO_2 reduction of each measure is shown in Table 4.1. However, for "No.8: Reduction of road construction," the amount of CO_2 increase resulting from traffic congestions will be calculated, and the measure will be evaluated based on this calculation.

Collecting and Processing Data

Data Used for Equations of the Amount of CO_2 Reduction Concerning each Measure

Table 4.2 shows the outlines of data used in equations (Table 4.1) for the amount of CO_2 reduction of each measure. The similar data of amount of CO_2 reduction for the 2005 and 2010 will be used in the calculations. Additionally, measures No.1, 2, 3, and 8 will be categorized to small- and large-size automobiles to calculate the amount of CO_2 reduction.

Table 4.1 Equation for each measures for the reduction of CO_2 emissions

No.	Equation
1	Amount of CO_2 reduction [t-CO_2] = diffusion rate of fuel-efficient automobiles [%] × number of running automobiles according to automobile type [automobiles] × mileage [km]/(fuel efficiency of automobiles to be updated [km/l] – fuel efficiency of fuel-efficient automobiles [km/l]) × CO_2 emission factor per fuel consumption [kg-CO_2/kWh]
2	Amount of CO_2 reduction [t-CO_2] = diffusion rate of ecological driving [%] × number of running automobiles according to automobile type [automobiles] × mileage [km]/fuel efficiency [km/l] × reduction effect [%] × CO_2 emission factor per fuel consumption [kg-CO_2/kWh]
3	Amount of CO_2 reduction [t-CO_2] = annual reduction of automobile travel [km] × active automobiles [automobiles]/fuel efficiency [km/l] × CO_2 emission factor per fuel consumption [kg-CO_2/kWh]
4	Amount of CO_2 reduction [t-CO2] = annual number of CO_2 reduction – active passenger automobiles [automobiles] × mileage [km]/fuel efficiency [km/l] × CO_2 emission factor per fuel consumption [kg-CO_2/kWh]
5	Amount of CO_2 reduction [t-CO_2] = annual number of passenger automobiles on work days [automobiles] × shift rate from passenger automobiles to public transportation [%] × number of passengers in automobile [people] × mileage [km]/fuel efficiency [km/l] × CO_2 emission factor per fuel consumption [kg-CO_2/kWh]
6	VICS diffusion promotion: Amount of CO_2 reduction [t-CO_2] = number of automobiles introduced to VICS [automobiles] × mileage [km] × primary unit of CO_2 reduction [kg-CO_2] ETC utilization promotion: Amount of CO_2 reduction [t-CO_2] = (primary unit of CO_2 reduction of non-ETC automobiles [kg-CO_2] – primary unit of CO_2 reduction of ETC automobiles [kg-CO2]) × number of running automobiles according to automobile type [automobiles] × ETC introduction rate to automobiles [%] × road extension [km]
7	Amount of CO_2 reduction [t-CO_2] = active number of automobiles with trip distance under 5 km [automobiles] × mileage [km]/fuel efficiency [km/l] × shift rate toward bicycle use [%] × CO_2 emission factor per mileage [kg-CO_2/kWh]
8	Amount of CO_2 increase [t-CO_2] = number of running automobiles according to automobile type [automobiles] × mileage [km/fuel consumption [kl] × CO_2 emission factor per fuel consumption during traffic congestions [kg-CO_2/kWh] × congestion hours [hours] × CO_2 emission factor per fuel consumption [kg-CO_2/kWh] – number of active automobiles [automobiles] × mileage [km]/fuel consumption [kl] × congestion hours [hour] × CO_2 emission factor per fuel consumption [kg-CO_2/kWh]
9	Amount of CO_2 reduction [t-CO_2] = (power consumption of electric bulb type signals [w] – power consumption of LED type signals [w]) × number of signal lights [lights] × annual utilization time [hours] × CO_2 emission factor concerning signal lights [kg-CO_2/kWh]

Table 4.2 Outlines of data used for calculating the amount of CO_2 reduction concerning each measure

No.	Data	Source
1	Diffusion rate of fuel-efficient automobiles	Ministry of Land, Infrastructure, Transport and Tourism [19] automobile fuel-efficiency list
	Number of running automobiles according to automobile type	Road traffic census
	Mileage	
	Fuel efficiency of automobiles to be updated	Ministry of Land, Infrastructure, Transport and Tourism [19] automobile fuel-efficiency list
	Fuel efficiency of fuel-efficient automobiles	
	CO_2 emission factor per fuel consumption	Ministry of the Environment [20] guideline on greenhouse gas calculation methods
2	Diffusion rate of ecological driving	2005: Ministry of Land, Infrastructure and Transport [17] internet monitoring questionnaire survey, 2010: Sony assurance Inc. [27] surveys concerning ecological drivers
	Number of running automobiles according to automobile type	Road traffic census
	Mileage	
	Fuel efficiency	Ministry of Land, Infrastructure, Transport and Tourism [19] automobile fuel-efficiency list
	Reduction effect	Ministry of Economy, Trade and Industry/Ministry of Land, Infrastructure and Transport [15] calculation method guideline on CO_2 emission in the distribution field
	CO_2 emission factor per fuel consumption	Ministry of the Environment [20] guideline on greenhouse gas calculation methods
3	Annual reduction of automobile travel	Road traffic census
	Active automobiles	Ministry of the Environment [20] guideline on greenhouse gas calculation methods
	Fuel efficiency	Ministry of Land, Infrastructure, Transport and Tourism [19] automobile fuel-efficiency list
	CO_2 emission factor per fuel consumption	Ministry of the Environment [20] guideline on greenhouse gas calculation methods

(continued)

Table 4.2 (continued)

No.	Data	Source
4	Annual number of CO_2 reduction-active passenger automobiles	Ministry of the Environment [20] guideline on greenhouse gas calculation methods
	Mileage	Road traffic census
	Fuel efficiency	Ministry of Land, Infrastructure, Transport and Tourism [19] automobile fuel-efficiency list
	CO_2 emission factor per fuel consumption	Ministry of the Environment [20] guideline on greenhouse gas calculation methods
5	Annual number of passenger automobiles on work days	Ministry of the Environment [20] guideline on greenhouse gas calculation methods
	Shift rate from passenger automobiles to public transportation	
	Number of passengers in automobile	Kato [9]
	Mileage	Road traffic census
	Fuel efficiency	Ministry of Land, Infrastructure, Transport and Tourism [19] automobile fuel-efficiency list
	CO_2 emission factor per fuel consumption	Ministry of the Environment [20] guideline on greenhouse gas calculation methods
6	VICS diffusion promotion	
	Number of automobiles introduced to VICS	Survey by the automobile information and communication center "number of shipments for automobiles equipped with VICS"
	Mileage	Road traffic census
	Primary unit of CO_2 reduction	Ministry of the Environment [20] guideline on greenhouse gas calculation methods
	ETC utilization promotion	
	Primary unit of CO_2 reduction of non-ETC automobiles	Abe et al. [1]
	Primary unit of CO_2 reduction of ETC automobiles	
	Number of running automobiles according to automobile type	Road traffic census
	ETC introduction rate to automobiles	Survey by the Ministry of Land, Infrastructure, Transport and Tourism "utilization situation of ETC"
	Road extension	Abe et al. [1]

7	Active number of automobiles with trip distance under 5 km	Ministry of the Environment [20] guideline on greenhouse gas calculation methods
	Mileage	Road traffic census
	Fuel efficiency	Ministry of Land, Infrastructure, Transport and Tourism [19] automobile fuel-efficiency list
	Shift rate toward bicycle use	Ministry of the Environment [20] guideline on greenhouse gas calculation methods
	CO_2 emission factor per mileage	
8	Number of running automobiles according to automobile type	Road traffic census
	Mileage	
	Fuel consumption	Ministry of Land, Infrastructure, Transport and Tourism [19] automobile fuel-efficiency list
	CO_2 emission factor per fuel consumption during traffic congestions	Dohi et al. [4]
	Congestion hours	Road traffic census
	CO_2 emission factor per fuel consumption	Ministry of the Environment [20] guideline on greenhouse gas calculation methods
	Number of active automobiles	
9	Power consumption of electric bulb type signals	Ministry of the Environment [20] guideline on greenhouse gas calculation methods
	Power consumption of LED type signals	
	Number of signal lights	National Police Agency Survey "number of repaired traffic lights according to each prefecture"
	Annual utilization hours	Ministry of the Environment [20] guideline on greenhouse gas calculation methods
	CO_2 emission factor concerning signal lights	

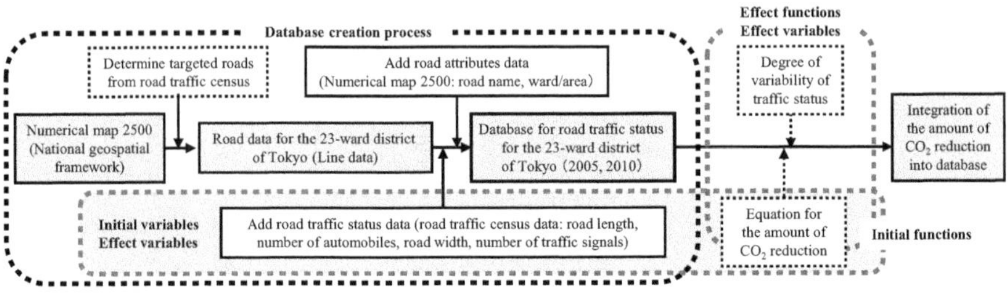

Fig. 4.1 Road database creation/usage process using GIS

Fig. 4.2 Major road network in the 23-ward district of Tokyo

Creation of Road Database Using GIS

Figure 4.1 shows road database creation/usage process using GIS. According to Fig. 4.1, first, the road database in digital map format, which is made up of 42,461 lines, will be created with the numerical map 2500 (national geospatial framework) as the base map. Next, road database will be created by extracting only the lines of the main roads (21 lines of expressways and 74 lines of national highways), which are subjects of the survey of the road traffic census, and integrating and adding each data in Table 4.2 as well as the amount of CO_2 reduction for 2005 and 2010 to each road. Figure 4.2 shows the major road network which consists of expressways, national highways, and principal prefectural roads in the 23-ward district of Tokyo. In the next section, using this road database, the amount of CO_2 reduction of each expressway and national highway unit for 2005 and 2010 will be calculated and visualized on the digital maps using GIS.

Calculation and Visualization of the Amount of CO_2 Reduction

Using both the initial functions and effect functions, the amount of CO_2 reduction of each road unit of each measure for 2005 and 2010 was calculated. As there are many targeted roads for this study, expressways with the high amount of CO_2 reduction were extracted and shown in Table 4.3a, 4.3b and Fig. 4.3. More specifically, the amount of CO_2 reduction at the two points is shown in Table 4.3a, 4.3b, and the results of "No.3: Reduction of automobile traffic" for large-size automobiles and "No.4: Acceleration of utilization of public transportation" are visualized on the digital maps using GIS in Fig. 4.3. These two measures belong to "B. Measures for traffic facilitation." Additionally, "No.7: Improvement of the environment for using bicycles" and "No.9: Conversion to LED lights" for main roads other than expressways are not shown in Table 4.3a, 4.3b or Fig. 4.3.

From Table 4.3a, 4.3b, concerning "No.3: Reduction of automobile traffic" and "No.4: Acceleration of utilization of public transportation" for both 2005 and 2010, it was made evident that the amount of CO_2 reduction for each expressway was higher than any other measures. Concerning "No.3: Reduction of automobile traffic," the amount of CO_2 reduction for most expressways was lower in 2010 than that in 2005, because the number of active automobiles for this measure just on most expressways decreased during the period of 2005–2010.

Additionally, when comparing the amount of CO_2 reduction for 2005 and 2010, the difference was varied according to each measure and road. However, in the case of "No.2: Promotion of eco-drive systems" for small-size automobiles, for most targeted roads, the amount of CO_2 reduction for 2010 was lower than that of 2005.

Table 4.3a Amount of CO_2 reduction for expressways in 2005 (t)

Road number	Road name	Road length [km]	No.1		No.2		No.3		No.4	No.5	No.6		No.8	
			Small-size automobiles	Large-size automobiles	Small-size automobiles	Large-size automobiles	Small-size automobiles	Large-size automobiles			ETS utilization promotion	VICS diffusion promotion	Small-size automobiles	Large-size automobiles
1	Kanetsu expressway	0.8	0.2	0.1	0.1	0.1	1388.7	761.4	111. 5	32.0	1.6	9.8	0.0	0.0
2	Inner circular route 1	3.9	2.5	2.0	1.5	1.1	15,785.6	10,695.6	11,345.0	363.7	3.9	116.8	377.7	76.8
3	Inner circular route 2	3.8	2.8	3.1	1.8	1.7	18,416.9	17,114.5	13,075.0	984.4	5.1	147.7	430.8	120.1
4	Inner circular route 3	3.7	2.9	2.1	1.8	1.1	18,859.3	11,334.3	13,629.9	1338.3	4.8	136.0	693.8	125.1
5	Inner circular route 4	2.2	2.0	1.1	1.2	0.6	12,435.3	6055.5	9564.5	1020.0	5.2	86.2	479.7	70.1
6	Expressway route 1 Haneda sen	12.6	4.0	2.5	2.7	1.3	27,363.8	13,524.2	17,869.3	951.3	2.0	190.1	0.0	0.0
7	Expressway route 1 Ueno sen	4.3	0.7	0.3	0.5	0.2	4831.0	1554.5	3217.4	492.1	1.0	31.5	0.0	0.0
8	Expressway route 1 Meguro sen	5.7	2.1	0.6	1.3	0.3	13,088.8	3008.9	9752.9	1005.8	1.9	82.4	0.0	0.0
9	Expressway route 3 Shibuya sen	11.6	6.8	8.6	4.3	4.6	43,641.6	46,935.2	31,220.2	1647.9	4.2	365.7	0.0	0.0
10	Expressway route 4 Shinjuku sen	13.4	9.2	5.5	5.9	3.0	59,857.2	30,298.6	111.5	1956.6	4.1	417.5	0.0	0.0

11	Expressway route 5 Ikebukuro sen 1	9.0	6.4	6.7	3.9	3.6	39,846.8	36,806.9	30,179.5	1939.3	4.7	319.0	711.6	197.2
12	Expressway route 5 Ikebukuro sen 2	9.4	5.1	5.3	3.4	2.9	35,026.8	29,113.7	22,704.0	1632.2	3.8	272.4	838.1	209.0
13	Expressway bay coastline 1	10.9	6.7	12.2	4.4	6.5	45,319.3	66,772.1	30,117.4	1.821.2	5.1	424.2	1108.7	490.1
14	Expressway bay coastline 2	8.2	2.1	5.6	1.4	3.0	14,337.7	30,450.0	9624.9	765.9	2.5	157.2	0.0	0.0
15	Expressway route 6 Mukoujima sen	9.6	4.3	5.5	2.8	3.0	28,308.7	30,368.5	19,425.6	1291.7	3.3	237.0	738.1	237.5
16	Expressway route 7 Komatsugawa sen	10.4	4.4	2.3	2.7	1.2	27,667.7	12,505.5	20,708.5	1165.3	2.4	189.3	0.0	0.0
17	Expressway route 9 Fukagawa sen	5.3	1.7	2.2	1.1	1.2	11,413.6	12,178.5	7930.3	943.3	2.4	95.4	0.0	0.0
18	Expressway route 11 Daiba sen	4.4	2.0	1.8	1.2	1.0	12,460.1	9942.5	9383.1	1240.4	2.9	95.9	0.0	0.0
19	Central circular route 1	12.1	4.7	13.7	3.3	7.4	33,263.1	75,163.6	20,729.8	1204.1	4.1	375.8	1235.5	837.5
20	Central circular route 2	6.8	3.2	6.0	2.1	3.2	21,881.7	32,590.8	14,089.7	1409.5	4.0	205.7	605.7	270.6
21	Central circular route 3	5.5	1.7	1.8	1.1	1.0	10,778.6	9991.2	30,179.5	858.4	2.1	86.4	711.6	0.0

Table 4.3b Amount of CO_2 reduction for expressways in 2010 (t)

			No.1		No.2		No.3				No.6		No.8	
Road number	Road name	Road length [km]	Small-size automobiles	Large-size automobiles	Small-size automobiles	Large-size automobiles	Small-size automobiles	Large-size automobiles	No.4	No.5	ETS utilization promotion	VICS diffusion promotion	Small-size automobiles	Large-size automobiles
1	Kanetsu expressway	0.8	16.8	11.8	0.1	0.1	1143.7	600.8	84.5	21.1	2.8	18.6	0.0	0.0
2	Inner circular route 1	3.9	249.0	133.2	1.8	0.7	3915.2	1566.3	11,284.1	1016.3	8.2	263.5	0.0	0.0
3	Inner circular route 2	3.8	249.5	117.3	1.8	0.7	6708.4	2358.6	11,303.2	391.6	8.3	259.1	674.3	237.1
4	Inner circular route 3	3.7	249.7	132.6	1.9	0.7	3218.4	1278.7	11,314.3	391.9	8.7	264.0	700.1	278.2
5	Inner circular route 4	2.2	163.0	84.7	1.2	0.5	7067.1	2745.2	7386.2	255.9	9.5	171.7	473.5	183.9
6	Expressway route 1 Haneda sen	12.6	451.3	203.4	3.3	1.1	2684.2	904.7	20,449.2	1068.9	4.5	466.1	0.0	0.0
7	Expressway route 1 Ueno sen	4.3	54.2	13.4	0.4	0.1	1202.1	222.7	2455.7	85.1	1.5	52.7	0.0	0.0
8	Expressway route 1 Meguro sen	5.7	169.4	26.8	1.3	0.2	3238.8	383.3	7674.0	407.4	3.4	160.0	0.0	0.0
9	Expressway route 3 Shibuya sen	11.6	652.7	605.0	4.8	3.4	19,550.3	13,556.3	29,575.2	2376.8	8.1	767.9	0.0	0.0
10	Expressway route 4 Shinjuku sen	13.4	858.4	382.4	6.4	2.1	10,910.5	3635.7	38,895.3	3125.9	8.1	885.3	2104.0	701.1

11	Expressway route 5 Ikebukuro sen 1	9.0	16.8	297.4	3.7	1.6	4895.9	2189.2	22,543.3	1296.3	7.2	535.9	0.0	0.0
12	Expressway route 5 Ikebukuro sen 2	9.4	249.0	397.1	4.0	2.2	11,302.4	6243.1	24,366.1	2599.7	7.8	602.1	0.0	0.0
13	Expressway bay coastline 1	10.9	249.5	991.2	4.6	5.5	2714.6	3243.9	28,113.6	2259.4	9.5	855.4	1364.6	1630.6
14	Expressway bay coastline 2	8.2	249.7	422.0	1.7	2.3	3578.7	4982.8	10,272.3	1049.8	4.9	330.6	0.0	0.0
15	Expressway route 6 Mukoujima sen	9.6	163.0	347.5	2.8	1.9	3809.6	2582.4	17,374.3	1369.5	5.7	448.7	0.0	0.0
16	Expressway route 7 Komatsugawa sen	10.4	451.3	140.2	3.0	0.8	11,821.2	3023.6	18,583.6	1493.5	4.8	410.2	0.0	0.0
17	Expressway route 9 Fukagawa sen	5.3	54.2	177.5	1.2	1.0	2446.8	2047.8	7187.6	249.0	4.5	195.8	0.0	0.0
18	Expressway route 11 Daiba sen	4.4	169.4	96.2	1.0	0.5	4744.3	2451.7	6310.8	218.6	4.3	153.9	0.0	0.0
19	Central circular route 1	12.1	652.7	874.4	3.3	4.8	5505.9	8102.2	20,140.8	2148.9	6.7	662.4	1201.5	1768.0
20	Central circular route 2	6.8	858.4	421.3	2.4	2.3	6620.0	6314.8	14,970.0	1597.2	7.6	423.4	859.7	820.1
21	Central circular route 3	5.5	16.8	228.6	1.4	1.3	8552.6	7838.4	8455.7	1042.8	5.2	236.3	0.0	0.0

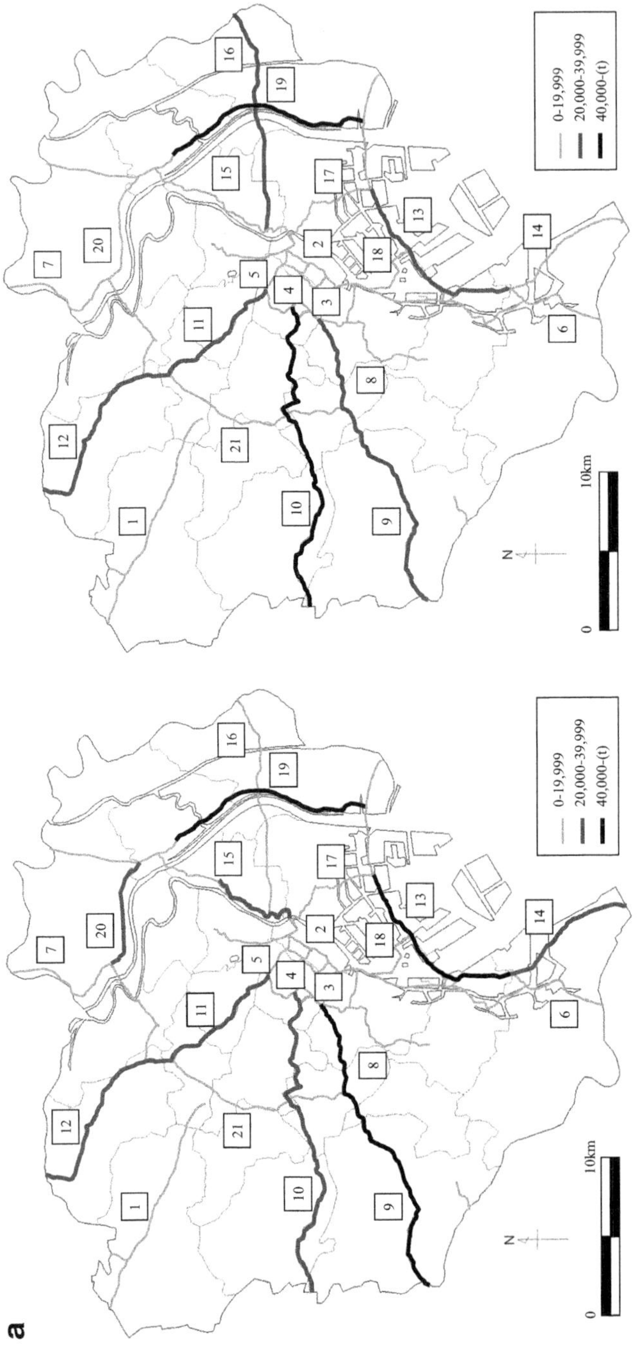

Fig. 4.3 (**a**) Visualization of the amount of CO_2 reduction for expressways in 2005 (*left side:* No.3 for large-size automobiles, *right side:* No.4)

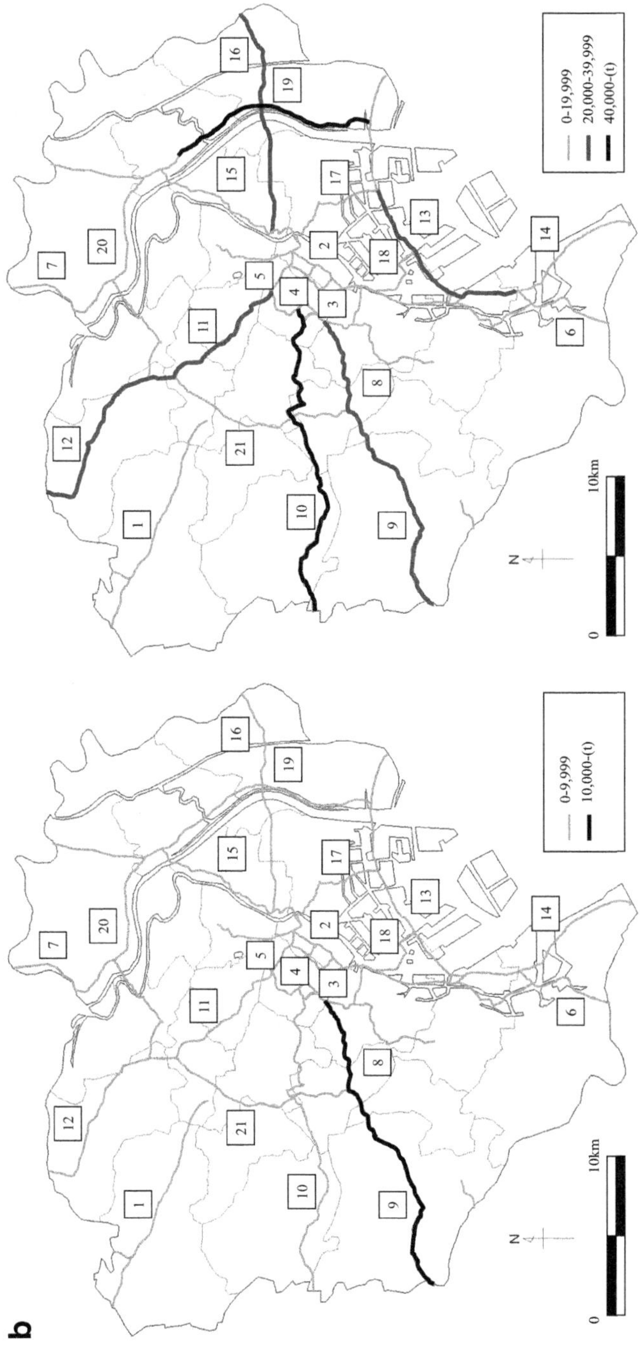

Fig. 4.3 (**b**) Visualization of the amount of CO_2 reduction for expressways in 2010 (*left side*: No.3 for large-size automobiles, *right side*: No.4)

Evaluation of Measures for the Reduction of CO_2 Emissions

Outlines of Evaluation

In order to evaluate the measures for the reduction of CO_2 emissions, the amount of CO_2 reduction calculated according to each main road line (21 lines of expressways and 74 lines of national highways) for the initial stage (2005) and medium stage (2010) of measures implementation will be compared, and evaluation will be conducted by grasping the implementation effect of each measure using statistical analysis. In this study, in order to grasp the implementation effects of measures for the reduction of CO_2 emissions concerning the period between 2005 and 2010, a t-test will be conducted if the R language can be used. R is a programming language of an open-source free-software for statistical analyses.

Evaluation

Table 4.4 shows whether the implementation of each measure for the reduction of CO_2 emissions was effective or not based on the t-test results. As shown in Table 4.4, other than the "No.2: Promotion of eco-drive systems" for small-size automobiles, all were $P < 0.1$, and it was evident that there was a significant difference between the initial and medium stages of measure implementation for the amount of CO_2 reduction. In short, besides the measure mentioned above, it was evident that the implementation of measures for the reduction of CO_2 emissions was effective. Regarding the measure mentioned above, the average difference of the amount of CO_2 reduction at the initial and medium measure implementation stage was −0.018, and the absolute value was 0.018. Therefore, as there is no difference between the amounts of CO_2 reduction at these two points, it can be said that the implementation for the above measure was ineffective. One point that can be given as a reason is, as mentioned in the previous section, the decrease in the amount of CO_2 reduction by the above measure in most targeted roads between 2005 and 2010. Additionally, there was no major increase in the diffusion rate of small-size automobiles with the results being 71.9% in 2005 and 79.8% in 2010. The fuel consumption was 15 km/l for small-size automobiles and 6 km/l for large automobiles, so it is evident that CO_2 emission per car for the former was little to begin with.

Conclusions

This study focuses on global warming which is one of the most critical environmental issues with the purpose of proposing a method for the evaluation of measures for the reduction of CO_2 emissions from automobile traffic. More specifically, with the

Table 4.4 The effectiveness of the implementation of each measures for the reduction of CO_2 emissions

Measure		Implementation effectiveness	p value
No.1: Updating to fuel-efficient automobiles			
	Large-size automobiles	Effective	1.0e-10
	Small-size automobiles	Effective	2.4e-07
No.2: Promotion of eco-drive systems			
	Large-size automobiles	Ineffective	0.28
	Small-size automobiles	Effective	4.3e-05
No.3: Reduction of automobile traffic			
	Large-size automobiles	Effective	7.1e-07
	Small-size automobiles	Effective	8.1e-05
	No.4: Acceleration of utilization of public transportation	Effective	1.2e-11
	No.5: Commutation traffic management	Effective	2.2e-16
No.6: Promotion of intelligent transport systems (ITS)			
	Large-size automobiles	Effective	2.2e-16
	Small-size automobiles	Effective	2.9e-09
	No.7: Improvement of the environment for using bicycles	Effective	1.2e-07
No.8: Reduction of road construction			
	Large-size automobiles	Effective	2.0e-02
	Small-size automobiles	Effective	1.9e-03
No.9: Conversion to LED lights		Effective	2.2e-16

23-ward district of Tokyo, which has the highest amount of CO_2 emission in Japan, as the evaluation area, and after the measures for the reduction of CO_2 emissions from automobile traffic were systemized, the initial functions of the initial measure implementation stage (2005) and the effect functions of the medium stage (2010) were set. Then, the amount of CO_2 reduction at these two points was calculated according to each road unit by means of the functions, and it was visualized on the digital maps using GIS. Further, evaluation was conducted by comparing the amount of CO_2 reduction of the two points and grasping the implementation effect of each measure using statistical analysis. According to the evaluation results, excluding "No.2: Promotion of eco-drive systems" for small-size automobiles, all measure implementations were effective.

In this study, limiting the evaluation area to the 23-ward district of Tokyo as mentioned in section "Framework and Method of Evaluation", the evaluation of measures for the reduction of CO_2 emissions using open big data with generally high versatility was conducted. Such data is provided to the public and is relatively easy to obtain by not only specialists and researchers but also by the general public. Therefore, if various data is obtainable by the same conditions as for this study, evaluation methods of this study can be applied in other metropolitan areas. In the same way, with functions used to calculate the amount of CO_2 reduction of each measure, by updating the observation data and technology data, the evaluation of measures for the reduction of CO_2 emissions can be continuously conducted. Therefore, the evaluation method of this study can be applied to other areas with the same standards as in this study from the spatial aspect, and continuous improvement of evaluation precision can be expected with information updates from the temporal aspect.

Regarding future research challenges, first, a more detailed and specific evaluation by utilizing a more detailed road traffic condition data and updating information according to the technological advancement for measures for the reduction of CO_2 emissions must be considered. Second, as in the case of the 23-ward district of Tokyo, in other metropolitan areas with high CO_2 emission, evaluation by applying the evaluation methods proposed in this study must also be conducted.

References

1. Abe, S., Nonaka, Y., Katsuragi, M., Ushiyama, H.: CO_2 emission effect by adopting ETC. Proc. Ann. Confer. Jap. Soc. Civil Eng. **60**(4), 613–614 (2005)
2. Ando, R., Nishihori, Y., Ouchi, D.: Development of a system to promote eco-driving and safe-driving. In: Balandin, S., Dunaytsev, R., Koucheryavy, Y. (eds.) Smart Spaces and Next Generation Wired/Wireless Networking, 6294 of the Series Lecture Notes in Computer Science, pp. 207–218. Springer, Berlin Heidelberg (2010)
3. Brink, P.t.: Mitigating CO_2 emissions from cars in the EU (regulation (EC) no 443/2009). In: Pallemaerts, M. (ed.) The new Climate Policies of the European Union: Internal Legislation and Climate Diplomacy, pp. 179–201. Asp-Academic & Scientific Pub, Brussels (2012)
4. Dohi, M., Sone, S., Takimoto, M., Ogawa, T., Namikawa, Y.: Grounds for the calculation of motor vehicle emission factors using environmental impact assessment of road project etc. (Revision of FY 2010). Technical Note of National Institute for Land and Infrastructure Management, 671, p. 289 (2012)
5. Hickmana, R., Ashirub, O., Banister, D.: Transport and climate change: simulating the options for carbon reduction in London. Transp. Policy. **17**(2), 110–125 (2010)
6. Intergovernmental Panel on Climate Change: The 5th Assessment Report. p. 58 (2013)
7. Ito, K., Kato, H., Shibahara, N.: Estimating the provision length of minimum LC-CO_2 mode in trunk lines for large reduction of CO_2 emission from regional passenger transport in Japan. Infrastruct. Plan. Rev. **67**(5), 45–67 (2011)
8. Kang, J., Zhao, T., Ren, X., Lin, T.: Using decomposition analysis to evaluate the performance of China's 30 provinces in CO_2 emission reductions over 2005-2009. Nat. Hazard. **64**(2), 999–1013 (2012)
9. Kato, K.: A study on the regulated and actual seating capacity of automobiles. In: Proceedings of the 57th Annual Conference of JSSD, pp. 57–59 (2010)
10. Lin, T.P.: Carbon dioxide emissions from transport in Taiwan's national parks. Tour. Manag. **31**(2), 285–290 (2010)
11. Masaka, M., Kato, K., Kondo, M., Okushima, M.: Transport policy toward low-carbon society focusing on health aspect in provincial city. Infrastruct. Plan. Rev. **68**(4), 400–411 (2012)
12. Matsuhashi, K., Kudo, Y., Moriguchi, Y.: Measures for mid and long-term drastic reduction of CO_2 emissions from automobile traffic. Global Environ. Res. **12**, 179–189 (2007)
13. Matsumoto, S., Tozawa, T., Hiraoka, T., Yamabe, S., Kawashima, H.: Fundamental study on effect of eco-driving behavior of preceding vehicle on following vehicle. Infrastruct.Plan. Rev. **67**(5), 1024–1028 (2011)

14. Matsumoto, T., Iwashita, T.: Future estimation of CO_2 emission reductions from automobiles based on high-resolution population projection and characteristics of vehicle use. Environ. Sci. **26**(6), 562–570 (2013)
15. Ministry of Economy, Trade and Industry/Ministry of Land, Infrastructure and Transport: Calculation method guideline on CO_2 emission in the distribution field. p. 6 (2007)
16. Ministry of Land, Infrastructure and Transport: Present situation of CO_2 emission from road traffic. p. 26 (2005)
17. Ministry of Land, Infrastructure and Transport: Internet monitoring questionnaire survey. http://www.mlit.go.jp/kisha/kisha06/00/000814_.html (2006) Accessed 22 Jan 2016
18. Ministry of Land, Infrastructure, Transport and Tourism: CO_2 emission from transport sector. http://www.mlit.go.jp/sogoseisaku/environment/sosei_environment_tk_000007.html (2016a) Accessed 15 Apr 2016
19. Ministry of Land, Infrastructure, Transport and Tourism: Automobile fuel efficiency list. http://www.mlit.go.jp/jidosha/jidosha_fr10_000019.html (2016b) Accessed 22 Jan 2016
20. Ministry of the Environment: Guideline on greenhouse gas calculation methods. p. 108 (2003)
21. Ministry of the Environment: Kyoto Protocol target achievement plan. p. 132 (2005)
22. Ministry of the Environment: Guidelines for measures to mitigate heat island (revised version in FY 2013). http://www.env.go.jp/air/life/heat_island/guideline/h24.html (2013) Accessed 22 Jan 2016
23. Ministry of the Environment: Manual of the calculation and report of greenhouse gas emission (II) Calculation method of greenhouse gas emission. p. 234 (2016)
24. Moona, G., Leeb, Y., Choib, K., Jeong, D.: Emission characteristics of diesel, gas to liquid, and biodiesel-blended fuels in a diesel engine for passenger cars. Fuel. **89**(12), 3840–3846 (2010)
25. Morita, H., Kaneoka, Y., Kato, H., Shibahara, N., Hayashi, Y.: Effect on CO_2 emissions of low-carbon-technology considering residents' activity schedule. Proc. Environ. Sanitary Eng. Res. **69**(5), 33–43 (2013)
26. Oda, T.: Estimation of CO_2 emissions from automobiles and its reduction with traffic signal control. J Inst. Electric. Eng. Jap. **130**(9), 600–603 (2010)
27. Sony Assurance Inc: Surveys Concerning Ecological Drivers. p. 15 (2010)
28. Ueda, T., Yamamuro, H., Taniguchi, M.: Effects on automobile CO_2 emissions and duration behavior from emerging cyberspace the case of shopping activity. J. City Plan. Inst. Jap. **46**(3), 763–768 (2011)
29. Wang, H., Fu, L., Bi, J.: CO_2 and pollutant emissions from passenger cars in China. Energy Policy. **39**(5), 3005–3011 (2011)

Chapter 5
Using Social Media Data to Infer Urban Attitudes About Bicycling: An Exploratory Case Study of Washington DC

Justin B. Hollander and Yaqi Shen

Abstract Biking as a travel mode has become more and more common and popular recently. However, some problems occured in the development of cycling. This chapter explores the use of microblog data in the form of sentiment analysis and statistical analysis to determine if relationships exist between how bikeable a place is and talking about on the microblog Twitter. The results demonstrate that there is relationship between peoples' attitudes, bicycling facilities, and physical environment factors. We also provide suggestions about some good strategies of developing cycling for bicycling planners and policymakers by using the results indicated in this study.

Keywords Urban attitudes • Bicycling • Social media • Sentiment analysis • Statistical analysis

Introduction

Recently, nonmotorized modes of travel, particularly bicycling, have become more common and popular. Bicycle use increased 39 percent in the U.S. from 2001 to 2011, because riding a bike saves on the cost of living in urban areas, and, also, bikes are available readily and are convenient to ride anywhere and anytime compared to public transportation [3]. Cities such as San Francisco, Chicago, and Washington DC have seen huge increases in cycling [3]; however, cycling plans in urban areas have some problems, and the development of such plans has not kept pace with the massive rise in cyclists.

Through a case study in Washington, DC, this chapter evaluates the ways in which bicycling facilities and city cycling plans influence people's attitudes and behaviors. The study includes a discussion of the Bicycling Score in Washington,

J.B. Hollander (✉) • Y. Shen
Tufts University, Department of Urban and Environmental Planning & Policy, Medford, MA, USA
e-mail: justin.Hollander@tufts.edu

A. Karakitsiou et al. (eds.), *City Networks*, Springer Optimization and Its Applications 128, DOI 10.1007/978-3-319-65338-9_5

an analysis of people's attitudes from Twitter data collected from March 3 to April 10, 2015, and an analysis of the relationship between cyclists' attitudes, cycling facilities, and city bike plans.

Finally, the chapter provides a detailed discussion and evaluation of the current research on cyclists' attitudes about bicycling.

The following research questions were explored:

1. How is bicycling ability represented in Washington DC?
   - What is the Bicycling Score in Washington, DC, and how is the index used?
   - What factors enter into the index, such as cycling infrastructures, hilliness, desirable amenities, and road connectivity?
2. What are people's attitudes (positive or negative) about bicycling in Washington?
3. Is there a relationship between people's attitudes, the bicycling index (cycling infrastructures, hilliness, desirable amenities, and road connectivity), and locations where people talk about bicycling?

The chapter also discusses the implications of these research questions with respect to bicyclists, bicycling facilities, and bicycling planning within the Washington DC area. We will also draw conclusions from the Washington data for application to the rest of the USA and provide some useful guidelines for future planning about bicycling based on human and social considerations.

Twitter data collected from March 3 to April 10, 2015 provide the most recent data and thus represent people's attitudes well. Current attitudes about bicycling contributed to the analysis of the research questions and identified problems better.

To answer the research questions above, the following software was used: ArcMap, SentiStrength, and STATA. ArcMap was used to produce maps of current bikeability in Washington DC and helped calculate the Bicycling Score there as well. In addition, the data created by ArcMap contributed to other statistical analyses. SentiStrength was used to estimate the sentiment score for each Twitter post, while STATA provided the ability to conduct statistical analyses of the sentiment scores and other biking data calculated by ArcMap. The methods with which these software programs were used are described in detail in the next sections.

We begin with a brief literature review that discusses several case studies of people's attitudes about bicycling in urban areas and the ways in which bikeability affects those attitudes. Next we offer an overview of current bikeability in Washington DC, followed by the methodologies employed in the chapter for data collection, determination of sentiment scores, and the relationship between sentiment scores and factors associated with biking. We conclude with a review of the limitation of the study and offer ideas for future relevant studies.

Literature Review

Currently, because of the increasing benefits associated with bicycling [6], more US municipalities are beginning to promote and develop policies and plan to make bicycling a safer and easier form of transport. Transportation planners have begun to realize that public participation in planning bicycle projects is an important part of the process, specifically with respect to the demand for bicycle facilities, network planning, suitability modeling, and other related issues [13]. In the next few sections, we will argue that data collected from Twitter can be an effective method of assessing public attitudes. Using social networks is helpful in tracking people's attitudes on many objects and acquiring feedback and suggestions. An analysis of Twitter data specifically could contribute to future bicycle planning.

Empirical Study of People's Travel Attitudes and Urban Design Features

There is a consensus within planning and urban design policy that the design of a sustainable urban environment would encourage people to reduce the use of automobiles and choose more sustainable modes of travel. However, based on one case study in the UK, sustainable features of urban design did not cause people to change their behavior, although such features may change their attitudes about walking and bicycling in the long-term [19]. Some sustainable features of urban design relevant to people's attitudes and behaviors, including secured bike storage, high connectivity of neighborhoods to nearby areas, natural surveillance, high-quality public realm, and issues of heavy traffic, will be discussed.

Facts from Survey of People's Attitudes and Behaviors About Bicycling

Data from the 2012 National Survey of Bicyclist and Pedestrian Attitudes and Behavior are worth mentioning: 46% of 7509 respondents have bicycle paths available within a quarter mile of where they live, and 39% have bicycle lanes available on roads within a quarter mile of where they live [17]. Among the 1350 respondents who do not have bicycle paths nearby, 40% never ride on bike paths. Further, among the 1176 people who do have bicycle paths nearby, only 12% of them always ride on bicycle paths, and another 11% never use these paths [17]. Moreover, the same report indicated that 38% of 1551 respondents ride a bike more often by comparison to a year ago, while 20% ride a bike less often than in 2011 [17]. These facts and case studies show that, although people's bicycling behaviors are changing, there still is no clear pattern.

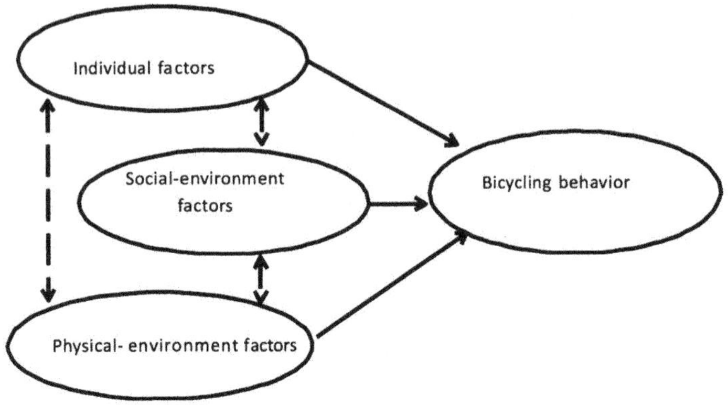

Fig. 5.1 Conceptual model [11]

Conceptual Model from Previous Case Studies in the USA

There is a conceptual framework that is used widely in research on physical activity, specifically in the field of public health [16]. This framework is composed of three factors—individual, social-environmental, and physical-environmental—that are used to explain people's behavior and attitudes [11]. Individual factors (e.g., attitudes) about the physical- environmental factors in transportation infrastructures and land-use patterns both offer strong support for the arguments in the next sections in this chapter.

Figure 5.1 shows that these three factors are hypothesized to affect bicycling behaviors directly [11]. Individual factors contribute to the motivation to bicycle, and social and physical environments are key factors in determining the quality of bicycling conditions that may enable and encourage people to bicycle [9, 10]. These factors are interrelated, and all influence bicycling behaviors. For example, bicycle infrastructure, one of the physical-environmental factors, influences an individual's attitudes and behaviors with respect to bicycling; because bicycle infrastructure affects travel time, safety, enjoyment, and other related bicycling experiences, it has an important influence on whether or not people decide to bicycle. In turn, increasing individual bicycling would help generate a supportive physical environment.

While individual factors are important in explaining bicycling behaviors [11], aspects of the physical environment also are important, including the distance to destinations as determined by land-use patterns in transportation bicycling and the presence of a network of off-street bicycle paths for both transportation and recreational purposes [11].

All of these are useful pieces of evidence that directed the research and analysis in the next few sections.

Approaches to Measuring People's Attitudes About Bicycling and Other Urban Issues

Because people's attitudes about bicycling and other urban issues are complex, an analysis of those attitudes is necessary and helpful in understanding people's perceptions. Moreover, social networking provides good tools for exploring people's attitudes, because of how much they share. We will use data from Twitter, because Twitter has 320 million monthly active users and supports more than 35 languages [21]. Thus, this large number of users will provide a sufficient sample size for exploring and analyzing attitudes. In the article "You are what your friends eat," the authors use social media data to track attitudes and explore changes in eating behavior and body weight [1]. This is a good example and provides evidence to show that social media are a powerful tool to identify people's attitudes and for research.

People's attitudes may be measured by either qualitative or quantitative methods. Qualitative methods are powerful tools to explore such complex issues because they allow investigation of an individual's own explanations of their behavior and attitudes ([2, 5]). Qualitative methods include participant observation, direct observation, unstructured interviews, and case studies. Quantitative approaches have the advantage of measuring the responses of many subjects to a limited set of questions and thus allow comparison and statistical aggregation of the data [14].

Qualitative Case Study

One qualitative study addressed travelers' attitudes about transport [2]. They used in-depth interviews of 24 individuals who ranged from 18 to 70 years of age and conducted a comparative analysis of the respondents' demographics (e.g., work status, car status, and income level).

Quantitative Case Study

In contrast, a quantitative study by Grabow [8] presented predictors from urban design and used them to model commuting behavior. The author developed the "active transportation index" (ATI) by using data from surveys and geographic information system (GIS) sources.

A brief description of these studies will be helpful in demonstrating the methodology employed in this chapter. The chapter combined both of these methods and included data collection, quantitative analysis by SPSS, and spatial analysis by GIS.

Literature Findings and Conclusions

Thus far, the literature has shown that the benefits of bicycling cannot be present. Data from Twitter will be useful in obtaining feedback from current bicyclists and suggesting options that will facilitate bicycle planning.

Further, individual, social-environmental, and physical-environmental factors all affect people's attitudes, which provided strong support for the purpose of this chapter, to evaluate how bicycling facilities and city cycling planning influence people's attitudes and behaviors through a case study in Washington, DC.

Moreover, the literature related to the various methodologies used to assess attitudes and urban issues was helpful in the selection of the methods used in this chapter. Ultimately, we concluded that a mixed quantitative and qualitative design would be an ideal way to address the hypotheses and perform the analyses.

Overview of Current Bikeability in Washington, DC

The primary focus of this study was an exploration of the relationship between people's attitudes about biking and bicycling facilities and city cycling planning in Washington DC. In preparation for the remainder of the analyses, we first conducted a bikeability analysis in Washington DC using GIS data.

The data for this analysis were derived from four main resources that contributed to six individual maps. The main goals of this analysis were to obtain an overview of bikeability in Washington DC in a map format that makes the results easy to interpret. Further, this analysis identified the best aspects of six indices separately and represented the highest biking score area as maps as well. Moreover, parts of these maps were used in data collection and the statistical analyses.

The physical environment is an important variable in explaining bicycling behaviors, particularly the distances to destinations as determined by land-use patterns for transportation bicycling, and the network of off-street bicycle paths for bicycling for both transportation and recreational purposes [11]. According to the empirical evidence from the opinion survey, and travel behavior analyses, four main domains combined to identify bikeability: bicycle facilities, street connectivity, topography, and neighborhood land use [22].

Thus, based on this evidence and the references, I used six physical environment factors in this analysis to evaluate bikeability in Washington DC: bike trails, signed bike routes, Capital Bikeshare locations, hilliness, destination density, and intersection density. This resulted in individual maps for the six factors, after which I calculated an overall score (scale of 1–5, with 5 the highest) to identify the best biking areas in Washington.

The reasons for using the Capital Bikeshare locations should be mentioned here. The Capital Bikeshare location is organized by the Capital Bikeshare (CaBi) system, which has operated since 2010 as a public-private partnership with Alta Bicycle

Share, and is a bicycle-sharing system that provides service in Washington DC and in some cities in Virginia and Maryland. This system has more than 300 stations and 2500 bikes, all of which are owned by local governments [4]. This system improved the development of bicycle transportation and enhanced people's attitudes about biking as well. Therefore, I used the Capital Bikeshare locations as one of the factors to evaluate bikeability in Washington DC.

Primary Methods for Constructing Score

The primary steps used to produce different maps differed slightly and are summarized in the following table (Table 5.1).

Results and Analyses of Score Map

In evaluating the biking scores, we used all six factors in the following formula: Biking Score = Destination Density Score * (100/6) + Intersection Density Score * (100/6) + hilliness score *(100/6) + Signed Bike Routes Score*(100/6) + Capital

Table 5.1 Steps in producing score maps

Factors	Methods
Destination density score	Point data from reference USA were processed with the display XY data and point density tools to rank proximity to existing destination
Intersection density score	Line data from M drive in GIS lab operated by Tufts University were calculated with the intersect tool, and then results were filtered with selection when more than two lines intersected (ICOUNT >= 3); the layers created in the previous steps were then processed with the point density tool to rank proximity to existing intersection points
Hilliness score	Spot elevation data were processed with the IDW interpolation and slope tools to rank hilliness
Signed bike routes score	A line shapefile was processed with the line density tool to rank proximity to existing signed bike routes
Capital Bikeshare	Point shapefiles of Capital Bikeshare locations were processed with the point density tool to rank proximity to existing Capital Bikeshare locations
Bike Trail score	A line shapefile of bike trails was processed with the line density tool to rank proximity to existing bike trails
Overall biking score	Raster calculator was used to calculate the biking score overall by the six factors described above, which were given equal weights (100/6)

Bike Share Location Score*(100/6) + Bike Trail Score*(100/6). The best biking areas were adjacent, and biking scores between 3 and 5 were located primarily in central Washington.

In order to investigate people's attitudes about bicycling in urban areas in Washington DC, we used the following method, based on the Tufts Urban Attitudes lab (UAL), a research team affiliated with the Department of Urban and Environment Policy and Planning at Tufts University.

Data Collection

In order to collect and filter the microblog data, we used a computer program developed by the UAL to collect data from the Twitter Streaming API. This program collects and indexes a continuous stream of Tweets available publicly, which total less than 1% of all Tweets generated at a given location (Hollander et al. [12]). Geotagged Tweets were collected from March 3 to April 10, 2015 (total of 13 .CSV files). Figure 5.2 shows a small set of sample Tweets as displayed within UAL that

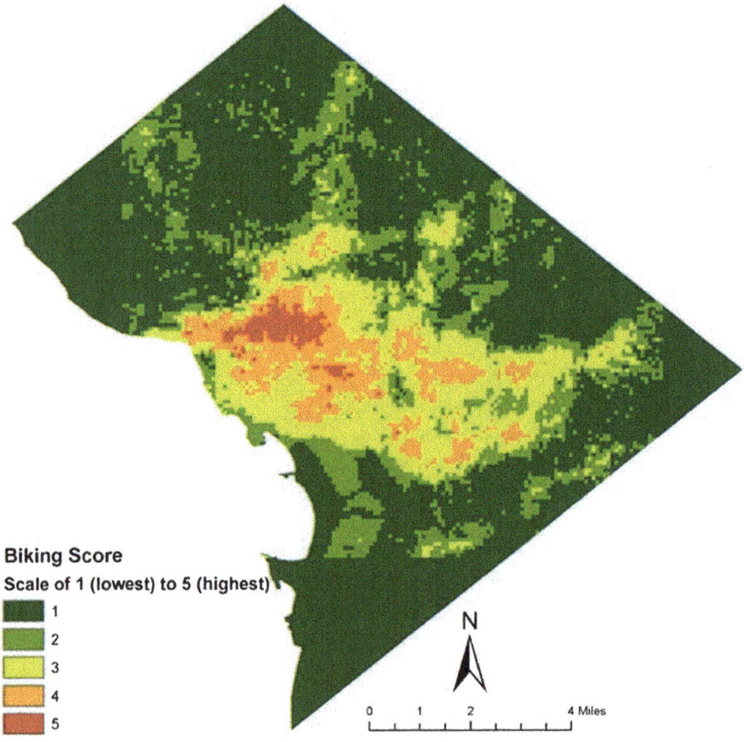

Fig. 5.2 Biking scores

A	B	C	D	E	F	G
5.73E+17	2802200413	@heartoutlander @beulahcrusoe @stenbergmika @Heughliots @heughanize	-77.086826	38.889995	2015-03-03 15:32:36	
5.73E+17	996017294	Back |||	-77.000631	38.925059	2015-03-03 15:32:38	
5.73E+17	190787153	Aint going tell u twice I'm tell u once	-76.917868	38.853918	2015-03-03 15:33:05	
5.73E+17	349921132	Nancy Pelosi is in the house. #Israel #IranDeal #jpost	-77.009266	38.88853	2015-03-03 15:33:16	
5.73E+17	180458590	Este sue▯o me tiene de muy mal humor y CERO tolerancia y CERO paci	-77.026458	38.99114	2015-03-03 15:33:16	
5.73E+17	375383005	@georgetown @RobertMGroves if we want to reach all survivors confi	-77.070697	38.907674	2015-03-03 15:33:19	
5.73E+17	1387436876	#AAP shld consider replacing the Jhadu with the scorpion as their s	-77.039532	38.905299	2015-03-03 15:33:22	
5.73E+17	190787153	@RafielVistaGWO: What's understood don't gotta be explained _ _	-76.917899	38.85392	2015-03-03 15:33:47	
5.73E+17	2802200413	@beulahcrusoe @stenbergmika @Heughliots @heughanizers @FansOSamHeug	-77.086826	38.889995	2015-03-03 15:33:48	
5.73E+17	2641339091	Friendship height we here's today.!!! http://t.co/XdZk8gH1JK	-77.085308	38.959943	2015-03-03 15:33:50	
5.73E+17	311267542	Im So Over These Big Ass Prenatal Pill	-76.978076	38.845022	2015-03-03 15:34:04	
5.73E+17	380290690	Because cat Sabi swim no mean say E be catfish this Wan sef na pun	-76.99386	38.926592	2015-03-03 15:34:08	
5.73E+17	128467838	Never get sloppy drunk but alcohol is problem solving!	-77.011489	38.895257	2015-03-03 15:34:10	
5.73E+17	2587789764	#Nursing #Job alert: RN / Registered Nurse / Travel RN job | Supple	-77.052129	38.904237	2015-03-03 15:34:15	
5.73E+17	450113186	Cannot stand it when people in college brag about their grades #not	-76.996555	38.9341	2015-03-03 15:34:20	
5.73E+17	59785902	_ @qweenpush: http://t.co/VHUcoitZiK_ me	-77.020233	38.924031	2015-03-03 15:34:23	

Fig. 5.3 Sample Tweets from Washington, DC

were exported to Excel in the .CSV file format. Column A is the Twitter ID, column B is the user's ID, column C is Tweets, columns D and E display longitude and latitude, and column F is the time the Tweets were posted (Fig. 5.3).

All 13 .CSV files included more than 400,000 rows of data, and thus, it was necessary to select only those Tweets related to "biking." Therefore, all data were filtered by the following key words: bike, bicycling/cycling, biking, bicycle, cyclists/bicyclists, bike facilities, bike infrastructure, and bikeability/bikeabilities. Ultimately, 366 data points were obtained. Further, we collected the same number of randomly selected data points (366) from the original data, except the 366 Tweets about biking selected previously. This helped compare people's attitudes about biking to that about other subjects.

The location of all 366 bike-focus sample Tweets and 366 random sample Tweets was determined by longitude and latitude; Twitter has a setting that users can turn on or off to post the geolocations of their Tweets. Thus, we used ArcGIS to geolocate all Tweets (see Fig. 5.4). The green dots show the bike-focus samples, and the red dots are random samples. From Fig. 5.4, we cannot observe any strong patterns of distribution for both bike-focus samples and random samples, and these Tweets posted everywhere in Washington DC. Figures 5.5 and 5.6 show the example of the bike-focus Tweets and random sample Tweets in excel file. In Fig. 5.5, we can see people talked about "bike" in many ways, such as someone complained about drivers and someone expressed their love of biking.

Sentiment Analysis

Once we collected all of the "biking" Tweets, we conducted a sentiment analysis using SentiStrength, which uses a set of Tweets and processes the dataset line by line to match words in a sentiment dictionary, after which it assigns either a positive or negative score to each Tweet.

SentiStrengh is available in ten languages, among them English, Arabic, French, Greek, Italian, Persian, Polish, Portuguese, and Swedish, and it is possible to apply

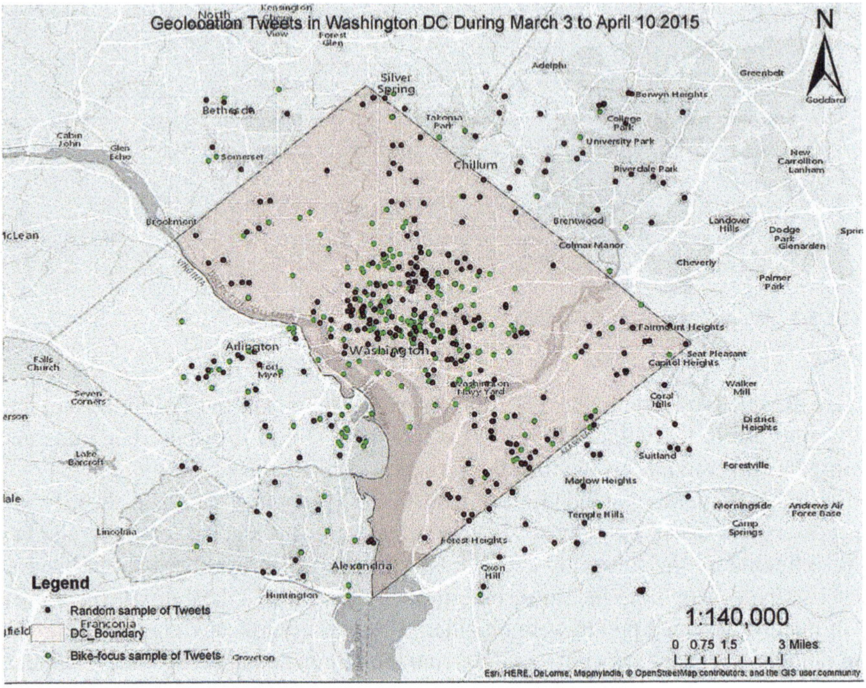

Fig. 5.4 Geolocation of Tweets in Washington, DC (random sample and bike-focus sample of Tweets)

Tweets in each language to the same scale. This software also can apply certain emoticons, such as “:(“, “:),” “^_^,” and “<3.” This lexicon has been used in a number of peer-reviewed articles and multiple studies and projects and thus is respected academically [7, 15, 20].

For each Tweet, SentiStrength [18] reports negative emotion scores from 1 (not negative) to 5 (extremely negative) and positive emotion scores from 1 (not positive) to 5 (extremely positive). For the purposes of this study, I used positive, negative, and overall emotion scores and analyzed them separately; the overall score was equal to the sum of the positive and negative scores. Figure 5.7 shows an example of the sentiment score for a single Tweet; the overall score for this sample would be 1, because the positive emotion score was 2, and the negative emotion score was −1. When a large number of data are processed, the software creates a .txt file with the same concept as the example. To begin the sentiment analysis, we used SentiStrength to process the data needed and exported them to a .txt file. Then, we modified it as an Excel file (see Fig. 5.8) to prepare for the next statistical analyses.

While this method is the best way to process large amounts of data, it does have several weaknesses. Firstly, although this software can be applied in ten languages, the number of words in each lexicon in the software differs. For example, there is a total of 1964 words in the Portuguese lexicon, of which 1627 words (83%) are

5.72787E+17	299467435	Things I'm going to do in spring: read outside. Picnic. Ride my bike to school like it's the 90's all over again.	38.951828	-77.002712
5.72899E+17	282196492	Moeen Khan talking about #RediscoveringPakistan-untold tale: Motorbike journey through Pakistan http://t.co/Tgyz3SOhJF	38.943397	-77.068232
5.72899E+17	282196492	Moeen Khan talking about #RediscoveringPakistan-untold tale.at#PakEmbassy.Motorbike journey through Pak http://t.co/MECRHOAKTC	38.943397	-77.068232
5.73195E+17	498852873	Why is Arlington even considering cuts to walking/biking improvements? The story coming up at 4:30 on @wamu885news #bikeva	38.882074	-77.076369
5.73199E+17	384657755	Few things have made me as happy as riding my bike in Amsterdam did â¤ï¸	38.951653	-76.972017
5.73291E+17	215452671	Shattered. #BikeDC #AnacoastiaRiver @ Frederick Douglass National Historic Site https://t.co/YVgnxihpjp	38.863704	-76.984425
5.73854E+17	244293422	son i RL fell on ice while on a bike ðŸ˜¤ðŸ˜	38.887879	-76.956496
5.74335E+17	896556121	My @jamisbicycles I rode on the mighty #HillsOfAnacostia today! @BicycleSPACE @ewilliams0305 @moeherbs #bikedc http://t.co/qO6Dt4sCKV	38.850643	-76.9787
5.74364E+17	896556121	It was all @jamisbicycles riding the mighty #HillsOfAnacostia today! @BicycleSPACE #JamisBikesDC #bikedc http://t.co/FjprLb3Twy	38.850657	-76.978717
5.74971E+17	10400902	The first ever school pick-up by bike and the WeeHoo #BikeDC @ National Presbyterian School https://t.co/UBxvou0JSI	38.941594	-77.080913

Fig. 5.5 Example of bike-focus sample Tweets from Washington, DC

5.73E+17	189614954	These girls are really crying ðŸ˜‚ðŸ˜‚ðŸ˜‚ðŸ’€ðŸ’€ðŸ’€ðŸ’€ðŸ’€	38.862169	-76.973275
5.73E+17	223922535	Just posted a photo @ Lincoln Memorial https://t.co/kXzaDzkfQG	38.847133	-76.980556
5.73E+17	556126496	Hot Seller: International Spy Museum: The Permanent Exhibit @ International Spy Museum (Washington DC) http://t.co/JcnmKezbFR	38.936559	-77.05988
5.73E+17	2765611839	Dat juss blew me bruhh ðŸ˜¤ðŸ˜¤ðŸ˜¤	38.859618	-76.950499
5.73E+17	610185044	My math teacher is so much kix eh...it's a pity I don't come to this class more often lol	38.854822	-77.043833
5.73E+17	66711859	Missing my seniors. Kinda wishing I took mike up on his New York offer. Oh well. Spring Break will go on ðŸ˜’	38.8651	-77.0391
5.73E+17	2314156260	CGI: Business Objects Developer (#WashingtonDC) http://t.co/vKOQdm6ALE #IT #Job #Jobs #TweetMyJobs	38.975184	-76.954439
5.73E+17	174377381	@JuliannsMine Love you more Abby Boo! â¤ï¸	38.92022	-77.071227
5.73E+17	42111754	Not even a weather-related DC shutdown can stop my fitness flow! #MarchMotivation http://t.co/AVgrEY588j	38.920154	-76.99619
5.73E+17	29765306	Snapchat: yeahcri	38.941025	-77.038776

Fig. 5.6 Example of random sample Tweets from Washington, DC

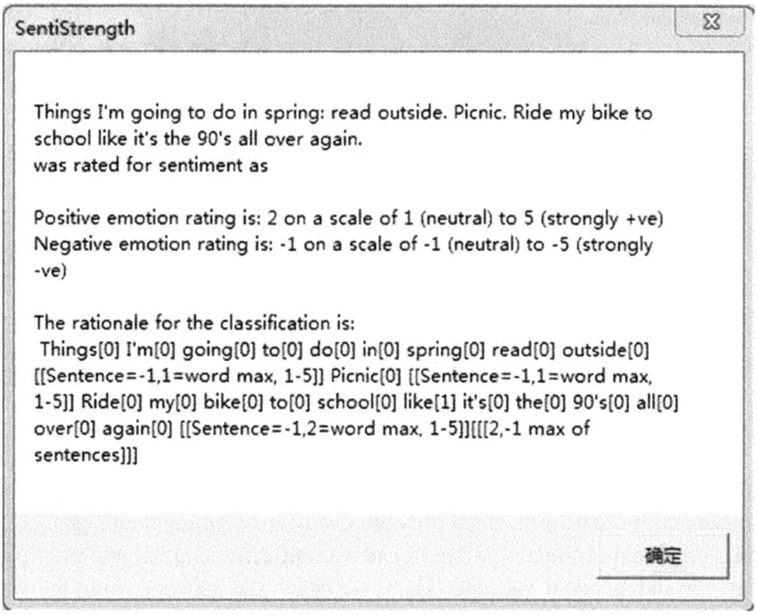

Fig. 5.7 Example of the sentiment analysis for a single Tweet

A	B	C	D	E	F	G
Twitter	TwitterUserID	Tweets	latitute	longitude	positive_e	negative_e
5.72787E+17	299467435	Things I'm going to do in spring: read outside. Picnic. Ride my bike to school like it's the 90's all ov	38.951828	-77.002712	2	-1
5.72899E+17	282196492	Moeen Khan talking about #RediscoveringPakistan-untold tale: Motorbike journey through Pakist:	38.943397	-77.068232	1	-1
5.72899E+17	282196492	Moeen Khan talking about #RediscoveringPakistan-untold tale.at#PakEmbassy.Motorbike journey	38.943397	-77.068232	1	-1
5.73195E+17	498852873	Why is Arlington even considering cuts to walking/biking improvements? The story coming up at ‹	38.882074	-77.076369	2	-1
5.73199E+17	384657755	Few things have made me as happy as riding my bike in Amsterdam did â¤í,	38.951653	-76.972017	2	-1
5.73291E+17	215452671	Shattered. #BikeDC #AnacoastiaRiver @ Frederick Douglass National Historic Site https://t.co/YVg	38.863704	-76.984425	1	-1
5.73565E+17	181009536	@bikeshare across from the Zoo! @yourtake @nbcwashington @WTOP @capitalweather @postl	38.93092	-77.056448	1	-1
5.73613E+17	250208461	@dwhitphoto snapping @NatCapVeloClub last night for @BikeArlington @ccbid and @CXHairs 1s	38.867999	-77.048497	1	-1
5.73854E+17	244293422	son i RL fell on ice while on a bike ðŸ˜¤ðŸ˜	38.887879	-76.956496	1	-1
5.74335E+17	896556121	My @jamisbicycles I rode on the mighty #HillsOfAnacostia today! @BicycleSPACE @ewilliams030!	38.850643	-76.9787	1	-1
5.74364E+17	896556121	It was all @jamisbicycles riding the mighty #HillsOfAnacostia today! @BicycleSPACE #JamisBikesD	38.850657	-76.978717	1	-1
5.74971E+17	10400902	The first ever school pick-up by bike and the WeeHoo #BikeDC @ National Presbyterian School htt	38.941594	-77.080913	1	-1
5.7565E+17	42873686	Army Navy Dr. is great to ride on and gets #BEaPAL messaging to traffic on I395. @BikeArlingtonâ€	38.856358	-77.08096	3	-1
5.75717E+17	5865822	Hey @anthonysiracusa wanna bring #NBS15 and #bikewalktn out to some hep #bikedc culture? h	38.856548	-76.977251	2	-1
5.75734E+17	10400902	Bike parking at #DCA is actually farther away than car parking #irony #BikeDC @ Ronald Reaganâ€	38.852342	-77.042369	1	-1
5.75793E+17	313603217	bike trail open again all the snow gone #cycling #wednesday	38.977742	-77.001216	1	-1
5.75794E+17	306457721	Dirt Bikes Jump Af Busting 12 Ass Coming Down 30th ðŸ˜,ðŸ˜,ðŸ˜,ðŸ˜,	38.85308	-76.965944	1	-2
5.76374E+17	2181447170	Good morning DC so excited the warm weather is coming bike riding summer breeze makes me	38.946536	-77.032853	3	-1
5.76376E+17	260810689	#BloomingdaleDC -1 of our fave neighbors in #EckingtonDC had her bike stolen Please keep eye oı	38.913193	-77.012231	2	-2
5.76461E+17	154387520	@jlisle @DDOTDC make it a bike lane problem solved! Then everyone could use it.	38.921139	-77.012127	1	-2

Fig. 5.8 Example of the sentiment analysis for Tweets after modification

negative [18]. By comparison, 2476 words were found in the English lexicon, and 1585 (64%) of those were negative [18].

Although these biking data all used English, some random data selected used other languages. Further, emoji (see Fig. 5.9) have become more and more popular in social media, but SentiStrength does not yet have a lexicon of emoji. In addition, the software can perform a sentiment analysis on only 122 emoticons; therefore,

Fig. 5.9 Example of emoji

it has limited ability to test the remainder of the emoticons used. Thus, these weaknesses might have influenced the outcome of the sentiment analysis.

Based on all data collected, we performed a sentiment analysis and used ArcMap to create the independent variable. Then, we began the analysis using the general-purpose statistical software, STATA.

Firstly, we used the summarizing tool to obtain detailed summary statistics and tables that were constructed to compare positive and negative sentiment scores across the various distances/values to determine whether there were noticeable differences in the scores across these categories. Then, we used the graphic tool to create histograms used to evaluate the sentiment scores visually for trends across the distance/value measures. Moreover, a series of chi-squared and Kruskal-Wallis tests was used to explore whether the differences were statistically significant. In addition, Spearman correlations were used to explore the statistical relationship between variables. Finally, a univariate linear regression model was performed to explore the linear relationship between the sentiment scores and the six physical environment variables

Discussion: Biking Data vs. Random Data

The purpose of comparing bike-focus sample Tweets and random sample Tweets was to determine whether people's attitudes about biking were below or above a standard level. Thus, because the original database was large, I selected another 366 Tweets from over 450,000 Tweets. The results of the data selected randomly from the original database represented the sentiment level of all Tweets. Table 5.2 shows the comparison between these two datasets. The mean score of the random sample overall was −0.44, and the mean of the biking sample was 0.26. Thus, people's attitudes about biking were positive and higher than the sentiment level of all Tweets.

Table 5.2 Sentiment analysis results

	Random sample Tweets	Bike-focus sample Tweets				
	Positive overall emotional score	Negative emotional score	Overall emotional score	Positive overall emotional score	Negative overall emotional score	Overall emotional score
Observation	366	366	366	366	366	366
Median	1	−1	0	1	−1	0
Mean	1.44	−1.49	−0.44	1.54	−1.28	0.26
Standard deviation	0.69	0.85	1.10	0.74	0.61	0.94

Table 5.3 Results of linear regression between dependent and independent variables

	p-value		
Independent variables	Positive score	Negative score	Overall score
Intersection density score	0.92	0.76	0.78
Destination density score	0.69	0.91	0.69
Hilliness score	0.68	0.04, coeff. = 0.82 R-squared = 0.0119	
Distance to signed bike routes	0.72	0.87	0.71
Distance to capital Bikeshare locations	0.01 Coeff. = −0.14 R-squared = 0.0280	0.31	0.12
Distance to bike trail	0.53	0.26	0.25

Results: Linear Regression

Table 5.3 shows the results of the regression between the sentiment scores and the six individual independent variables. The hypothesis was that there is no significant relationship between the dependent and independent variables. As the table shows, most *p*-values were higher than 0.05, and thus, the null hypothesis was accepted. However, two *p*-values less than 0.05 are highlighted in the table. The *p*-value between the hilliness and negative sentiment scores was 0.04, and thus, the null hypothesis was rejected in that case. When the hilliness score increased by 1, the negative sentiment score increased by 0.08, indicating that a higher hilliness score reduced people's negative attitudes about biking.

The r-squared were very low in this case, and they were around 0.01 and 0.03. It means the model explains only 1% and 3% of the response data around their means.

There was another significant relationship between the distance to Capital Bikeshare locations and the positive sentiment score. The *p*-value between them was 0.01, and thus, the null hypothesis was rejected. When the distance to a Capital

Bikeshare location increased by 1 unit, the positive sentiment score fell by 0.14. The greater the distance to a Capital Bikeshare location, the more negative people's attitudes about biking.

Conclusions, Limitations, and Recommendations for Future Research

The first research question asked, "How is bicycling ability presented in Washington DC?" To address the question, we used six indices to calculate bicycling scores in Washington: intersection and destination density, hilliness, bike trails, signed bike routes, and Capital Bikeshare locations. We were unable to provide specific scores for the whole city, but by using these six individual factors and processing ArcMap, we found that the best biking areas were adjacent and were located primarily in the central areas of the city.

The second research question was, "What are people's attitudes (positive or negative) about bicycling in Washington?" and the answer overall was "positive." As Chap. 5 mentioned, by comparing the biking data to random data, the sentiment scores for biking data were positive and higher than were those for random data. Based on the results presented in Chap. 5, I concluded that people's attitudes toward biking in Washington are positive.

The last research question asked, "Is there a relationship between people's attitudes, the bicycling index (cycling infrastructures, hilliness, desirable amenities, and road connectivity), and locations where people talk about bicycling?" Because the answer is complex, I will describe it specifically as follows, based on the six biking indices. Firstly, there was no clear pattern between people's attitudes and the distance to bike trails and signed bike routes. Further, the results showed no relationship between people's attitudes and different scores for intersection and destination densities.

However, the analyses did demonstrate some significant relationships. The linear regression showed relationships between the hilliness score and negative sentiment scores and distance to Capital Bikeshare locations and positive sentiment scores. When the hilliness score increased by 1, the negative sentiment score increased by 0.08, indicating that a higher hilliness score reduced people's negative attitudes about biking. Another set of data was related to the distance to Capital Bikeshare locations and positive sentiment scores. When the distance to Capital Bikeshare location increased by 1 unit, positive sentiment scores fell by 0.14, suggesting that greater distances to Capital Bikeshare locations reduced positive attitudes. Thus, the answers to the research questions were relatively conclusive.

These results could be improved further by optimizing the whole process. We discuss the limitations of this study in the next section.

Limitations in Data Collection

Some limitations in the process overall included the ways in which the data were collected, as well as the statistical analyses, either or both of which may have influenced the outcomes of the study.

The first limitation relates to data collection. Firstly, we collected Twitter data for only 5 weeks. Thus, the sample size was not sufficiently large to be representative. In addition, all of the data collected were from Twitter and contained words and emoticons only. Therefore, some pictures, videos, and emoji were not considered and applied in the sentiment analysis. Further, within the small sample size (366), some samples came from one Twitter user. Thus, if a user was overrepresented, some elements of his/her personality may have biased the data. For example, if a user was a negative person, he/she might be likely to complain about everything, and thus, his/her Tweets might skew the results negatively, given that the sample size was small already. Figure 5.10 is the example of selected Tweets from one

Twitter	TwitterUse	Tweets	positive emotion rating	negative emotion rating	Biking_Score
5.754E+17	46416755	This is a removable/lockable rear trunk! The design is sweet! #womenbike #NBS15 http:/	2	-1	4
5.754E+17	46416755	.@MCMHandles worked on advocacy in Boyle Heights- all the people selected for leading	1	-1	4
5.754E+17	46416755	If you are asking community leaders to do outreach for your Transpo project make sure y	1	-1	4
5.754E+17	46416755	To reach the Boyle Heights community @MCMHandles went to local businesses- not cor	1	-1	4
5.754E+17	46416755	Just had a presenter drop #KeepIt100 at the National Bike Summit. #nbs15 #womenbike	1	-2	4
5.754E+17	46416755	The entire room at #womenbike just chanted outdoor Afro #NBS15	1	-1	4
5.754E+17	46416755	@ambrown @qpdx definitely is- @peopleforbikes study says Women are more likely to l	1	-1	4
5.754E+17	46416755	All of these discussions at #NBS15 confirm for me: bike fun is the future of coalition buildir	2	-1	4
5.754E+17	46416755	@BikePortland lots of people still don't think that's true! And sadly many of them are the	2	-2	4
5.754E+17	46416755	@GoddardTara I'm not quite sure what your question is but what I mean is our biggest pi	2	-1	4
5.754E+17	46416755	@ambrown I agree we need more family friendly bike fun- but I think the advantage of b	2	-1	4
5.754E+17	46416755	@gerikkransky @GoddardTara should've come to my talk on Bike Fun building advocate	2	-1	4
5.754E+17	46416755	@GoddardTara I'm not saying replacing car trips- that's the. Broadening the bike commu	1	-1	4
5.754E+17	46416755	@GoddardTara @gerikkransky If we focus on only commute trips we lose those that car	1	-2	4
5.754E+17	46416755	@GoddardTara @gerikkransky we find women are more likely to participate in activities,	2	-1	4
5.754E+17	46416755	@GoddardTara @gerikkransky Utility=Fun isn't what I was saying. Bike Fun- community	2	-1	4
5.754E+17	46416755	@gerikkransky @GoddardTara To build women&minority bike leaders we first nee	2	-1	4
5.754E+17	46416755	@GoddardTara @gerikkransky Andando en Bicicletas en Cully and We all Can Ride and B	3	-1	4
5.754E+17	46416755	@GoddardTara @gerikkransky The #womenbike keynote last year talked about how in t	1	-1	4
5.754E+17	46416755	@gerikkransky @GoddardTara I didn't say Bike Fun is the only key to getting people on b	2	-1	4
5.754E+17	46416755	@gerikkransky @GoddardTara It's much easier to invite someone to a neighborhood bik	1	-1	4
5.754E+17	46416755	@GoddardTara @gerikkransky The #womenbike keynote last year talked about how in t	1	-1	4
5.754E+17	46416755	@gerikkransky @GoddardTara I didn't say Bike Fun is the only key to getting people on b	2	-1	4
5.754E+17	46416755	@gerikkransky @GoddardTara It's much easier to invite someone to a neighborhood bik	1	-1	4
5.754E+17	46416755	@gerikkransky @GoddardTara really? How about ABC members who lead Day of the de	1	-3	4
5.754E+17	46416755	@gerikkransky @GoddardTara It's the first step - fee want to advocate for bikes until the	1	-1	4
5.754E+17	46416755	@ladyfleur @cyclelicious here's the bike fun podcast! http://t.co/5ADss2R6MZ	2	-1	4
5.732E+17	498852873	Why is Arlington even considering cuts to walking/biking improvements? The story comi	2	-1	1
5.732E+17	498852873	Tune to @wamu885news at 4:30 for my story on Arlington County considering cuts to bik	2	-1	3
5.735E+17	498852873	Progressive Arlington may cut funding for bike/pedestrian programs: http://t.co/GLeFnU	1	-1	3
5.739E+17	498852873	.@WABADC says @DDOTDC not effectively implementing bike projects at a pace that m	1	-1	3
5.739E+17	498852873	.@WABADC's Greg Billing says DDOT can't get projects done. And now... they're attemp	1	-1	3
5.739E+17	498852873	@darsal @beyonddc I didn't realize there were bike lanes there.	1	-1	3
5.739E+17	498852873	Oversight hearing continues. Domsjo yet to testify. Lots of testimony from bike advocat	2	-2	3
5.739E+17	498852873	@darsal @beyonddc The number of miles of protected bike lanes is rather small. Can't re	1	-1	3
5.739E+17	498852873	.@marycheh says we still don't have comprehensive bike lane network. Key word: netw	1	-2	3
5.764E+17	498852873	@jeffreyanders19 @maustermuhle Protected bike lanes are the clincher for some riders	1	-1	4
5.764E+17	498852873	Look at all that traffic! #bikedc http://t.co/FRvg7kbvVW	1	-1	5
5.764E+17	498852873	Look who is riding a bike to work today. (Hint: @marycheh) http://t.co/RHwfoYteFJ	1	-1	5
5.764E+17	498852873	Is @DDOTDC moving fast enough to build a protected bike lane network? Tune in Mond	1	-1	5
5.743E+17	896556121	My @jamisbicycles I rode on the mighty #HillsOfAnacostia today! @BicycleSPACE @ewi	1	-1	1
5.744E+17	896556121	It was all @jamisbicycles riding the mighty #HillsOfAnacostia today! @BicycleSPACE #Jar	1	-1	1
5.772E+17	896556121	Mighty @PhilKinDC showing off his old school racing strip sweats on @BicycleSPACE #City	1	-1	1
5.86E+17	896556121	Today @DCPoliceDept #bikedc patrol officer showed me proper way to carry my bike up	2	-1	1
5.861E+17	896556121	Wal-Mart on H St in #bikedc no customer parking for bike riders! No bike racks in parking	2	-1	1
5.743E+17	896556121	It's all good @ewilliams0305! Always great riding mighty #HillsOfAnacostia w/ you &am	3	-1	2
5.743E+17	896556121	Frederick Douglas Bridge a #bikedc @DDOTDC disaster! Today's @BicycleSPACE #HillsOf	1	-2	2
5.746E+17	896556121	On @BicycleSPACE #CityExplorers ride! @MOMsOrganicMrkt food stop! @OrMoBike @	1	-1	2
5.856E+17	896556121	Walking enriches your life! Just found 1989 Lincoln Penney on the National Mall! @allwal	1	-1	3
5.767E+17	896556121	Rain or shine! Its no jive! I am riding the mighty #HillsOfAnacostia today! 8a @BicycleSPA	1	-1	4
5.768E+17	896556121	Stoked! to see @bikeshare carrying @jamisbicycles! Nice! #JamisBikesDC #bikedc http:/	2	-1	4
5.858E+17	896556121	My @GiantBicycle riding across the giant Frederick Douglass Bridge! Wrong Mayor on sig	1	-2	4

Fig. 5.10 Example of bike-focus sample Tweets from one user

user. We picked user 46,416,755 as the example here. The user 46,416,755 posted 27 Tweets during 5 weeks about biking, and the mean of positive score from these Tweets is 1.52, and the mean of negative score from these Tweets is −1.19.

Based on this evidence, the attitude of user 46,416,755 tended to be positive and had possible influence on results.

Future Recommendations

Although the study had some limitations, some suggestions and recommendations can be offered. The first suggestion addresses future relevant studies. Based on the limitations described above, future studies may be able to overcome or minimize the limitations in this study. For example, adding more physical-environmental factors would address some of the potential biases in the statistical analyses. The sample size also should be expanded considerably. Moreover, identifying more relationships or including more factors in the analysis would help cycling planners and policymakers make decisions in the development of biking.

Higher hilliness scores were found to have a statistically significant relationship with peoples' negative attitudes. While this analysis was limited and many variables were not included, there may be an important relationship between topography and bike user attitudes that deserves further exploration by local government planners. But more than anything, the findings presented here illustrate the remarkable power of social media data to shed light on urban policy issues and help shape the way that planning can happen.

Acknowledgments We gratefully acknowledge the feedback and advice provided on earlier versions of this research by Dr. Sumeeta Srinivasan, assistance by Claudia Aliff and Isabelle Shutt with copyediting and formatting, and Hui Shen and Yan Zhang for their support throughout this project.

Bibliography

1. Adam, F., Chris, B., Annik, S.: You are what your friends eat: systematic review of social network analyses of young people's eating behaviours and bodyweight. BMJ. Glob. Health. 1–8 (2011). https://doi.org/10.1136/jech.2010.113936
2. Beirao, G., Cabral, J.A.S.: Understanding attitudes towards public transport and private cars: a qualitative study. Transp. Policy. **14**, 478–489 (2007)
3. Ben, A.: The rise of urban biking. In: The Nation Instigating Progress Daily. Retrieved from http://www.thenation.com/article/rise-urban-biking/ (2011)
4. Capital Bikeshare: About Capital Bikeshare. Retrieved from https://www.capitalbikeshare.com/about (2016)
5. Clifton, K.L., Handy, S.L.: Qualitative methods in travel behavior research. Paper presented at the international conference on transport survey quality and innovation, Kruger National Park, South Africa (2001)

6. Frank, L.D., Engelke, P.O., Schmid, T.L.: Health and Community Design: The Impact of the Built Environment on Physical Activity. Island Press, Washington, DC (2003)
7. Garas, A., Garcia, D., Skowron, M., Schweitzer, F.: Emotional persistence in online chatting communities. Sci. Rep. **2**, 402 (2012)
8. Grabow, M.L.: Active Transportation and Health: Predictors from Urban Design and Modeled Commuting Behavior, pp. 1–140. ProQuest LLC, Ann Arbor (2013)
9. Handy, S.L.: Urban form and pedestrian choices: a study of Austin neighborhood. Transp. Res. Rec. **1552**, 135–144 (1996)
10. Handy, S.L.: Walking, bicycling, and health. In: Healthy Equitable Transportation Policy. PolicyLink, Oakland (2009)
11. Handy, S.L., Xing, Y., Buehler, T.J.: Factors associated with bicycle ownership and use: a study of six small U.S. cities. Transportation. **37**, 967–985 (2010)
12. Hollander, J., Graves, E., Levanthal, T.: Using Big Data to Study Urban Sentiments: Twitter Data vs. Published Meeting Minutes. Tufts University (2014)
13. Molina, J.: The Case for Crowdsourcing in Bicycle Planning: an Exploratory Study, pp. 2–135. ProQuest LLC. Tufts University (2014)
14. Patton, M.Q.: Qualitative Evaluation and Research Methods. Sage, Newbury Park (1990)
15. Pfitzner, R., Garas, A., Schweitzer, F.: Emotional divergence influences information spreading in Twitter, ICWSM-12 (2012)
16. Sallis, J.F., Owen, N.: Ecological models of health behavior. In: Health Behavior and Health Education: Theory, Research, and Practice, vol. 4, pp. 465–486. San Francisco, CA, Jossey-Bass (2008)
17. Schroeder, P., Wilbur, M.: 2012 national survey of bicyclist and pedestrian attitudes and behavior, volume 2: findings report. (Report No. DOT HS 811 841 B). Washington, DC, National Highway Traffic Safety Administration (2013)
18. Sentistrength: About Sentistrength. Retrieved from http://sentistrength.wlv.ac.uk/#About (2016)
19. Susilo, Y.O., Williams, K., Morag, L., Dair, C.: The influence of individuals' environmental attitudes and urban design features on their travel patterns in sustainable neighborhoods in the UK. Transp. Res. D. **17**, 190–200 (2012)
20. Thelwall, M., Buckley, K., Paltohlou, G.: Sentiment in twitter events. J. Am. Soc. Inf. Sci. Technol. **62**(2), 406–418 (2011)
21. Twitter.com: About Twitter. Retrieved from https://about.twitter.com/company (2015)
22. Winters, M., Brauer, M., Setton, E.M., Teschke, K.: Mapping bikeability: a spatial tool to support sustainable travel. Environ. Plann. B. Plann. Des. **40**(5), 865–883 (2015)

Chapter 6
Sustainable Mobility

Fotini Kehagia

Abstract The EU's long-term outlook for transport in the EU and its associated emissions demonstrates that the 2050 decarbonization goals for the transport sector require not only incremental changes but a systematic change (EEA Transitions towards a more sustainable mobility system. EEA technical report, Copenhagen, 2016). The intensive problems that many urban areas are facing caused by the operation of the transport system such as traffic congestion, air pollution, degradation of the environment, etc. require a paradigm shift in the planning process. Banister (Transp. Policy 15:73–80, 2008) said that "transport planning has a crisis of identity and its future is uncertain, particularly as the increasing complexity of cities and societies make simple approaches to analysis, which views transport congestion as the problem and transport as the solution". Transport should have a basic role in achieving sustainable development. Towards this direction sustainable urban transport planning is a challenge.

Keywords Transportation • Sustainability • Economic growth • Urban mobility • Strategies

Transport and Sustainable Development: An Introduction

The concept of sustainable development is in danger of becoming everything
(Holden 2007)

Worldwide, cities evolve and change and increasingly face problems caused by transport and traffic. In Europe, a large majority of citizens live in an urban environment, with over 60% living in urban areas of over 10, 000 inhabitants. Urban mobility accounts for 40% of all CO2 emissions of road transport and up to 70% of other pollutants from transport [15]. Congestion in the EU is often located in and around urban areas and costs nearly EUR 100 billion, or 1% of the EU's GDP, annually [7]. The main challenges are the improvement of the performance of urban

F. Kehagia (✉)
Aristotle University of Thessaloniki, School of Engineering, Department of Civil Engineering, Thessaloniki, Greece
e-mail: fkehagia@civil.auth.gr

A. Karakitsiou et al. (eds.), *City Networks*, Springer Optimization and Its Applications 128, DOI 10.1007/978-3-319-65338-9_6

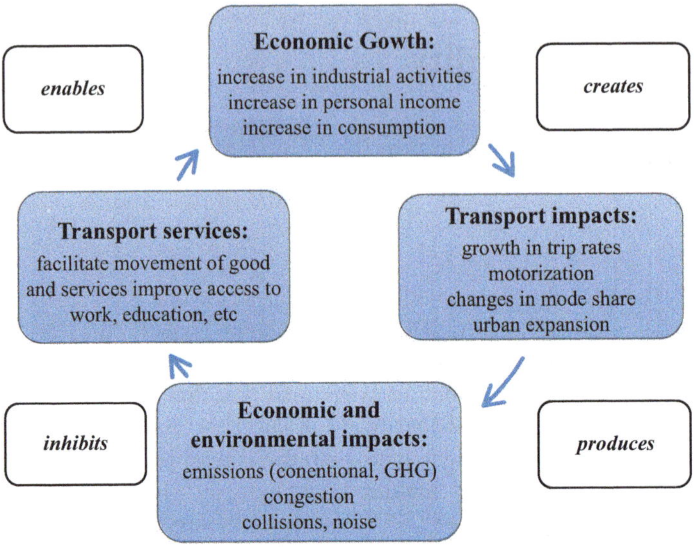

Fig. 6.1 The challenges of making mobility sustainable [39]

transport systems and the reduction of the negative impacts of transport activities on the climate, the environment and citizens' health, providing a more sustainable urban mobility.

Transport is fundamental to our economy and society. Moreover, it is an important factor in the context of sustainable development due to the pressure it places on the environment and its economic and social impacts. The basic principles of the definition of transport are given in Fig. 6.1, namely, whereas transport leads to economic growth, economic growth creates basic problems to the achievement of sustainable transport.

It has been three decades since the term "sustainable development" and "sustainability" appeared following the 1987 publication of the United Nations World Commission on Environment and Development (WCED) report, "Our Common Future", commonly referred to as the Brundtland Report. According to this definition, four different objectives have to be accomplished: safeguarding long-term ecological sustainability, satisfying basic human needs and promoting inter- and intragenerational equity [23, 37]. There has been extensive literature on the addressing of the meaning of the term which emerged from different disciplines such as economics, environmental studies, and sociology. The concept of sustainability has been interpreted in many different ways, often incompatible, but its principal element consists of an approach to development that looks to balance different, and often competing, needs against an awareness of the environmental, social and economic limitations we face as a society. Sustainable development has become the dominant paradigm of development in both developed and developing countries. The large number of definitions and interpretations which were made by scientists, politicians and businessmen had as objective to use it as a potential solution for the myriad of global, regional and local problems of societies in the late twentieth century.

The concept of sustainability has been interpreted in many different ways, often incompatible, but its principal element consists of an approach to development that looks to balance different, and often competing, needs against an awareness of the environmental, social and economic limitations we face as a society. Sustainable development has become the dominant paradigm of development in both developed and developing countries. The large number of definitions and interpretations which were made by scientists, politicians and businessmen had as objective to use it as a potential solution for the myriad of global, regional and local problems of societies in the late twentieth century. Despite the vagueness of the interpretation of the term, sustainable development as an ideal is as persistent a political concept as are democracy, justice and liberty [27]. Holden et al. [25] address the appropriate indicators of each of the four objectives – dimensions of sustainability. Moreover, they assign threshold values that should be met in order for development to be deemed sustainable. The indicators and the suggested thresholds for the future of the four dimensions of sustainable development are presented in Table 6.1.

Since transport is a key driver in sustainable development, a new approach to transport policy-making in order to address the challenges of the future is needed. More specifically, an integrated approach to tackle the environmental impacts of transport consistent with the socio-economic development policies is needed [33]. Thus, a sustainable transport system must have the ability to support the constantly improved modern lifestyle with optimal management of resources for construction

Table 6.1 Dimensions, indicators and suggested 2030 threshold values for sustainable development [25]

Dimension	Indicator	2030 Threshold
Safeguarding long-term ecological sustainability	Yearly per capita ecological footprint[a]	Maximum 2.3 gha/capita
Satisfying basic human needs	Yearly per capita GDP PPP[b]	Minimum USD3350 (2000) per capita yearly
Promoting intragenerational equity	Gini coefficient[c]	Maximum 0.40
Promoting intergenerational equity	The amount of renewable to total energy in primary energy production[d]	Minimum 27%

[a]The Ecological Footprint tracks humanity's demands on the biosphere by comparing its consumption against the Earth's regenerative capacity

[b]Gross Domestic Product Purchasing Power Parity (GDP PPP) is about basic human needs. A high level of this indicates that countries have sufficient means to provide its inhabitants with the necessary services to meet their basic needs

[c]The Gini index measures the distribution of either household income or consumption spending in a country

[d]The fraction of renewable energy to total primary energy is used as an indicator for promoting intergenerational equity

and operation of transport systems (criterion of economic viability), the improvement of life quality (energy conservation, air pollution reduction and ensuring health) without restricting access (criterion of environmental sustainability) and finally ensuring an affordable, time-reliable, secure and flexible movement for all members of society (social sustainability criterion).

Sustainable Transport

There can be no sustainable development without sustainable transportation

(Gudmundsson [22])

Sustainable transport is an essential component not only because transportation is a prerequisite to development in general but because the operation of transportation systems has significant impacts on sustainability contributing substantially to a wide range of environmental problems (Table 6.2).

Although there is a growing global interest in the concept of sustainable transport, there is not a universally accepted definition for the terms to guide politicians in solving transport challenges. Sustainable transportation systems are the systems that derived from the concept of sustainable development, and they include its basic principles [3]. defines it as "transport that meets the current transport and mobility needs without compromising the ability of future generations to meet these needs", while Pearce et al. [32] argue "transport and mobility with non-declining capital, where capital includes human capital, monetary capital and natural capital".

The initiative of the studies for sustainable transportation is derived from the Organisation for Economic Co-operation and Development (OECD) which in 1996 set in action the Environmentally Sustainable Transport (EST) project [29]. Since then, strategies, best practices and future visions of sustainable transport have been included in OECD's agenda. In 2002, in the framework of EST, the Environment Ministers of OECD member countries endorsed the guidelines for governments in order to develop and implement strategies towards sustainability in transport. In specific these guidelines provide a solution to making transport policy more sustainable, enabling economic development and enhancing quality of life without causing undue health and environmental impacts and depletion of finite resources [31].

Table 6.2 Transportation impacts on sustainability [28]

Economic	Social	Environmental
Traffic congestion	Inequity of impacts	Air and water pollution
Mobility barriers	Mobility disadvantaged	Habitat loss
Accident damages	Human health impacts	Hydrologic impacts
Facility costs	Community interaction	DNRR
Consumer costs	Community liveability	
Depletion of non-renewable resources	Aesthetics	

In 1996 in the publication of the World Bank entitled Sustainable Transport: Priorities for Policy Reform, the need for new challenges associated with transport policies to be faced in developing countries is presented [36]. It is acknowledged that transport is crucial to development, and World Bank operations have contributed to the creation of essential transport infrastructure in developing countries to improve access to jobs, education and health facilities and to facilitate domestic and international trade. New challenges facing transport sector have to be addressed:

- Changing standards for evaluating transport performance by travellers' needs
- Globalization of production and trade changes transport patterns
- Rapid motorization

In order the above mentioned challenges to be faced, it recommends the need to reform transport policies incorporating the idea of sustainability and to adopt policies that are more sustainable economically and financially, as well as environmentally and socially. Economic and financial sustainability requires that resources be efficiently used in the sector and that assets be properly maintained. Environmental and ecological sustainability requires that the external effects of transport are fully taken into account when public or private decisions are made that determine future development. Social sustainability requires that the benefits of transport improvements reach all sections of the community.

In the United States, many transportation agencies have shown an increasing interest in sustainable transportation. In 2001, the US Department of Transportation's Federal Highway Administration (FHWA) sponsored a study group that travelled to the four countries in Europe (Sweden, Germany, the Netherlands, the UK) to examine the way in which sustainable transportation issues are addressed [19]. Black [4] in Transportation Research Board Symposium on Sustainable Transportation in Baltimore conducts a systematic review of existing definitions on sustainable transportation. He argues that current transport system is no sustainable due to many parameters as diminishing petroleum reserves, global atmospheric impacts, local air quality impacts, fatalities and injuries, congestion, noise, low mobility, biological impacts and lack of equity. In 2011, FWA published a report titled *Transportation Planning for Sustainability Guidebook* to provide practices around the world and to examine how sustainability considerations could be better incorporated into transportation planning [20].

In Canada, Transports Canada (2007) thinks that a sustainable transportation system should be "safe, efficient and environmentally friendly" and in order to turn this system into reality, it requires integration of "economic, social and environmental considerations" into transportation policy-making [42].

The European Community has taken important steps towards the direction of a common transport policy in European countries. The Treaty of Rome in 1957 was the first step towards the generation of a Common European Transport Policy. The opening of transport market was the basic concept of the 1st White Paper of the Committee published in 1992. In 2001 the new White Paper "European Transport Policy for 2010: time to decide" was published [6]. One of the most important issues of White Paper is transport growth's gradual decoupling from economic development. Research programmes concerning alternative transport modes, transport multimodality, renewable energy sources, new technologies for improved energy consumption and safety are developed towards the direction of sustainable and anthropocentric transport systems.

European Conference of Ministers of Transport (ECMT), an intergovernmental organization, contributes to the creation of a European transport system based on the principles of sustainable development. It proposes a number of measures to be applied in all member countries [18]. An accurate definition of a sustainable transportation system is one that:

1. Allows the basic access and development needs of individuals, companies and society to be met safely and in a manner consistent with human and ecosystem health and promotes equity within and between successive generations
2. Is affordable, operates fairly and efficiently, offers a choice of transport modes and supports a competitive economy, as well as balanced regional development
3. Limits emissions and waste within the planet's ability to absorb them, uses renewable resources at or below their rates of generation and uses non-renewable resources at or below the rates of development of renewable substitutes while minimizing the impact on the use of land and the generation of noise

Holden et al. [25] have adapted the four main dimensions of sustainable development to the passenger transport sector, and they have identified suitable indicators and threshold values (Table 6.3). The total energy consumption per capita for passenger transport (all modes) is an indicator of ecological sustainability because all types of energy (renewable and non-renewable) cause a threat to long-term sustainability. In accordance with the Brundtland Report's low-energy scenario, the maximum threshold level was set at 5.6 kWh per capita per day. All people should have access to affordable and appropriate means of transport in order for basic transport needs to be satisfied (for work, health or education). Holden et al. [25] set 9.2 km daily as the minimum level, assuming that people having access to motorized travel above this level would be able to meet their basic transport

Table 6.3 Dimensions, indicators and suggested 2030 threshold values for sustainable passenger transport [25]

Dimension	Indicator	2030 Threshold
Impacts of transport activities must not threaten long-term ecological sustainability	Daily per capita energy consumption for passenger transport	Maximum 5.6 KWh per capita per day
Satisfying basic transport needs	Daily per capita travel distance by motorized transport	Minimum 9.2 km per capita per day
Promoting intragenerational transport equity	Public transport accessibility level (PTAL)	Minimum PTAL 3
Promoting intergenerational transport equity	The amount of renewable to total energy used for transport	Minimum 15%

needs. Intragenerational transport equity means that access to transport does not vary systematically across population groups. The challenge for such accessibility standards are needed to assess the performance of a transport network in terms of its success or failure to provide minimal levels of accessibility to all population groups. PTAL is a detailed and accurate measure of the accessibility of a point in the public transport network, taking into account walk access time and service availability. The PTAL is categorized from 1 to 6, where 6 represents an excellent level of accessibility and 1 a very poor level. Intergenerational transport equity requires that future generations be able to meet their transport needs. The amount of renewable to total energy used for transport is an indicator that future generations' transport needs can be met using alternative types of modes without the need for fossil fuel energy.

These threshold values for sustainable passenger transport define a four-dimensional space which Holden et al. call the "sustainable transport space" (STS). The performance of some countries addressing the challenge to satisfy two of the above thresholds is presented in Fig. 6.2. Sustainable transport is defined as the area in the lower right quadrant where the maximum and minimum requirements are met – i.e. per capita energy consumption is below 5.6 kWh/day and per capita travel distances is above 9.2 km/day. As it can be seen, the countries of Romania and Slovakia are included in the STS area, while the performance of Armenia and Albania (in the lower left quadrant) is out of STS, showing that the latter should focus on strategies towards increasing motorized travel per capita. However, this should be done taking into consideration appropriate policies referring to the

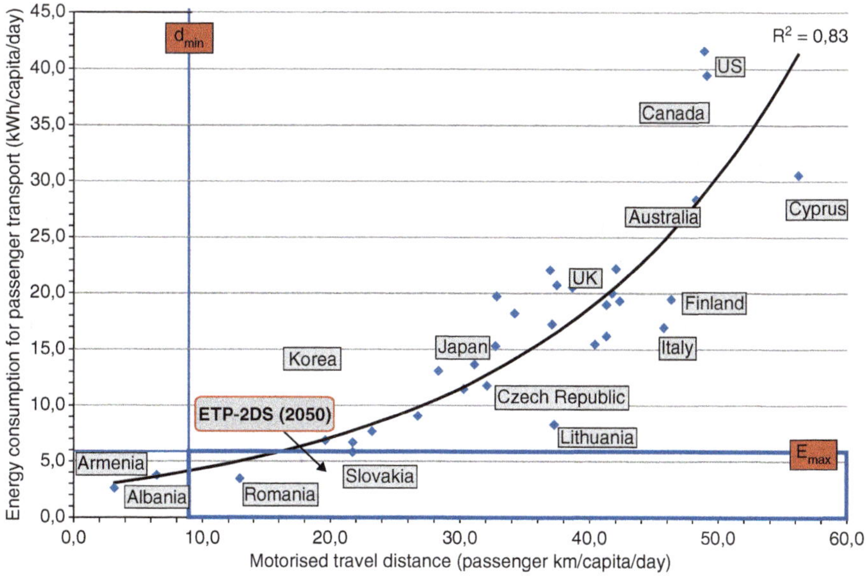

Fig. 6.2 The sustainable transport space (Data from [25])

sustainable land-use planning and to the promotion of public transport systems to ensure that the increased motorized travel remains at the lowest possible level of energy consumption. In developed countries, achieving sustainable transport creates different challenges. For example, Lithuania and the UK have comparable travel distances per capita but different energy consumption per capita, and consequently different strategies have to be implemented towards sustainability.

Sustainable Urban Mobility

Transport as Derived Demand or as Valued Activity?

(Banister 2008)

Urban Mobility and New Perspectives in Transport Planning

Urban areas constitute the living environment of the vast majority of the population. The quality of life in these areas should be as high as possible to support the growth and employment. Urban mobility is of growing concern to citizens. According to "Action Plan on Urban Mobility" [8], nine out of ten EU citizens believe that the traffic conditions in their area should be improved. The choices that people make in the way they travel will affect economic well-being of citizens, future urban development and urban environment.

Socio-economic factors are related with the transport mode used, as the increase in GDP leads to growth in car ownership (Fig. 6.3). However, in some wealthy cities like Hamburg and Helsinki, decoupling of these two parameters appears to have occurred.

Urban residents' mobility choices are strongly linked with density. In denser areas there is a higher prosperity to opt for alternative modes other than private cars (Fig. 6.4). The shorter trip distances associated with high-density areas naturally lend themselves to more walking and cycling. Higher-density areas also involve high concentration of activities, thereby allowing public transport to efficiently connect the locations where the origins or destination trips are concentrated.

Worldwide, car travel has increased over the past 20 years. Between 1995 and 2012, the rate of car ownership per 1000 inhabitants rose in all countries in Europe.

A recent survey of residents of 75 EU cities explored which mode of transport people use most often on a typical day [11, 38]. This survey does not cover the commuting zone and so it does not include people working in the city, but living outside the city. The results of this survey by mode of transport shows that the mean share of car use is below 30%, with Lefkosia having more than 70% and Paris less than 10%. In almost all the cities surveyed, at least 20% of the residents rely on public transport, while for 21 cities it was the main mode for more than half of the residents. The only city, where walking is the main mode of transport for the

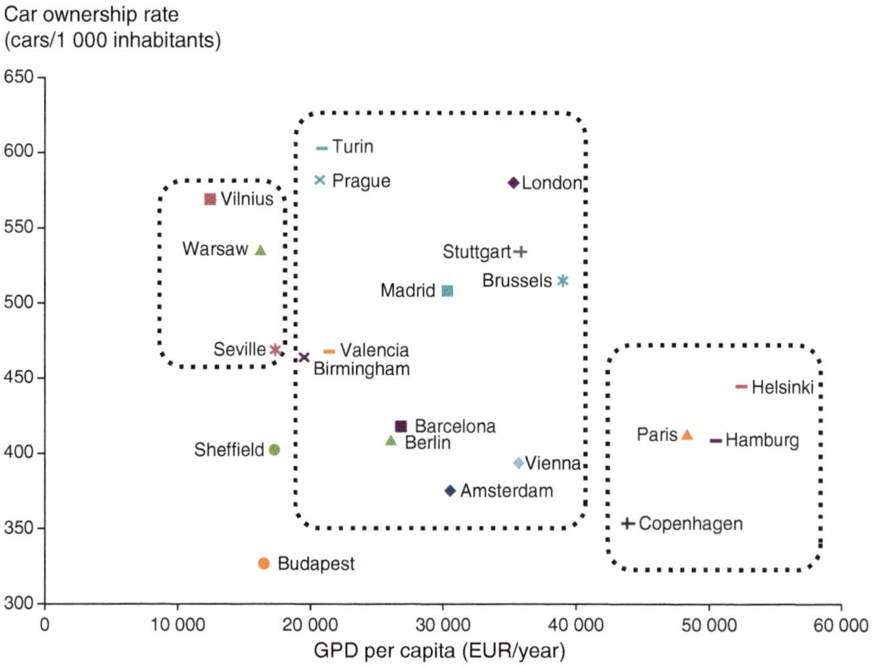

Fig. 6.3 GDP in relation to car ownership growth (EEA TERM 2013)

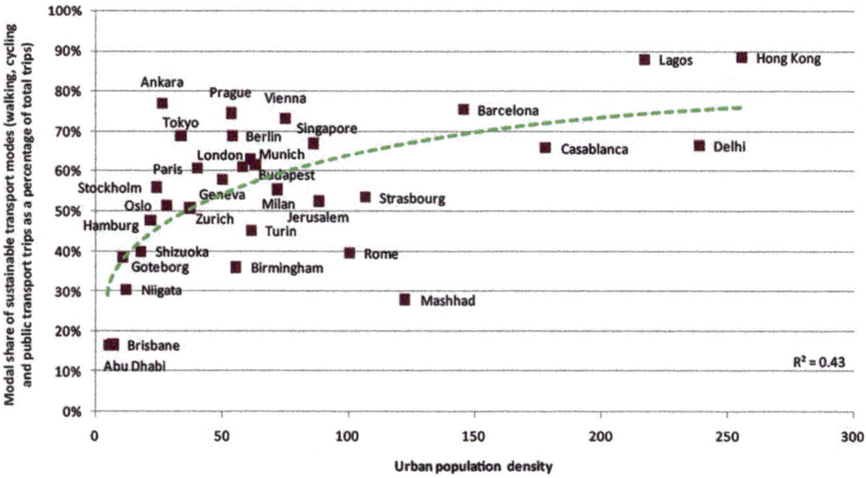

Fig. 6.4 Share of total daily trips undertaken by sustainable transport modes – walking, cycling and public transport – out of the total number of daily trips compared by urban population density in metropolitan area [38]

majority of residents, is Paris. Nevertheless, in two out of three cities, at least 25% of the population walk in most places. The survey showed that in half of the cities surveyed, cycling did not reach more than a10% modal share, in fourteen it reached a share of more than 25% and in three (Amsterdam, Copenhagen and Groningen) the share was even over 50%.

The developing and implementing policies of sustainable urban mobility is a great challenge. It provides a balanced service of conflicting and potential additional environmental, social and economic needs of the functioning of the city. Rethinking urban mobility involves optimizing the use of all the various modes of transport and organizing "co-modality" between the different modes of collective transport5 (train, tram, metro, bus, taxi) and the different modes of individual transport (car, motorcycle, cycle, walking). It also involves achieving common objectives in terms of economic prosperity and managing transport demand to guarantee mobility, quality of life and environmental protection [7].

In recent years, there was a shift from the conventional approach of the transport planning system, where key priorities were to increase mobility and minimize the travel time (physical dimensions in terms of mobility) to the new sustainable mobility approach which considers social dimensions in terms of accessibility and people. Priorities in this new approach are the increased access to various activities, the multimodal development of transport with upgraded hierarchy of the mobility of pedestrians and bicyclist and the ensuring of reasonable travel times leading to a reliable transport system which is now valued not only with economic but also with environmental and social criteria. The accessibility is secured not only by the appropriate development of transportation system but also of the land-use planning, where compactness and functional diversity contribute significantly towards its achievement. Banister [2] argues that the key policy objective becomes the reasonable travel time than travel time minimization. In these terms, the conventional planning strives to minimize travel time and thus to speed up traffic, whereas sustainable mobility attempts to realize reasonable and reliable travel times which may require slowing down movement. Transport policy measures can reduce levels of car use through the promotion of walking and cycling and the development of the new transport hierarchy by reallocating space to public transport, through parking controls and road pricing making thus easier to use public transport. The main differences between the conventional and new alternative approach in transport planning are presented in Table 6.4.

Sustainable Urban Transport Policies and Strategies

WBCSB in "The Sustainable Mobility Project" (2001) defies sustainable mobility as "the ability to meet the needs of society to move freely, gain access, communicate, trade and establish relationships without sacrificing other essential human or ecological values today or in the future" [39, 41].

Table 6.4 Contrasting approaches to transport planning [2]

The conventional approach – transport planning and engineering	An alternative approach – sustainable mobility
Physical dimensions	Social dimensions
Mobility	Accessibility
Traffic focus, particularly on the car	People focus, either in (or on) a vehicle or on foot
Large in scale	Local in scale
Street as a road	Street as a space
Motorized transport	All modes of transport often in a hierarchy with pedestrian and cyclist at the top and car users at the bottom
Forecasting traffic	Visioning on cities
Modelling approaches	Scenario development and modelling
Economic evaluation	Multicriteria analysis to take account of environmental and social concerns
Travel as a derived demand	Travel as a valued activity as well as a derived demand
Demand based	Management based
Speeding up traffic	Slowing movement down
Travel time minimization	Reasonable travel times and travel time reliability
Segregation of people and traffic	Integration of people and traffic

Urban transport systems are integral elements of the European transport system and, as such, constitute an integral part of the Common Transport Policy under Articles 70–80 EC Treaty. In addition, other EU policies (cohesion policy, environment policy, health policy, etc.) cannot achieve their objectives without taking into account urban specificities, including urban mobility. EU-funded initiatives, often supported by the Framework Programmes for research and technological development, have helped to develop innovative approaches which stimulate authorities at local, regional and national level to adopt the long-term integrated policies. The role of public authorities in providing the planning, the funding and the regulatory framework is essential. The main benchmark European policies in sustainable urban policies are:

1992 The Green Paper where the Commission of the European Union (EC) launched the concept sustainable mobility as the challenge to initiate a public debate on the issue of transport and the environment [17].

Sustainable mobility is a mobility in accordance with the principles and requirements of sustainable development. This introduces the two concepts: mobility and sustainable development. In order to understand the concept of sustainable mobility, it is necessary to understand both of these basic concepts. With the objective of contributing to a critical discourse, the overall perspective has again to be critical. Both concepts are complex and subject to large differences in understanding. Such differences can originate in perspectives and traditions given by various scientific disciplines. However, they can also have a more fundamental basis through variances in value systems and preferences. The complexity only increases when the two are combined into one: sustainable mobility.

1992 The document on the future development of the Common Transport Policy: A Global Approach to construction of a community framework for sustainable mobility published by EC highlighted that achieving sustainable transport is a matter of reducing traffic intensity than transport volumes [12].

2007 The Green Paper "Towards a new culture for urban mobility" adopted by EC in order to set a new European agenda for urban mobility while respecting the responsibilities of local, regional and national authorities in this field [7].

It refers that urban mobility should make possible the economic development of towns and cities and in the meanwhile should secure the quality of life of their inhabitants and the protection of their environment. The main challenges that European towns and cities have to meet as part of an integrated approach for urban mobility are:

- Free-flowing towns and cities
- Greener towns and cities
- Smarter urban transport
- Accessible urban transport
- Safe and secure urban transport

However, in order the challenge facing urban areas in the context of sustainable development to be met, the need of creating a new "urban mobility culture" is imperative. A joint effort, setting up partnerships, will make it possible to encourage the search for innovative and ambitious urban transport solutions, new planning methods. Education, training and awareness rising have an important role to play towards this direction.

2009 The Action Plan in urban mobility is published by EC, providing a coherent framework for EU initiatives in the area of urban mobility. It proposes short- and medium-term practical actions to be launched progressively addressing specific issues related to urban mobility in an integrated way [8].

2011 In the White Paper 2011, entitled "Roadmap to a Single European Transport Area – Towards a competitive and resource-efficient transport system", the EC defines a long-term vision until 2050 for the transport sector [9]. It provided a solid ground for upcoming policy debates and actions and a roadmap of 40 concrete initiatives for the next decade to build a competitive transport system that will increase mobility, remove major barriers in key areas and fuel growth and employment. Specifically, for the urban context, different strategy policies are suggested involving land-use planning, pricing schemes, efficient public transport services and infrastructure for nonmotorized modes and charging/refuelling of clean vehicles in order to address the sustainable mobility. The White Paper sets out specific urban targets:

- Conventionally fuelled cars will be phased out in cities by 2050, and their use will be halved by 2030.
- A target of CO2-free city logistics in major urban centres by 2030.
- Reducing road accident fatalities by 2030 by half and to zero by 2050.

Moreover, it is suggested that cities above a certain size should be encouraged to develop Urban Mobility Plans that should be fully aligned with Integrated Urban Development Plans aiming to address current and future transport needs sustainably.

2013 The Urban Mobility Package is published together with towards competitive and resource-efficient urban mobility [10]. With the Urban Mobility Package, the Commission reinforces its supporting measures in the area of urban transport by:

Sharing experiences, showcasing best practices and fostering cooperation

- Providing targeted financial support
- Focusing research and innovation on delivering solutions for urban mobility challenges
- Involving the member states and enhance international cooperation

The sustainable transport planning approach, contrary to the priorities of the conventional approach which are the encouragement of the use of private vehicles and the construction of additional road infrastructure, focuses on the promotion of the alternative means of transport, i.e. walking, bicycle and public transport, and sets as a primary objective the provision of mobility and information services as well as the better management of the existing networks [40].

Holden [23, 24] argues that there are three main approaches for developed countries to enter to sustainable passenger transport, efficiency, alternation and reduction, terms that can be characterized, respectively, as "travel more efficiently", "travel differently" and "travel less". Additionally, he presents fourteen theses regarding the roles of technology, public transport, green attitudes and land-use planning in achieving sustainable mobility.

Options to improve urban transport have been analysed in terms of three key approaches by Dalkan and Brannigan [5]:

- **Avoid** – the need to travel to access goods and services through efficient urban planning, communication technology, consolidation activities and demand management.
- **Shift** – where appropriate, people and goods moved towards more sustainable modes such as walking, cycling and public transport rail.
- **Improve** – the environmental performance of vehicles with the adoption of low-emission vehicle technologies and more efficient operation of vehicles.

The implemented strategies that aim at improving the performance of urban transport and reduce their environmental impacts have to be in a context of integrated approach taken into consideration the characteristics of each individual city [14, 15]. suggests strategies based on the three key approaches of Dalkan and Brannigen (Table 6.5).

Table 6.5 Strategies towards urban sustainability

Avoiding the need to travel	Supporting modal shift	Improving modal efficiency
Land use and planning	Increasing the share of walking and cycling	New mobility services
Information and communications (ICT)	Developing the use of public transport	Traffic management and integration
Access management	Alternatives to road freight	Driver behaviour
Consolidating supply and demand		Regulation and pricing

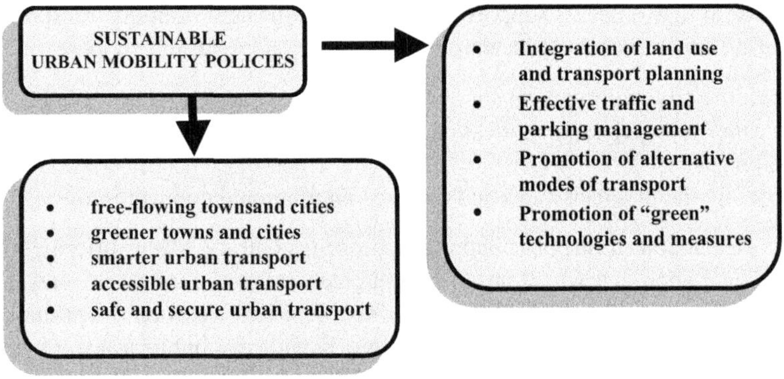

Fig. 6.5 Sustainable urban mobility strategies [34]

Four principles for the new approach of sustainable urban transport and development were presented by Banister [2]:

- Reducing the need to travel – substitution
- Transport policy measures – modal shift
- Land-use policy measures – distance reduction
- Technological innovation – efficiency increase

Sustainable transport can be achieved with a strong combination of the above principles, based on regulation, land-use development including planning and regulations, use of technology and information.

Policies in achieving sustainable urban travel have to be examined in a context of integrated approach taking into consideration land-use planning, effective traffic and parking management, reliable transport system, infrastructure for alternative modes of transport (cycling, walking) and promotion of "green" technologies [34] (Fig. 6.5).

However, many barriers to the implementation of an integrated approach of sustainable mobility are common and make the progress to be very slow. Banister [1] describes some of them:

- Application of general planning and car parking standards and prices which are
- inappropriate and unachievable in a town centre/high street context
- Reluctance of people to use public transport and cycle/walk, even for local trips
- Strong desire of urban residents to use their cars
- Fragmentation of the organization, integration and management of public transport
- Uncertainty over the funding of public transport and nonmotorized modes
- Separation of planning and transport functions within local authorities

The most important and widespread barriers are however the lack of political will and funding. Measures to overcome barriers are the raising awareness about the effectiveness of integrated approaches and creating individual culture to support the measures through active involvement and action [14].

Monitoring Sustainable Mobility in Cities

It was Agenda 21, the action plan adopted in 1992 at the United Nations Conference on Environment and Development in Rio, that first calls on countries as well international organizations to develop indicators as a tool to monitor sustainable development. In specific the implementation of indicators can provide a solid basis in order the complex and challenging idea of sustainable development can be determined and measured. Indicators are defined as statistics or quantitative measures designed to identify significant trends, point out problems, track the progress over time towards a specific vision-objective, contribute to the priority-setting and inform both the experts and the public about a complex phenomenon in simple way [13, 30].

Urban sustainability is a multidimensional concept that includes environmental, economic and political dimensions, and its assessment is a major challenge for political, transport and environmental authorities. The use of urban sustainability indicators constitutes internationally an important tool to assess sustainability and monitor the progress. The Institute for Environment and Sustainability of the EC Joint Research Centre after an extensive literature review developed a comprehensive indicator system in order to assess the sustainability of the transport activities. The WBCSD in the paper Mobility 2030: Meeting the challenges to sustainability proposes a framework of 11 different sets of indicators, measuring progress towards a set of seven "goals" to "improve the outlook" of sustainable mobility adopting policies in this direction.

Gillis et al. [21] examined sustainable mobility indicators across literature by the principles of neutrality and transferability. They gave a set of 22 indicators that cover different aspects of sustainable mobility that are applicable in different social and economic contexts. All these indicators are positioned in a four-dimensional space a = (g,q,e,m) (g-global environment, q-quality of life, e-economic success, m-mobility system).

Karagiannakidis et al. [26] present the results of literature review of main sustainable urban mobility indicators. The plurality of indicators found in the literature and the considerable number of sustainable mobility indicator initiatives are the result of many parameters including the high interest in sustainability issues, the great complexity of the transport system and the specific features of each urban area. However, specific criteria have to be met in order for an indicator to be selected. Tafidis et al. [35] examined different criteria of indicators based on extended literature review. The main character is that an indicator should provide useful information concerning the performance in terms of social, economic and environmental sustainability (relevance to sustainability), while it has to be capable of illustrating even the slight changes. Indicators should be relevant to the policies, objectives and goals that are expected to measure and to illustrate the impact of transport-related policies. The structure of an indicator should be simple and transparent in order to be easily understandable either to experts, policymakers and other stakeholders or to the public. Moreover, an indicator should enable comparisons both between different urban areas and time periods. The latter but extremely significant criterion for selecting an indicator consists of the affordability, i.e. the necessary cost and time in order to collect the original required data and subsequently estimate the indicator's value. Affordability is considered to be dependent on the data availability, the data frequency and the data reliability (accuracy), and it comprises in most cases the weakest point during the selection process, as a result of the limited abilities in gathering data of many local authorities.

Pitsiava-Latinopoulou [34] proposes a comprehensive set of indicators that covers different aspects of sustainable mobility taking into account the four strategies, outlined previously (Fig. 6.5), for urban mobility. The main objectives identified were to integrate the transport policies and the land-use planning, reduce the use of private vehicle trips by promoting the use of public transport and of "green modes" walking cycling as well as the use of new technologies, improve the environmental quality and promote the urban economy (Table 6.6).

Good Practices

In European cities, potential solutions for sustainable mobility need to be drawn into a consistent and coherent mobility plan, integrated with other city plans and policies [14]. The 2013 Urban Mobility Package sets out a concept for Sustainable Urban Mobility Plans (SUMPs) that has emerged from a broad exchange between stakeholders and planning experts across the EU. New approaches to urban mobility planning are emerging as local authorities seek to develop strategies that can

Table 6.6 Sustainable urban mobility indicators [34]

Integration of land-use and transport planning	Effective traffic and parking management	Promotion of alternative modes of transport	Promotion of green technologies and measures
GDP per capita	Traffic volume	PT network coverage	Share of population exposed to values of Lden above 60 dB(A) or Lnight above 55 dB(A)
Population density	Modal split	PT modal share	Annual CO2 emissions per capita
Land-use mix	Occupancy rate of passenger vehicles	PT reliability, comfort, safety and security	Annual emissions of air pollutants (CO, NOx, SO2, VOC) per capita
Ratio of total number of jobs to the total population living in the city	Number of trips	Number of park and ride lots	Average age of vehicle fleet
Network connectivity	Total vehicle-km travelled	Share of household income devoted to PT	Share of vehicle fleet by engine technology and type of fuel
Share of population living within 500 m from PT stops	Average speed of passenger vehicles	Modal share of nonmotorized modes (walking, bicycle)	Share of heavy vehicles powered with CNG
Share of population living within 500 m from basic services	Average PT speed	Total length of PT, pedestrian and bicycle networks	Average concentrations of air pollutants and exceedances of air quality standards
	Number of road accidents	Pedestrian and bicycle volumes	
	Travel cost and average fuel consumption per vehicle	Total length of roads with traffic calming measures	
	Traffic noise levels	Walkability index	
	Emissions of air pollutants	Number of bicycle parking spaces	
		Number of road accidents involving vulnerable users	

stimulate a shift towards cleaner and more sustainable transport modes. The process for developing a SUMP was clearly set out in the guidelines developed by the Eltis Plus project. SUMP consists of four key stages covering analysis of the existing situation, setting improvement goals, developing a clear set of actions and implementing strategy. In developing these stages, they need to consider all aspects of mobility, both passenger and freight, and the wider economic development of the city [16]. Many of Europe's towns and cities are leading the way in addressing these issues. In each of the main policy areas, examples of highly successful approaches can be identified. Nevertheless, these good practices should not be taken for implementation as they are by every town or city, but they should be adopted to its individual specific characteristics and its individual needs (Table 6.7).

Conclusions

The need for transformations in mobility systems demands the "avoid, shift and improve framework" that is based on three directions: to rethink the need of mobility avoiding unnecessary trips, to shift to a more environmental-friendly transport mode and to improve the efficiency of transport modes. Implementing new technology, improving public transportation, increasing individuals' green attitudes, promoting sustainable land-use planning and implementing information and communication technology are all important elements towards the achievement of sustainable mobility. A comprehensive indicator system based on a number of different criteria such as policy relevance, continuity, compatibility, sensitivity as well as data availability, frequency and reliability is a robust instrument in order to assess the sustainability of the transport activities and the effectiveness of the implemented policies. However, as Holden [23] argues, changes towards sustainable mobility must start with the transformation of the attitudes and values of a large majority of people. Individual attitude in travel choices has to be changed as sustainable mobility is not just a goal but a change in direction of our lives.

Acknowledgments I would like to thank Professor Pitsiava-Latinopoulou Magdalini of Civil Engineering Department of Aristotle University of Thessaloniki whose experience in urban transport planning contributed to this work.

Table 6.7 Good practices in European cities

	With the elegantly designed, modern, highly functional and environmentally friendly tram system, Strasbourg has made a name for itself as a pioneer in urban transport. The city was one of the first in France to opt for a return to a transport system which had been abandoned in the second half of the nineteenth century and, in so doing, heightened its international reputation. 55.5 km of commercial tracks, 6 routes, 69 stations, 300,000 passengers daily
	Since 1962 Copenhagen's largest shopping area has been centered around Strøget in the heart of the city. Strøget is far and away the most famous street in Copenhagen. The walking street is 1111 meters long, making it Europe's longest pedestrian street. The street is a wonderful place to go shopping or simply to go for a pleasant stroll, taking in the sights and sounds of this vibrant city.
	Vienna has a well-developed public transport network with reliable, clean and convenient service. Buses, trains, trams and underground lines will take you almost anywhere in the city in no time at all.
	Vauban is a neighbourhood to the south of the town centre in Freiburg, Germany. The transport is primarily by foot or bicycle. Vauban limits car use through parking-free residential streets, spatially and fiscally separated parking and filtered permeability to prevent through traffic. Attractive alternatives include: Frequent rail-based transit system and extensive, high quality non-motorized transport infrastructure.

References

1. Banister, D.: Unsustainable Transport. City Transport in the new Century. Routledge, New York (2005)
2. Banister, D.: The sustainable mobility paradigm. Transp. Policy. **15**, 73–80 (2008)
3. Black, W.R.: Sustainable transportation: a U.S. Perspective, Journal of Transport Geography. **4**, 151–159 (1996)
4. Black, W.R.: 2005. Sustainable transport definitions and responses. In: Transportation Research Board, 2005. TRB conference proceedings 37: integrating sustainability into the transportation planning process, Washington, DC
5. Dalkan, H., Brannigan, C.: Transport and climate change: module 5e. In: Sustainable Transport: A Sourcebook for Policy-Makers in Developing Cities. Deutsche Gesellschaft für Technische Zusammenarbeit (GTZ), Eschborn Germany (2007)
6. EC: European transport policy for 2010: time to decide. White paper from the office for official publications of the European Communities (2001)
7. EC: Towards a New Culture for Urban Mobility. Green paper from the office for official publications of the European Communities (2007)
8. EC: Action plan on urban mobility. Opinion of the European economic and social committee on the communication from the commission to the Euro-pean parliament, the council, the European economic and social committee and the committee of the regions (2009)
9. EC: Roadmap to a single European transport area. Towards a competitive and resource efficient transport system. White paper from the office for official publications of the European Communities (2011)
10. EC: Urban mobility package. Office for official publications of the European Communities (2013)
11. EC: European cities 2016. Cities leading the way to a better future (2016)
12. EC (European Commission): The future development of the common transport policy: a global approach to the construction of a community framework for sustainable mobility. White paper from the commission to the European council (1992)
13. EEA: EEA core set of indicators: guide. EEA technical report, Copenhagen 2005
14. EEA: A closer look at urban transport. EEA technical report, Copenhagen 2013
15. EEA: Transitions towards a more sustainable mobility system. EEA technical report, Copenhagen 2016
16. Eltis: The urban mobility observatory. www.eltis.org
17. EUCOM: Green paper on the impact of transport on the environment. A community strategy for “Sustainable mobility”. Commission of the European community. COM (192) 46 final. Brussels (1992)
18. European Conference of Ministers of Transport (E.C.M.T.): Sustainable Transport Policies. OECD, Prague (2000)
19. Federal Highway Administration (FHWA): Sustainable Transportation Practices in Europe. Federal Highway Administration, Washington, DC (2001)
20. Federal Highway Administration (FHWA): Transportation Planning for Sustainability Guidebook. U.S. Federal Highway Administration, Washington, DC (2011)
21. Gillis, D., Demanjski, I., Lauwers, D.: How to monitor sustainable mobility in cities? Literature review in the frame of creating a set of sustainable mobility indicators. Sustainability. **8**, 29 (2015)
22. Gudmundsson, H., Hall, R., Marsden, G., Zietsman, J.: Sustainable Transportation. Indicators, Framworks and Performance Management. Springer, Berlin (2016)
23. Holden, E.: Achieving Sustainable Mobility: Everyday and Leisure-Time Travel in the EU. Ashgate, Aldershot (2007)
24. Holden, E., Linnerud, K.: The sustainable development area: satisfying basic needs and safeguarding ecological. Sustain. Dev. **15**(3), 174–187 (2007)

25. Holden, E., Linnerud, K., Banister, D.: Sustainable passenger transport: back to Brundtland. Transp. Res. A. **54**, 67–77 (2013)
26. Karagiannakidis, D., Sdoukopoulos, A., Gavanas, N., Pitsiava-Latinopoulou, M.: Sustainable urban mobility indicators for medium-sized cities. The case of Serres, Greece. 2nd Conference on Sustainable Urban Mobility, Volos, Greece (2014)
27. Lafferty, W.M.: Governance for Sustainable Development. The Challenge of Adapting Form to Function. Edward Elgar, Cheltenham (2004)
28. Litman, T., Burwell, D.: Issues in sustainable transportation. Int. J. Global Environmental Issues. **6**(4), 331–347 (2006)
29. OECD: OECD proceedings: towards sustainable transportation. Conference organized by the OECD, Vancouver 1996
30. OECD: Indicators for the integration of environmental concerns into transport policies. Working group on the state of the environment (1999)
31. OECD: Guidelines towards Environmentally Sustainable Transport. OECD, Paris (2002)
32. Pearce, D.W., Markandya, A., Barbier, E.: Blueprint for a Green Economy. Earthscan, London (1989)
33. Pitsiava-Latinopoulou, M., Basbas, S., Christopoulou, P.: Sustainable transport systems: trends and policies. WIT Trans. Built Environ. **89**, 13–22 (2006)
34. Pitsiava-Latinopoulou, M.: Indicators for sustainable mobility in urban areas. Newsletter, Hellenic Institute of Transportation Engineers. **182**, 13–17 (2012)
35. Tafidis, P., Sdoukopoulos, A., Pitsiava-Latinopoulou, M.: Sustainable urban mobility indicators: policy versus practice in the case of Greek cities. 3rd conference on sustainable urban mobility 2016, Volos 2016
36. The World Bank: Sustainable Transport: Priorities for Policy Reform. World Bank, Washington DC (1996)
37. The World Commission on Environment and Development, United Nations: Our Common Future (1987)
38. UITP, Advancing Public Transport: Mobility in cities database. Synthesis report (2015)
39. WBCSD: The Sustainable Mobility Project, Mobility 2030: Meeting the challenges to sustainability (2004)
40. WBCSD: Mobility for development. Report of the mobility for development project (2009)
41. World Business Council for Sustainable Development (WBCSD): Mobility 2001–world mobility at the end of the twentieth century and its sustainability (2001)
42. Zhou, J.: Sustainable transportation in the US: a review of proposals, policies and programs since 2000, Frontiers of architectural research (2012) 1, pp. 150–165 (2012)

Chapter 7
Sustainable Operations of Closed-Loop Logistics Chain from an Economic and Environmental Performance Perspective

B.Y. Liu and H.D. Yang

Abstract With increasingly scarce global resources, gradually worsening waste emissions, and rising environmental protection awareness, an increasing number of countries enacted strict regulations to protect the environment; in addition, the issues on recycling and reusing used products have caught worldwide attention. Adapting to environmental protection and consumer requirements has compelled manufacturers and distributors to formulate forward and reverse logistics networks (RLNs) simultaneously. Moreover, the increasingly transparent contradiction between the supply and demand of resources coincides with the enterprises' experience of growing cost pressures, for which recycling and remanufacturing gradually became important techniques to reduce the production cost. Under this background, the closed-loop logistics (CLL) has received significant attention from the academia and industry. In this chapter, we defined the related concepts with CLL; proposed the theoretical framework, including the CLL network structure and sustainable operations of logistics; and analyzed the effect of CLL on manufacturing, services, and people's lives with a special focus on the main sustainable operations, namely, recycling and remanufacturing. The results revealed that implementing CLL management became a strategic choice for many enterprises, with its vital significance to decrease waste emissions, protect the environment, reduce production costs, improve economic efficiency, enhance competitiveness, promote enterprise technology innovation, and strengthen environmental protection.

Keywords Closed-loop logistics chains • Reversed logistics • Waste emission • Economic performance • Environmental protection

---

B.Y. Liu (✉) • H.D. Yang
School of Economics and Management, Fuzhou University, Fuzhou, 350116, China
e-mail: jasperseu@fzu.edu.cn; yanghaidong428@163.com

A. Karakitsiou et al. (eds.), *City Networks*, Springer Optimization and Its Applications 128, DOI 10.1007/978-3-319-65338-9_7

What Is Closed-Loop Logistics?

Given limited energy and resources, sustainability has recently become an important topic in environmental protection. The green supply chain (GSC) management is now suggested as an efficient tactic to achieve this goal. Based on the three Rs, namely, recycling, reuse, and recovery in GSC management, a green company strives to prevent any wastage of materials from the life cycle of a product. Therefore, closed-loop logistics (CLL) is required to facilitate the 3R processes.

Although an increasing number of scholars have focused on studying CLL and proposed many types of definitions and research framework since the twenty-first century, the academia currently has no authoritative definition on CLL.

From the product life cycle perspective, CLL refers to the entire process from raw material (parts) procurement, transportation, processing and manufacturing, distribution, consumption, recycling, and remanufacturing to resale and reuse. The latter achieves the circulation of material and function, maximizes the value of the products by recycling and reusing, and minimizes the impact on the environment. Figure 7.1 shows the material flow of a CLL. CLL refers to activities associated with the transformation and the flow of goods and services with their information from the sources of the materials to the end users.

CLL is essentially a new concept and a novel supply chain management perspective that gestated in the social ideological trend of sustainable development and circular economy; it also plays a highly significant role in environmental protection. CLL is a type of logistics chain in the inverse direction of supply chain, based on and bound to the traditional forward supply chain. Furthermore, CLL considers social resources, environmental requirements, and economic interests to realize resource recycling and subsequently maximize resource savings and reduce waste emissions.

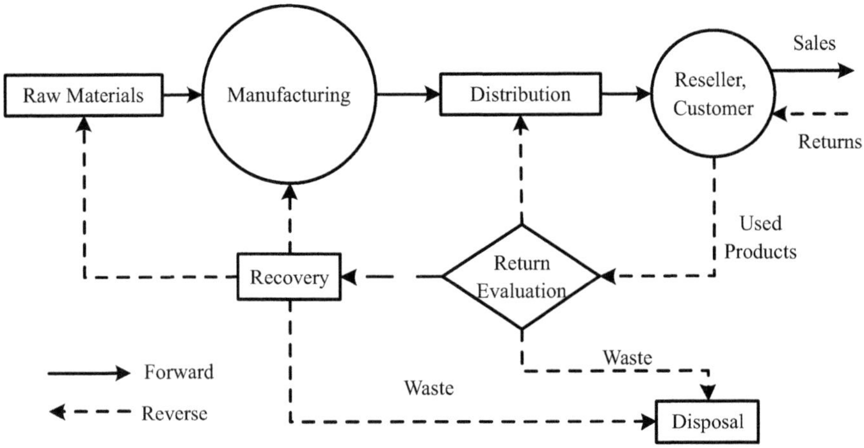

Fig. 7.1 Material flow of CLL [62]

Fleischmann et al. [20] emphasized that a closed-loop supply chain may include traditional manufacturers and retailers with logistics service providers in the forward channel; moreover, this chain may also contain specialized parties of secondary material dealers and material recovery facilities in the backward channel. One of the key aspects in closed-loop management is the simultaneous improvement of both economic and environmental performance throughout the chain by establishing long-term relationships between buyers and suppliers (Zhu and Cote 2004). Hence, building a stable CLL in a chain is necessary when companies aim at minimizing the total cost and the involved environmental impacts.

Therefore, apart from the conventional logistics described by a network of suppliers, manufacturing sites, distribution centers (DCs), and customer locations through logistics, an important module on dismantlers or recyclers is incorporated into a supply chain network for a CCL problem. These dismantlers handle the recovered resources into many different types for further use or disposal [7]. If the recycled resources can be used again at dismantler sites, then the resource should be shipped to a manufacturer for reproduction; otherwise, useless resources must be sent to a landfill [59].

Network Structure and Sustainable Operations in CLL

Technically, CLL consists of two parts, namely, forward and reverse logistics. For the forward logistics (as a conventional logistics), distributors will deliver the final products to the customers after manufacturing to satisfy their demands. In this part, the position of customers is typically the end of the process.

For the reverse logistics, the flow of used products is processed from customers back to dismantlers to perform sorting or disassembling for recovery, reuse, or disposal ([7, 30, 48], Schultmann et al. 2009). CLL management aims to ensure the production of the least material waste by following the conservation law of material life cycles.

In general, the following three issues must be considered for a network planning problem: validity of the model, efficiency of the solution, and applicability. The configuration of forward and reverse logistics network (RLN) should be integrated for a CLL problem.

In most previous studies, the designs of forward and reverse logistics have been separately considered. However, the forward logistics network (FLN) configuration is thoroughly impressed by the RLN and vice versa because they share a number of resources, such as transport and warehouse capacity. Moreover, returns information should be integrated with forward logistics data to achieve optimum planning and to reduce costs. Therefore, the design of FLN and RLN should be integrated to avoid the suboptimalities caused by a sequential design [42]. The CLL framework is shown in Fig. 7.2. The integration of the forward and reverse logistics model is called the CLL model.

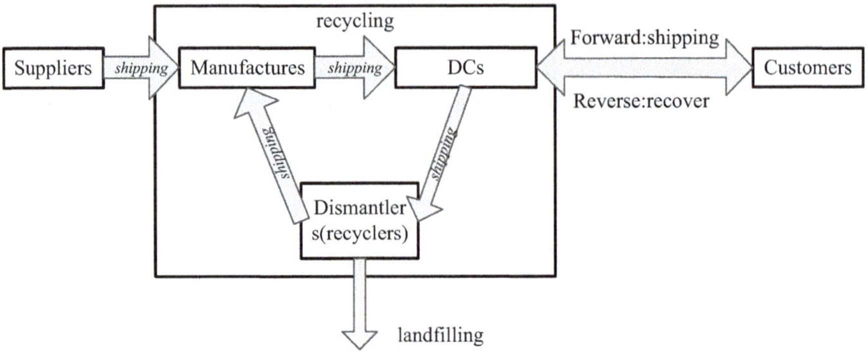

Fig. 7.2 CLL framework

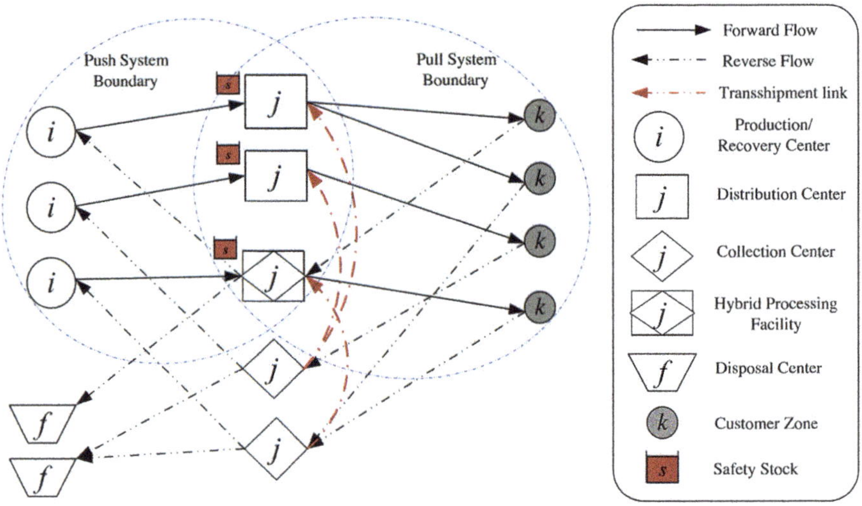

Fig. 7.3 Proposed integrated FLN//RLN [39]

An integrated forward and reverse logistics network (IFRLN) generally consists of supplying raw materials from suppliers, converting these materials to end products, shipping them to proper DCs, and delivering to customer zones; subsequently, the process proceeds to collecting used products from customer zones and finally recovering or remanufacturing and performing proper disposal. Instead of only separately considering forward processing facilities (i.e., DCs) and backward processing facilities (i.e., collection centers), Keyvanshokooh et al. [39] considered hybrid processing facilities wherein both distribution and collection centers are established at the same locations as Fig. 7.3. Compared with separate distribution or collection centers, hybrid processing facilities suggest other advantages consisting of cost savings and pollution reduction because of the sharing of material handling equipment and infrastructures [43]. In this integrated network, hybrid distribution–

collection facilities are considered to accomplish additional cost savings. Whether the hybrid processing facility is used or not depends on the trade-off of fixed opening and variable costs. Thus, such a logistics network turns the use of hybrid processing facilities into a decision variable.

In CLL, sustainable operations include two core disposition decisions, namely, remanufacturing and recycling [19]. Recycling means material recovery, and it is an attractive option for products, where returns have minimal economic value due to technological obsolescence or poor-quality products. Remanufacturing is a value-added operation with the potential for high profitability among disposition decisions.

Recycling in CLL

The recent rapid development of information technology and social economy increased product upgrades, especially on electrical and electronic equipment (EEE), which simultaneously resulted in massive waste products [66, 72]. The failure to collect or properly discard these waste products may pose a serious threat to the environment and lead to the failure to obtain the economic salvage value. For example, the entire global waste from EEE would reach approximately 65 million tons by 2017, which will increase by 33% compared with 49 million tons in 2012. Over 700 chemical materials are identified in the Waste Electrical and Electronic Equipment (WEEE), and half of them are hazardous to human health, whereas WEEE may derive 80–1500 g of gold and 350–1850 g of silver per ton, valued at several thousand US dollars [67]. To realize the economic value of WEEE and correspondingly reduce its environmental harm, relevant legislations for WEEEs' implementation and an improvement in consumers' environmental protection consciousness emerged as significant practices.

Given to various reasons, especially those pertaining to environmental concerns and economic potential associated with used products and resulting business options, many enterprises, such as IBM, Dell, GM, HP, Kodak, Xerox [65], and ReCellular, Inc. [29], started recycling and disposing WEEEs themselves or outsourced such tasks to third-party logistics providers [12, 15, 26, 38], such as FedEx, ASTRA, and GENCO [41].

To achieve sustainable development, many governments have taken measures to stimulate the collection of used products, especially for WEEE. To illustrate, a number of legislations with respect to product recovery, such as the Paper Recycling Directive, End-of-Life Vehicle Directive, and WEEE Directive, among others, has been implemented in the European Union [24].

Through the Extended Producer Responsibility (EPR) law, the government financially incentivizes manufacturers to design environment-friendly products and hold them responsible for collecting and recovering products at the end of the product life. EPR integrates a product's life cycle environmental costs with its production costs when deciding on the market price of products. In Europe, producers take full financial responsibility of collecting used vehicles, batteries, packaging, and all WEEEs. The Paper Recycling Directive, End-of-Life Vehicle Directive, WEEE Directive in Europe, and Specified Household Appliance Recycling (SHAR) are examples of such legislations. The WEEE Directive (Directive 2002/96/EC) evolved into a European law in 2003, which contains mandatory requirements on collection, recycling, and recovery for all types of electrical goods, with a minimum rate of 4 kilograms per head of population per annum [22]. Some EU member states applied economic instruments, including taxes, to create incentives for residents to separate recyclables from regular waste streams. A deposit–refund system provides one type of such incentives to ensure maximum reuse and recycling. Denmark implemented one of the oldest deposit–refund systems called "Dansk Retursystem" for beverage bottles and mineral water bottles [54]. The Netherlands also adopted a deposit–refund system for collecting PET bottles (>0.5 L) with special collection machines in supermarkets. For PET bottle collection, a deposit–refund system is in place, where PET bottles are collected using a different channel. A trade-off occurs between the cost of using the refund collection system and the collection rate. WEEE-like legislation has also been introduced in Canada, Japan, China, and many states in the United States [55]. Thus far, 25 US states have passed a legislation mandating statewide e-waste recycling (http://www.electronicstakeback.com/promote-good-laws/state-legislation/). Producers in Japan are also responsible for recycling cars and WEEE, whereas the Chinese government has enacted similar legislations and started implementing the WEEE Recycling Management Regulation and adopted subsidization to the manufacturer to stimulate collection.

Considering the government's important role in promoting the recycling of used products, government intervention has also received significant attention from the academia. Atasu and Subramanian [2] investigated the implications of collective and individual producer responsibility models of product take-back laws for WEEE. Plambeck and Wang [53] found that adopting the "fee-upon-disposal" policy motivates manufacturers to design for recyclability. Aksen et al. [1] developed models to guide the government in subsidizing manufacturers, whereas Atasu et al. [4] discussed effective conditions for government to legislate WEEE recycling and remanufacturing policy. Atasu and Wassenhove [3] proved that enforcing collection and recycling of WEEE is necessary from the operations perspective. Rahman and Subramanian [56] found that the government legislation can motivate computer recycling operations. Govindan et al. [27] found that CLSC members are not only directed by the value of the WEEEs but also by regulations. Atasu et al. [5] compared two forms of product take-back legislation with manufacturer- and state-operated systems, and Wang et al. [67, 68] proposed a reward–penalty mechanism.

The three types of collectors are those dominated by producers, retailers, or designated third parties. These three typical collection methods are illustrated in Fig. 7.4. According to Chao et al. [11], the manufacturer generally performs the collection when the quantities to be returned are large and especially when transportation costs are low (Fig. 7.4a); however, in many cases, collection centers are required to maximize the recovery of products or materials. Third-party collectors (Fig. 7.4b) or distributors (Fig. 7.4c) may serve as collection centers when the

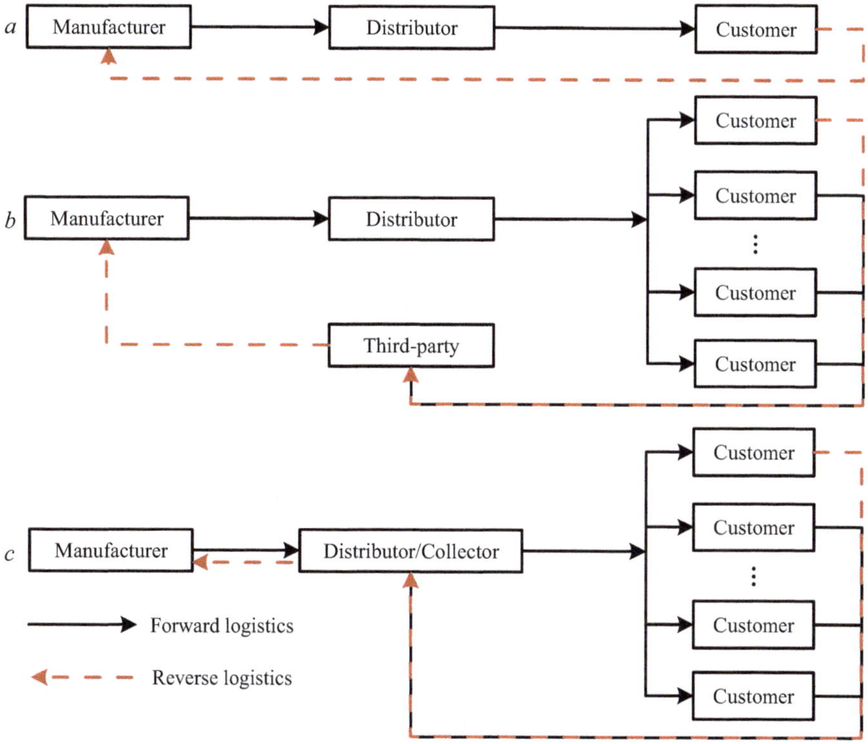

Fig. 7.4 Collection dominated by (**a**) manufacturers, (**b**) third parties, and (**c**) distributors [23]

number of customers is large or the quantities to be collected from each customer are small. Savaskan et al. [58] proposed three decentralized models and found retailer-based collection to be highly effective; furthermore, they proposed a coordination mechanism if the retailer is incentivized to induce collection. Savaskan and Van Wassenhove [57] further extended the above model to multiple settings, where retailing is competitive, and showed that manufacturers should entrust the collection of used products to retailers under certain conditions.

Owing to the inconvenience of recycling, consumers retain or abandon a great deal of end-of-life products instead of recycling them; their reluctance to act means that recyclable dealers are forced to wait [34, 66, 71]. A promising method to this challenge could be achieved by utilizing an online recycling channel. Recyclable dealers, such as Changhong Green Group Company Limited and Shanghai Xin Jin-qiao Environmental Protection Company Limited, who previously collected WEEEs through traditional recycling channels, recently began collecting e-wastes through online recycling channels. This service conveniently assists consumers in dealing with WEEEs by relaxing the traditional recycling constraint caused by physical distance and space. Such service also facilitated the recyclable dealers' accurate attainment of appropriate WEEEs while reducing collection and transportation costs [18]. In addition, under the context of sharing economy and based on Internet of things, crowdsourcing logistics may become a new collection method today.

Remanufacturing in CLL

Remanufacturing is the process of restoring used products to a "like-new" functional state. This process offers many benefits for original equipment manufacturers (OEMs), such as savings in labor, material costs, and energy consumption [31]. The cost for remanufacturing is 40%–60% of that for manufacturing a new product [16]. For example, in the 2008/2009 financial year, Fuji Xerox Australia remanufactured over 230,000 equipment parts, which led to a $6 million cost savings compared to sourcing new parts [9]. Apart from material conservation, remanufacturing requires only 15% of the energy used to manufacture a new product, thereby significantly reducing greenhouse gas emissions [25].

In recent years, carbon emissions have received significant attention among world leaders and scientists because it is commonly believed to be one of the main contributors to global warming [50, 51]. The growing consensus on the significance of emission reduction has been reflected in actions and in legal frameworks to drive action further forward, to meet the challenge, and to address broad issues on sustainability [47]. From the Kyoto Agreement (Graham et al. 2015; David et al. 2015) to the Paris Agreement [13], most countries have accepted the premise that an urgent need has risen to set carbon reduction targets and put regulations into action in order to decarbonize the economy [21, 37]. The United States, the European Union, and many other countries have implemented various instruments ranging from taxes, permits, and voluntary incentives for a decarbonized economy. Therefore, current carbon emission credits are becoming a type of an increasing important resource and production factor for companies. Such resource has been one of the critical factors of production decision-making. In a carbon-constrained circumstance, decisions on manufacturing and remanufacturing should be consequently made while considering both economic benefits and environmental costs.

In practice, two different remanufacturers exist depending on whether they manufacture new products. The processes of deciding on manufacturing and remanufacturing for these different remanufacturers with different environmental regulations and determining the type of appropriate regulations to promote remanufacturing for policymakers are extremely important problems. We first summarize existing regulations; then, we analyze how these regulations affect production decision and how to make decisions for independent remanufacturers (IRs) that only generate remanufactured products and for OEMs that produce new and remanufactured products.

Carbon Emission Regulations

Over the last few decades, global warming has received increasing attention; with an aim to protect the environment, many governments and organizations have agreed to adopt regulation policies to reduce carbon emissions to at least half by 2050 [36]. Investigating the related literature showed a number of carbon emission regulation policies, such as carbon cap, cap and trade, carbon taxes, subsidy, and carbon offset, which have been promulgated to help impose restrictions on emissions of carbon dioxide and other greenhouse gases by governments worldwide. Governments set a specific limit or mandatory capacity on the amount of carbon emissions, called carbon cap, for a particular company [17]. Under a carbon tax mechanism, a firm is charged with a fee for each unit of emission [50]. The cap and trade mechanism (also known as the carbon trading scheme) allows firms to sell or purchase emission allowances in the market [73]. The scheme both pressures and incentivizes companies to achieve appropriate environmental initiatives [73]. To encourage industry to reduce emissions voluntarily, some governments also provide a carbon subsidy according to the former's capability to reduce carbon emissions [44]. The carbon offset scheme allows organizations to invest in environmental projects in order to balance out their own carbon footprints [6, 50].

Among these mentioned regulations, the carbon cap and trade is regarded as one of the widely implemented programs based on eco-economic theory [14]. For example, the Kyoto Protocol (which went into effect in 2005) established emission caps for each country and allowed emission trading [10]. Furthermore, the carbon cap and trade mechanism is more effective in carbon emission reduction [45, 46, 60]. Early in 2013, global carbon emission trading volume reached 54.9 billion dollars [76]. Under the carbon trade mechanism, firms are assigned a carbon emission cap for a certain period, in which they choose to purchase or sell carbon permits through the carbon market according to their own actual carbon emission levels, such as the European Climate Exchange and Chicago Climate Exchange [33, 63, 74].

Decision-Making for Independent Remanufacturers (IRs) with Different Carbon Emission Regulations

Owing to a dramatic increase in volume, such as that in the automotive industry, the global number of end-of-life vehicles (ELVs) will easily exceed 100 million by 2020 [61], and OEMs can no longer effectively and efficiently process numerous returned used products within their own capacity. Considering that many products are unsuitable for large-scale remanufacturing, most durable products, such as automobiles, laser printers, and cell phones, are frequently redesigned as old designs remain in use; even if one returned product can be restored to its original condition, that old product may no longer be in demand. Therefore, a remanufacturing strategy for such products should not aim at the entire product but at the parts that the products are assembled from [40]. Therefore, with the increased number of end-of-life products, IRs are increasingly developed in a number of countries. These IRs collect used products/components/parts to remanufacture and sell them to special markets. Automotive remanufacturers, industrial equipment remanufacturers, and tire retreaders are some of the numerous examples [28]. In North America, some OEMs (e.g., Dell, Hewlett-Packard, and IBM) outsourced their remanufacturing operations to third-party remanufacturers [52]. Moreover, the National Development and Reform Committee (NDRC) in China published 14 enterprises as pilot automotive component remanufacturers in 2008 to promote the development of remanufacturing industry [75].

These remanufacturers collect and remanufacture used products/components/parts, even if the process is usually more environment friendly than that in traditional manufacturing because the process involves recycling used products in lieu of the input of raw materials [32, 35]. As a traditional manufacturing process, remanufacturing operations and processes also affect the environment (e.g., through carbon emissions). Similar to traditional manufacturers, IRs must determine how to trade off the economic costs generated in the production process and environmental costs under carbon emission regulation policies.

As mentioned in section "Carbon Emission Regulations", although a number of carbon emission regulations have been promulgated to help impose restrictions on the emissions of carbon dioxide and other greenhouse gases, many major carbon-emitting nations (e.g., the United States, China, and Japan) still debate which regulatory policy they should adopt; hence, many manufacturers and remanufacturers are unaware of which future policy is ideal for them. Moreover, many firms do not know how to adapt to the policies and how to handle different magnitudes of incentives and penalties and the stringency of constraints, which can fluctuate over time. For example, carbon emission permit prices have exhibited considerable volatility. In the European Union, permit price increased from around 7 euros in January 2005 to over 30 euros in April 2006 and then crashed to below 10 euros within 3 days. The price subsequently rose again and stabilized above 15 euros, for approximately 4 months before decreasing to nearly 0 by mid-2007 [8]. These uncertainties would significantly influence the production decisions of firms.

Thus, firms and policymakers must figure out how to trade off environmental and financial benefits from carbon emission regulations, which currently play an important role in production decision-making. Remanufacturers must determine the economically optimal remanufacturing quantity and the corresponding maximum expected profit under different carbon emission policies; moreover, they must decide which used products are profitable to remanufacture depending on different quality conditions (i.e., yield rate). Policymakers must formulate policies and decide magnitudes of incentives, penalties, and stringency of constraints; moreover, they must investigate which policy is most conducive to promote the development of an environment-friendly remanufacturing industry.

Of the five existing major carbon emission policies that have been discussed in section "Carbon Emission Regulations", the first three policies are independent, whereas the fourth and fifth policies can be regarded as a special case of the second one. Moreover, they seem to be discussed less in the literature. Therefore, we must only investigate decision-making using the first three policies, namely, carbon cap, carbon tax, and cap and trade.

By analyzing the possible cost and decision process, three optimization models corresponding to the three common regulation policies, namely, mandatory carbon emissions capacity, carbon tax, and carbon cap and trade, are presented to determine the remanufacturing quantity that maximizes the total profits in the work of Liu et al. [45, 46]. Moreover, the numerical results show the following: (1) Different carbon emission policies affect the critical values of yield for remanufacturing but only have minor effects on the optimal remanufacturing quantity when the yield rate is higher than the critical values. Furthermore, significant effects occur on the total profits. These effects decrease with the increase in yield rate. (2) The carbon cap policy is most sensitive to yield rate, that is, the effect on the total profit caused by this policy varies drastically between different yield rates. The carbon tax policy varies the least with different yield rates; hence, it is the most stable policy to promote the development of the remanufacturing industry. (3) Based on the above analysis, we recommend that remanufacturers should aim to improve yield rate in order to maximize profit irrespective of the implemented carbon emission policy. In addition, we suggest that policymakers should give preference to carbon tax policy among these three policies; however, if either of the other two policies is performed, a remanufacturing discount (e.g., a high carbon emission cap or lower penalty) should be implemented to promote the development of remanufacturers.

Decision-Making for a Hybrid Manufacturing/Remanufacturing System with Carbon Regulations

Based on the study on IRs, many OEMs also simultaneously produce new and remanufactured products. GE, GM, Ford, Boeing, Bosch, Xerox, IBM, HP, Kodak, and Caterpillar are some examples of those who have made remanufacturing an integral part of their business strategy. Hence, this section also analyzes the decision-making for OEMs with environmental regulations.

Remanufacturing is generally more environment friendly than traditional manufacturing because the process involves recycling used products in lieu of the input of raw materials [32, 35]. In a more carbon-constrained circumstance, integrating carbon emission values into the production operation process is necessary for manufacturers engaged in both manufacturing and remanufacturing.

A few studies have analyzed the effect of carbon emission constraints on production decisions for remanufacturing. Considering the highly variable quality of used products, Yang et al. [69] studied an acquisition and remanufacturing problem in a multiproduct remanufacturing system under carbon emission regulations. Yenipazarli [70] investigated the effects of emission taxes on the optimal production and pricing decisions of an enterprise engaged in manufacturing and remanufacturing using a leader–follower Stackelberg game model and analyzed the different effects on tactical decisions between emission taxes and emission trading policies. Tornese et al. [64] characterized the carbon equivalent emissions associated with the pallet remanufacturing operations for two repositioning scenarios, such as cross docking and take-back. Miao et al. [49] addressed the problems of optimal pricing and manufacturing/remanufacturing decisions of the manufacturer with trade-ins under carbon regulations. Chang et al. [10] constructed two profit maximization models to make optimal manufacturing/remanufacturing decisions under a carbon cap and trade mechanism. The results indicated that in a hybrid manufacturing–remanufacturing system, the cap and trade regulation cannot induce the firm to choose the low-carbon remanufacturing technology if an independent demand market exists; however, the opposite is observed when a substitutable demand market is in place.

Wang et al. [68] addressed the issues of manufacturing/remanufacturing decisions under the capital constraint and/or carbon emission constraint for a manufacturer, where new and remanufactured products are sold to segmented markets using a downward substitution strategy. For the manufacturer, the optimal production policy is found to significantly depend on the interaction effects of capital and carbon emission constraints. The analysis and results show the following: (1) Capital constraint can constantly encourage manufacturers to remanufacture used products at a high-quality level; however, only when capital constraint is considered will the carbon emission constraint distinctly affect the quality level of used products. (2) The carbon cap only has significant influences on the manufacturing/remanufacturing quantities of capital-constrained manufacturers, and the production policies of capital-constrained manufacturers are less sensitive to carbon price variation. (3) Based on the above analysis, the present study suggests that manufacturers should strive to enhance yield rate in order to achieve maximum profit when only considering the capital constraint; if the carbon emission constraint is also considered, manufacturers should produce other remanufactured products. The following are some proposed suggestions for policymakers: they should provide an ideal financing environment to capital-constrained manufacturers, which can promote the development of these firms, and they should implement different carbon trading policies between large- and small- and medium-sized manufacturers, i.e., a higher carbon cap for small- and medium-sized manufacturers, which tend to encounter capital dilemmas.

Conclusions

The popularity of environmental protection and sustainable development has encouraged the increasing importance of CLL in recent years. CLL is one of the most essential keys in relation to the cost incurred by companies, and its overall cost has become an urgent concern for the supply–demand chain management of companies.

This chapter introduced the essence of CLL and analyzed its network structure and sustainable operations during the CLL process with special focus on discussing the main sustainable operations, namely, recycling and remanufacturing. The results demonstrated the following: first, forward logistics and reverse logistics should be integrated in a CLL network; second, apart from the three traditional collection methods, recycling, which depend on online platforms and the public, using the Internet would be a popular approach considering its economic benefits and sustainable development; and finally, environmental regulations and government intervention play important roles in promoting the development of CLL.

References

1. Aksen, D., Aras, N., Karaarslan, A.G.: Design and analysis of government subsidized collection systems for incentive-dependent returns. Int. J. Prod. Econ. **119**(2), 308–327 (2009)
2. Atasu, A., Subramanian, R.: Extended producer responsibility for E-waste: individual or collective producer responsibility? Prod. Oper. Manag. **21**(6), 1042–1059 (2012)
3. Atasu, A., Wassenhove, L.N.: An operations perspective on product take-back legislation for E-waste: theory, practice, and research needs. Prod. Oper. Manag. **21**(3), 407–422 (2012)
4. Atasu, A., Wassenhove, L.N., Sarvary, M.: Efficient take-back legislation. Prod. Oper. Manag. **18**(3), 243–258 (2009)
5. Atasu, A., Özdemir, Ö., Van Wassenhove, L.N.: Stakeholder perspectives on E-waste take-back legislation. Prod. Oper. Manag. **22**(2), 382–396 (2013)
6. Barari, S., Agarwal, G., Zhang, W.C., Mahanty, B., Tiwari, M.K.: A decision framework for the analysis of green supply chain contracts: an evolutionary game approach. Expert Syst Appl. **39**(3), 2965–2976 (2012)
7. Baumgarten, H., Butz, C., Fritsch, A., et al.: Supply chain management and reverse logistics-integration of reverse logistics processes into supply chain management approaches. In: Electronics and the Environment. IEEE International Symposium on IEEE, pp. 79–83 (2003)
8. Benz, E., Trück, S.: Modeling the price dynamics of CO2 emission allowances. Energy Econ. **31**(1), 4–15 (2009)
9. Bulmuş, S.C., Zhu, S.X., Teunter, R.: Capacity and production decisions under a remanufacturing strategy. Int. J. Prod. Econ. **145**(1), 359–370 (2013)
10. Chang, X., Xia, H., Zhu, H., et al.: Production decisions in a hybrid manufacturing–remanufacturing system with carbon cap and trade mechanism. Int. J. Prod. Econ. **162**, 160–173 (2015)
11. Chao, G.H., Iravani, S.M.R., Savaskan, R.C.: Quality improvement incentives and product recall cost sharing contracts. Manag. Sci. **55**(7), 1122–1138 (2009)
12. Choi, T.M., Li, Y., Xu, L.: Channel leadership, performance and coordination in closed loop supply chains. Int. J. Prod. Econ. **146**(1), 371–380 (2013)

13. Cooper, M.: Renewable and distributed resources in a post-Paris low carbon future: the key role and political economy of sustainable electricity. Energy Res. Soc. Sci. **19**, 66–93 (2016)
14. Dales, J.H.: Pollution. University of Toronto Press, Toronto (1968)
15. De Giovanni, P., Reddy, P.V., Zaccour, G.: Incentive strategies for an optimal recovery program in a closed-loop supply chain. Eur. J. Oper. Res. **249**(2), 605–617 (2016)
16. Dowlatshahi, S.: Developing a theory of reverse logistics. Dermatol. Int. **30**(3), 143–155 (2000)
17. Fahimnia, B., Farahani, R.Z., Marian, R., Luong, L.: A review and critique on integrated production–distribution planning models and techniques. J. Manuf. Syst. **32**(1), 1–19 (2013)
18. Feng, L.P., Kannan, G., Li, C.F.: Strategic planning: design and coordination for dual-recycling channel reverse supply chain considering consumer behavior. Eur. J. Oper. Res. **260**(2), 601–612 (2017)
19. Ferguson, M.: Strategic and tactical aspects of closed-loop supply chains. Foundations and trends® in technology. Inf. Oper. Manag. **3**(2), 101–200 (2010)
20. Fleischmann, M., Bloemhof-Ruwaard, J.M., Dekker, R., et al.: Quantitative models for reverse logistics: a review. Eur. J. Oper. Res. **103**(1), 1–17 (1997)
21. Galinato, G.I., Yoder, J.K.: An integrated tax-subsidy policy for carbon emission reduction. Resour. Energy Econ. **32**(3), 310–326 (2010)
22. Georgiadis, P., Besiou, M.: Environmental and economical sustainability of WEEE closed-loop supply chains with recycling: a system dynamics analysis. Int. J. Adv. Manuf. Technol. **47**(5–8), 475–493 (2010)
23. Giannetti, B.F., Bonilla, S.H., Almeida, C.M.V.B.: An emergy-based evaluation of a reverse logistics network for steel recycling. J. Clean. Prod. **46**, 48–57 (2013)
24. Giri, B.C., Sharma, S.: Optimizing a closed-loop supply chain with manufacturing defects and quality dependent return rate. J. Manuf. Syst. **35**, 92–111 (2015)
25. Giutini, R., Gaudette, K.: Remanufacturing: the next great opportunity for boosting US productivity. Bus. Horiz. **46**(6), 41–48 (2003)
26. Govindan, K., Popiuc, M.N.: Reverse supply chain coordination by revenue sharing contract: a case for the personal computers industry. Eur. J. Oper. Res. **233**(2), 326–336 (2014)
27. Govindan, K., Khodaverdi, R., Jafarian, A.: A fuzzy multi criteria approach for measuring sustainability performance of a supplier based on triple bottom line approach. J. Clean. Prod. **47**, 345–354 (2013)
28. Guide Jr., V.D.R.: Production planning and control for remanufacturing: industry practice and research needs. J. Oper. Manag. **18**(4), 467–483 (2000)
29. Guide, V.D.R., Wassenhove, L.N.: Managing product returns for remanufacturing. Prod. Oper. Manag. **10**(2), 142–155 (2001)
30. Guide, V.D.R., Wassenhove, L.N.: Business Aspects of Closed-Loop Supply Chains. Carnegie Mellon University Press, Pittsburgh (2003)
31. Guide Jr., V.D.R., Srivastava, R., Spencer, M.S.: An evaluation of capacity planning techniques in a remanufacturing environment. Int. J. Prod. Res. **35**(1), 67–82 (1997)
32. Gungor, A., Gupta, S.M.: Issues in environmentally conscious manufacturing and product recovery: a survey. Comput. Ind. Eng. **36**(4), 811–853 (1999)
33. He, P., Zhang, W., Xu, X., Bian, Y.: Production lot-sizing and carbon emissions under cap-and-trade and carbon tax regulations. J. Clean. Prod. **103**(15), 241–248 (2015)
34. Huang, P., Zhang, X., Deng, X.: Survey and analysis of public environmental awareness and performance in Ningbo, China: a case study on household electrical and electronic equipment. J. Clean. Prod. **14**(18), 1635–1643 (2006)
35. Ilgin, M.A., Gupta, S.M.: Environmentally conscious manufacturing and product recovery (ECMPRO): a review of the state of the art. J. Environ. Manage. **91**(3), 563–591 (2010)
36. International Energy Agency. Office of Energy Technology, R & D., & Group of Eight (Organization). Energy technology perspectives. Int. Energy Agency. (2008)
37. Jaehn, F., Letmathe, P.: The emissions trading paradox. Eur. J. Oper. Res. **202**(1), 248–254 (2010)

38. Kannan, D., Govindan, K., Shankar, M.: India: formalize recycling of electronic waste. Nature. **530**(7590), 281–281 (2016)
39. Keyvanshokooh, E., Fattahi, M., Seyed-Hosseini, S.M., et al.: A dynamic pricing approach for returned products in integrated forward/reverse logistics network design. Appl. Math. Model. **37**(24), 10182–10202 (2013)
40. Kongar, E., Gupta, S.M.: A multi-criteria decision making approach for disassembly-to-order systems. J. Electron. Manuf. **11**(02), 171–183 (2002)
41. Krumwiede, D.W., Sheu, C.: A model for reverse logistics entry by third-party providers. Omega. **30**(5), 325–333 (2002)
42. Lee, D.H., Dong, M.: A heuristic approach to logistics network design for end-of-lease computer products recovery. Transport. Res. Part E Log. Transp. Rev. **44**(3), 455–474 (2008)
43. Lee, D.H., Dong, M.: Dynamic network design for reverse logistics operations under uncertainty. Transp. Res. Part E Log. Transp. Rev. **45**(1), 61–71 (2009)
44. Li, J., Du, W., Yang, F., Hua, G.: The carbon subsidy analysis in remanufacturing closed-loop supply chain. Sustainability. **6**(6), 3861–3877 (2014)
45. Liu, L., Chen, C., Zhao, Y., et al.: China's carbon-emissions trading: overview, challenges and future. Renew. Sust. Energ. Rev. **49**, 254–266 (2015a)
46. Liu, B., Holmbom, M., Segerstedt, A., et al.: Effects of carbon emission regulations on remanufacturing decisions with limited information of demand distribution. Int. J. Prod. Res. **53**(2), 532–548 (2015b)
47. Long, T.B., Young, W.: An exploration of intervention options to enhance the management of supply chain greenhouse gas emissions in the UK. J. Clean. Prod. **112**, 1834–1848 (2016)
48. Lu, Q., Christina, V., Stuart, J.A., et al.: A practical framework for the reverse supply chain. In: Electronics and the Environment. ISEE 2000. Proceedings of the 2000 IEEE International Symposium on. IEEE, pp. 266–271 (2000)
49. Miao, Z., Mao, H., Fu, K., et al.: Remanufacturing with trade-ins under carbon regulations. Comput. Oper. Res. (2016) https://doi.org/10.1016/j.cor.2016.03.014
50. Palak, G., Ekşioğlu, S.D., Geunes, J.: Analyzing the impacts of carbon regulatory mechanisms on supplier and mode selection decisions: an application to a biofuel supply chain. Int. J. Prod. Econ. **154**, 198–216 (2014)
51. Park, J.Y., Kug, J.S., Bader, J., Rolph, R., Kwon, M.: Amplified Arctic warming by phytoplankton under greenhouse warming. Proc. Natl. Acad. Sci. **112**(19), 5921–5926 (2015)
52. Patel, G.S.: A Stochastic Production Cost Model for Remanufacturing Systems. MS thesis, The University of Texas-Pan American, USA (2006)
53. Plambeck, E., Wang, Q.: Effects of e-waste regulation on new product introduction. Manag. Sci. **55**(3), 333–347 (2009)
54. Pro-Europe: Packaging waste legislation in Denmark. http://pro-e.org/Denmark (2009)
55. Quariguasi Frota Neto, J., Walther, G., Bloemhof, J., et al.: From closed-loop to sustainable supply chains: the WEEE case. Int. J. Prod. Res. **48**(15), 4463–4481 (2010)
56. Rahman, S., Subramanian, N.: Factors for implementing end-of-life computer recycling operations in reverse supply chains. Int. J. Prod. Econ. **140**(1), 239–248 (2012)
57. Savaskan, R.C., Van Wassenhove, L.N.: Reverse channel design: the case of competing retailers. Manag. Sci. **52**(1), 1–14 (2006)
58. Savaskan, R.C., Bhattacharya, S., Van Wassenhove, L.N.: Closed-loop supply chain models with product remanufacturing. Manag. Sci. **50**(2), 239–252 (2004)
59. Schultmann, F., Zumkeller, M., Rentz, O.: Modeling reverse logistic tasks within closed-loop supply chains: an example from the automotive industry. Eur. J. Oper. Res. **171**(3), 1033–1050 (2006)
60. Stavins, R.N.: A meaningful US cap-and-trade system to address climate change. Harv. Environ. Law Rev. **32**, 293–371 (2008)
61. Tian, J., Chen, M.: Sustainable design for automotive products: dismantling and recycling of end-of-life vehicles. Waste Manag. **34**(2), 458–467 (2014)

62. Tonanont, A., Yimsiri, S., Jitpitaklert, W., et al.: Performance evaluation in reverse logistics with data envelopment analysis. In: IIE Annual Conference. Proceedings. Institute of Industrial and Systems Engineers (IISE): p. 764 (2008)
63. Toptal, A., Özlü, H., Konur, D.: Joint decisions on inventory replenishment and emission reduction investment under different emission regulations. Int. J. Prod. Res. **52**(1), 243–269 (2014)
64. Tornese, F., Carrano, A.L., Thorn, B.K., Pazour, J.A., Roy, D.: Carbon footprint analysis of pallet remanufacturing. J. Clean. Prod. **126**, 630–642 (2016)
65. Üster, H., Easwaran, G., Akçali, E., et al.: Benders decomposition with alternative multiple cuts for a multi-product closed-loop supply chain network design model. Nav. Res. Logist. **54**(8), 890–907 (2007)
66. Wang, Z., Zhang, B., Yin, J., et al.: Willingness and behavior towards e-waste recycling for residents in Beijing city, China. J. Clean. Prod. **19**(9), 977–984 (2011)
67. Wang, W., Zhang, Y., Zhang, K., et al.: Reward–penalty mechanism for closed-loop supply chains under responsibility-sharing and different power structures. Int. J. Prod. Econ. **170**, 178–190 (2015)
68. Wang, Y., Chen, W., Liu, B.: Manufacturing/remanufacturing decisions for a capital-constrained manufacturer considering carbon emission cap and trade. J. Clean. Prod. **140**, 1118–1128 (2017)
69. Yang, C.H., Liu, H., Ji, P., et al.: Optimal acquisition and remanufacturing policies for multi-product remanufacturing systems. J. Clean. Prod. **135**, 1571–1579 (2016)
70. Yenipazarli, A.: Managing new and remanufactured products to mitigate environmental damage under emissions regulation. Eur. J. Oper. Res. **249**(1), 117–130 (2016)
71. Yin, J., Gao, Y., Xu, H.: Survey and analysis of consumers' behaviour of waste mobile phone recycling in China. J. Clean. Prod. **65**, 517–525 (2014)
72. Yu, J., Williams, E., Ju, M.: Analysis of material and energy consumption of mobile phones in China. Energy Policy. **38**(8), 4135–4141 (2010)
73. Zakeri, A., Dehghanian, F., Fahimnia, B., Sarkis, J.: Carbon pricing versus emissions trading: a supply chain planning perspective. Int. J. Prod. Econ. **164**, 197–205 (2015)
74. Zhang, B., Xu, L.: Multi-item production planning with carbon cap and trade mechanism. Int. J. Prod Econ. **144**(1), 118–127 (2013)
75. Zhang, T., Chu, J., Wang, X., Liu, X., Cui, P.: Development pattern and enhancing system of automotive components remanufacturing industry in China. Resour. Conserv. Recycl. **55**(6), 613–622 (2011)
76. Zhang, Y., Kuang, J.C., Wang, C.: Status quo analysis of Chinese and foreign carbon trading market. Foreign Energy Sources. **002**, 3 (2014). https://doi.org/10.4337/9781781001882.00008

Chapter 8
Bridging Borders: Integrating Data Analytics, Modeling, Simulation, and Gaming for Interdisciplinary Assessment of Health Aspects in City Networks

Jayanth Raghothama, Elhabib Moustaid, Vinutha Magal Shreenath, and Sebastiaan Meijer

Abstract The health perspective in urban science brings new methodological challenges to planning of city networks. Due to the system of systems nature of healthcare, new methods are needed to facilitate disciplinary integration and management of models and models-of-models. Participation of stakeholders and policy makers demands the uptake of new methods and a new perspective on the use of interfaces and boundary objects. In this chapter, the authors discuss evidence from five projects that use gaming, simulation, modeling, and data analytics in unconventional ways for design of large-scale urban systems to provide a methodological path forward for overcoming traditional engineering approach issues.

Keywords City health • Large-scale urban systems • Gaming • Simulation • Modeling • Data analytics

Introduction

The emergence of a science of cities provides the foundation for unifying a variety of disciplines that seek to study and explain contemporary urban phenomena. As contemporary urbanism throws up unprecedented challenges for urban research, management, and planning, the science of cities has moved away from thinking of cities as places in space to the idea of cities as systems of networks and flows [1].

J. Raghothama (✉) • E. Moustaid • V. Magal Shreenath • S. Meijer
School of Technology and Health, KTH Royal Institute of Technology, Huddinge, Sweden
e-mail: jayanthr@kth.se; elhabib@kth.se; vinutha@kth.se; smeijer@kth.se

A. Karakitsiou et al. (eds.), *City Networks*, Springer Optimization and Its Applications 128, DOI 10.1007/978-3-319-65338-9_8

The idea of a city as a network of flows, people, information, and goods has been advocated before. Scholars have argued that location is central to how networks of relationships form and hold the city together and that cities are networks of information, constantly producing and being subject to messages [26]. The idea of cities as networks of systems necessitates a shift of focus from space and place toward interactions and to see locations as emerging from interactions of these networks.

This new science of cities has been aided in large part by the emergence of large-scale sensing instruments providing rich streams of data, the increase in computing power, and increased sophistication in modeling techniques. These constructs enable city governments to better grapple with the size, scale, and complexity of urban spaces [46].

Urban planning and design, however, has traditionally been concerned with physical rather than social, political, or economic means of planning and management. Applying this new science of cities together with coexisting urban planning and policy making processes raises many questions about the future of urban governance. Among many issues regarding the new science of cities, transparency of computer-enabled governance, security, and privacy are of most concern. However, it also tends to ignore what is often a very messy reality of planning that of wicked problems which "have no definitive description, involve value judgments, and take place in unique contexts that make it difficult to accurately test solutions" [14, 39, 46].

There is a need to resolve the conflict between the ambiguity inherent in planning processes and the formal, data- and model-driven description of the city generated by this new science. This resolution needs to account for the roles of planners and citizens in planning processes. The failures of technocratic approaches to city planning resulted in a shift in focus toward community-oriented development [5]. This led to the insight that the tacit and implicit expertise of people about the systems they govern and experience is integral to the effective management and planning of cities, a lesson that ought not to be lost in the new wave of data-driven and automated governance. This resolution also needs to account for inherent structural uncertainties in models and data and a suspect predictive ability [2].

The criticism in the 1960s and 1970s of comprehensive urban modeling efforts prompted a rethink and scaling back of ambition. These efforts failed mostly for lack of sound theory, input data, and difficulties in application [28]. With the rise of increased computational power and the Internet, scholars again began developing complex models for urban planning.

In this paper, we propose a methodology to resolve this conflict through a variety of ways, exploring modeling, data-driven approaches, participatory simulations, and a combination of all of them. Through case studies, we explore different mechanisms that combined together can provide an effective framework that can integrate planning processes and scientific methods.

What Makes a City Healthy?

According to World Bank data, 54% of the world population today lives in cities, compared to 33% on 1960. This is a result of both massive immigration to urban areas and indigenous growth of populations in cities. Those figures are even expected to rise as Africa and Asia shift toward more urbanization. The growth of urban population has changed the nature of cities and the challenges they face. Cities need efficient housing, transportation, sanitation, healthcare, and economic and social activities. In today's cities, those challenges are more than even connected and affect each other in a massive way [10].

Ensuring quality of life in cities to all people is a continuous and comprehensive challenge. The satisfaction of demand to basic levels of services such as sanitation, water supply, heat, and waste disposal is a necessary aspect of making a city healthy. In 1990, about 600 million people lived in cities in life-threatening conditions [3]. The reasons for such a situation are a combination of several factors, such as poor sanitations, inadequate housing, non-efficient waste disposal, pollution, and inadequate local government's policies [21]. Those conditions were at the origin of terrible physical and social environments that led to development of chronic and infectious diseases at the level of individuals. City management needs to not only deal with growing population but complex environments and structures that results from such growth.

The provision of basic daily services is deemed necessary but not sufficient to claim quality of life and healthy cities. Making a city healthy is a matter of creating sustainable environments in which populations experience good living conditions. A healthy city is one that provides the conditions of existence of an environment that can ensure the well-being of the population and the processes to sustain it. The WHO, after seeing the rise in the number of cities and megacities and the effects of poor city management on population health, has picked on the notion of healthy cities launching the Healthy Cities project [45]. Our definition of healthy cities goes hand in hand with WHO's notion of Healthy Cities. WHO does not define a healthy city by outcomes, but rather by processes. Lack of diseases is less considered a healthy aspect of a city than a process that can prevent all form of physical and social non-healthiness and disease. One can define then a healthy city as a city that provides its citizens an equitable, inclusive, empowering, supportive, nonexploitative social environment. It also has to provide a clean safe physical environment and a healthy growing, diverse, vital economy. A healthy city is both a city that takes care of the sick and that prevents all sorts of illnesses, physical or moral, for the non-ill. Health is a consequence of many aspect systems in the city functioning in harmony [4, 13]. These systems include sanitation, waste disposal, housing, food provision, transportation, social networks, and green and individual empowering infrastructures [11]. Metrics measuring the efficiency of these systems often evaluate the quality of life in cities. Such metrics include, for example, air and water pollution indexes, quality and availability of healthcare, safety, traffic, or even purchasing power.

As the well-being of the whole population is the ultimate objective of healthy cities, it is crucial to go beyond the averages on the whole population to ensure not only equal but equitable conditions to the city population [29]. That requires taking a special care of demographic shifts and economic volatility and puts in place systems that can adapt to them to recognizable trends, such as elderly population growth, as well as sudden changes such as economic crises [43].

Detecting trends in demographic shifts, economic and social trends are important to define the future of healthy cities. The aging population sets new challenges on how to take care of the elderly in Europe and how to make their use of the city as healthy as possible, including provision of care, access to transportation, social activities, and even home care [43]. The developments in technology also changes the way population perceives their individual health and provide new tools for provision of healthcare. Technology and information are also changing many aspect systems that are influencing health in cities, such as transportation and work habits.

Health in cities is not an isolated challenge but a shared responsibility with a multitude of actors and disciplines, as conceptualized in Fig. 8.1. Health finds itself as a driver, a component, and an objective of city flow networks of goods, information, and people. Coordinating, collaborating, and bridging different aspects of city management, city use, and city growth is today an imperative to make a city healthy [12].

Who Makes a City Healthy?

Vlahov et al. [48] identify actors within different areas (government, market, civic society) whose interactions are determinant for the well-being in cities. Governments are responsible for providing basic services relating to activities that promote health. City authorities are responsible for providing good quality of basic sanitation services, access to water and electricity grids, as well as the settings for

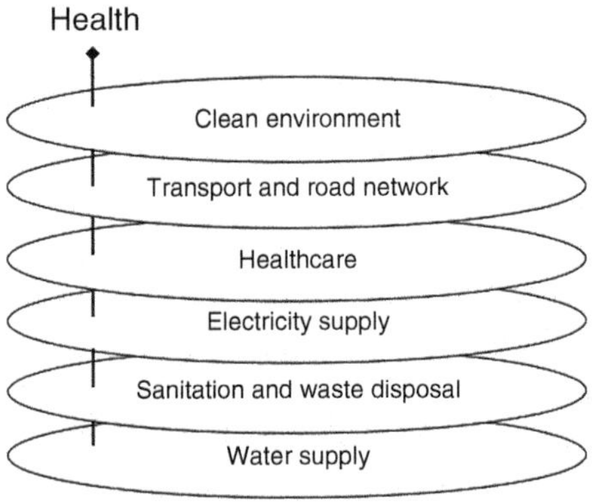

Fig. 8.1 System of systems in cities

having an inclusive social environment that promotes values of inclusion and well-being. Public education, transportation, and adequate housing to all segments of populations are also areas that need clear governance in which city authorities and governments are involved [41]. Legislators are responsible for making laws that regulate the urban spaces and control or encourage market activities. Markets can drive an economy to provide the basis for a healthy lifestyle. Markets can affect housing prices, public transport availability, and food prices and quality. Besides the responsibility of economic and market actors, governments and authorities have the duty of regulating markets and control economical activities if they threaten public health.

Cities are not isolated from their local context in their countries or even global trends. While city authorities are acting on local level, they are easily affected by trends both at the national and global level. Cities can be affected by global trends such as newly economic trends, immigrants' influx, and climate change [48].

The multitude of actors in the cities and the need for each to operate within their own disciplines and to take into account others call for new approaches and tools to urban development. Cities cannot be approached as closed systems that can operate in total autonomy from the rest of the world. They are as well subject of developments going around them, outside the city [9].

Methodological Challenges

A central tenet common in the planning of cities is the use of analytical tools such as models, simulations, data analytics, and so on. Used for their perceived ability to predict some future states, models have gained in ubiquity and complexity over the last two decades. However, as mentioned earlier, as models represent increasingly complex systems, it becomes harder to validate them.

This then raises the question: How do we know a model is reliable enough, to depend on it for a particular task?

A core question of validation, which scholars have attempted to address in different ways, either by focusing on the constructs of the simulation model or the function, is expected to serve. Most of these techniques and processes were designed in engineering fields for validating models of technical systems that followed fundamental physical laws. Complex systems, on the other hand, are sociotechnical in nature, follow nonlinear dynamics in their interactions, and are characterized by open boundaries.

A central difficulty in validating such tools arises from the fact that our understanding about the system is characterized by structural uncertainty, which implies that it is hard if not impossible to describe the system in sufficient detail to validate it. A way to overcome this is to blend together a diverse mixture of models of different aspects and effects.

Derived from the difficulties around analytical tools, the complexity of systems and city networks and the ambiguities around social and political aspects are central challenges to the issue of city planning. These challenges need to be addressed in an integrated methodology or approach, ideally one that is capable of capturing multiple dimensions of both the system and its attendant challenges. In the following sections, we describe these central challenges in detail and approach to address them. We then describe five cases where parts of this approach were used to solve city planning problems, followed by a discussion on the results of these cases.

Disciplinary Integration

Studies of complex systems and interdisciplinary studies go hand in hand [33]. While multidisciplinary studies typically study the same phenomena from different disciplinary perspectives, they often tend to report in their own "home" discipline. An interdisciplinary study is where understanding and knowledge from different disciplines are integrated to conduct and report the study. Planning in a city is one such process that needs to integrate many disciplines.

The framing of the problem such that it relates to the relevant discipline is a challenge. This is highly related to the type of models that are built in that discipline to understand the problem. Often this situation leads to a proliferation of models from different disciplines studying the same phenomena [34].

Lack of standard methodologies for interdisciplinary research, especially in communicating the interactions within the system being studied, is another challenge. There is often a tension in presenting social interactions as physiological phenomena or in claiming more rigor in one discipline compared to the other [16, 33, 38]. Translation to terminology from different disciplines to neutral descriptions of phenomena being studied is an important aspect of moving forward with conducting and communicating interdisciplinary research.

Integration of results from different disciplines into an insight or a solution requires that the results are integratable. As mentioned earlier in this section, the definitions, the scope and scale of the study, and the outputs from different models need to be coherently combined under a concept, or there should be a clear way to switch between different concepts that are represented by different models.

Model Integration and Model of Models

A mechanism to achieve complete, realistic representations of a complex system in a model, without going through the onerous engineering and modeling effort of modeling it from scratch, is to build integrated models. Integrated models provide the benefit of structuring and linking aspect systems within one model framework. Technical, economical, sociological, information, and policy processes could potentially be linked within one framework, potentially providing a more complete and high fidelity representation of a complex system.

Multiple pathways to integration exist. A simple way would be to link existing disciplinary models in a sequence, with the output of one model acting as input to another. This approach is more suitable for linking subsystems rather than related aspect systems. While linking is relatively easy, it leads to an inextricable tangle of processes while obscuring the true dynamics of the system. Further, merely linking subsystems does not provide for exploring the many interactions and feedback loops present within systems.

A second and more promising approach based on parallel and distributed simulations is to integrate existing disciplinary models closely with each other, by integrating them semantically and temporally. The suite of models to be integrated would exchange information about events and processes within each model and synchronize events with each other. The information exchanged would be used by the other models to account for events and processes happening in the other aspect systems, necessitating the development of the interaction logic between these aspect systems. This approach results in a consistent, complete, and complex model system, where interactions and feedback loops are modeled allowing for the investigation of complex and nonlinear dynamics. However, as mentioned earlier, as more models are added and integrated within this suite, the harder validation and analytical understanding of the model and reference system becomes.

Taking advantage of networking and distributed simulation approaches and technologies, the models can be computationally distributed achieving greater computational performance. However, this approach trades off fidelity with comprehension: macro models are easier to integrate and understand, since the number of interactions at that scale is relatively few. Integrating micro models with each other while possible is harder and requires extensive low-level technical knowledge of both system (to be modeled) behaviors and simulation architectures.

Despite the advances and advantages of model integration and distribution, such technical integrations of models work best when representing the technical components of reference systems. As with any model, technical components are usually completely documented, are easier to describe, and have closed boundaries. Further, it is hard to incorporate the social, institutional, and political aspects of the system(s) in such an approach.

Participation and Gaming

Populist movements in urban planning started refuting the hegemony of the expert urban planner to make the process of planning more democratic and inclusive. Stemming from the need to address the specific needs of individuals, local communities started asserting their rights to organize their local environments to participate in and influence the urban planning process. Cities are envisioned as holistic systems consisting of multiple subsystems organized under the influence of multiple networked actors, such as contractors, residents, planners, individuals, and others [17, 20, 23, 24, 25, 26]. Derived from the need for participation, this approach is rooted in qualitative approaches which seek to build upon multiple narratives, values, and perspectives. This communicative approach accepted the uncertainties inherent in planning and sought to build plans through collective and communicative processes. From the 1990s, planning shifted from the technical-rational perspective toward this communicative-rational perspective [42]. This approach while embracing the political, social, and institutional aspects of planning remained largely divorced from the technical and engineering aspects of systems.

Theoretically and methodologically aligned with the communicative turn in policy making, simulation games provide experiential environments where players can experience the effects of their actions through feedback mechanisms. Games have been used for a variety of different purposes, such as learning, experiments, and policy testing and data collection. Games have been used since the 1960s as a method to bring together policy makers and stakeholders in a participatory event. Games provide ways to collectively decide on the problem formulation, on the system boundaries, and on the dynamics of the system that will be addressed. Then, policies can be formulated and tested in this simulated environment [7, 15].

Lately, games are usually constructed by combining sophisticated computer simulations with interaction and role play [6]. Simulations provide a realistic context to the policy being tested. This participatory dialogue can be augmented if the stakeholders (players) can interact with the complex computer simulation built by integrating models. Including all relevant stakeholders in this participatory dialogue expands the scope of the social and institutional factors of the aspect system(s) included in the exercise. Interfacing these stakeholders with the complex model simultaneously expands the scope of the technical factors of the aspect system(s) in the exercise. The stakeholders can then make changes to the model and experience the effects of their changes.

Such an environment creates a powerful, complex multilogue [7], a dialogue between the stakeholders about the political nature of the planning problem. It also creates a dialogue between the stakeholders, representing the political and social aspect and the model representing the technical aspect of the large complex system. However, this requires a shift in how stakeholders interact with and perceive simulations.

Interfaces

A multilogue with stakeholders and complex models of the reference system, where stakeholders can experience the effects of their changes, necessitates different interfaces to the model system. Stakeholders must be able to interact with the model themselves and interact with it live, ideally when the model is running and is able to provide immediate and timely feedback. This creates a feedback loop and a chain of cascading decisions by stakeholders that can be observed and formalized into a plan.

Such interfacing requires interaction with the model and visualization of output data from it. Visualization of output data can be achieved during runtime using various graphics, animation, and gaming technologies. Simulations and analytical tools can stream data to the visualization tool where it can be animated and rendered in real time and in 3D, creating highly realistic, highly dynamic visual representations of the model.

Making changes in the model is harder, the extent of which is determined by the simulation or tool used within. If the simulation supports interference and changes during runtime through an application programming interface (API), then stakeholders can make said changes, which the simulation can immediately factor in and adapt dynamically. For example, during traffic management scenario, an operator can choose to close a link in the simulation forcing all vehicles in the traffic simulation to reconfigure their routes.

The advantage of such an approach is that the visualization can be customized to the stakeholder, creating a unique perspective on the model for that stakeholder. The options available to each stakeholder could also be different, reflecting their agency in reality. Stakeholders can also choose parameter settings, thereby customizing the model according to their beliefs of how the system should work and, at the same time, communicating those beliefs to other stakeholders in the multilogue. Depending on the flexibility offered by the underlying models, this customization can be continuously extended allowing stakeholders to experiment with parameters, structure, and so on.

In the following section, we provide evidence from five different cases where a different aspect of this approach was evaluated. Each case considers a planning problem, where the problem was first studied using either simulation models or big data analysis. The resulting analytical tool and results were evaluated in a participatory setting with multiple stakeholders.

Evidence

A pluralistic approach to city planning and model building can address the challenges described previously. This approach integrates multidisciplinary models of different aspect systems (related systems to the system being addressed) into a single complex whole, creating a complex, complete model. Participatory methods are then used to design exercises and experiments where multiple stakeholders interact with each other and with the integrated model to solve a particular problem. Stakeholders interact with each other based on their reflective, strategic interests and with the model from their domain-based perspectives, creating in a single environment, a pluralistic perspective on a complex model of the technical system, and a participatory multilogue structured by feedback loops between actors and models. Below, we describe planning problems in different cities where parts of this approach were used. Figure 8.2 provides an illustration of the different use cases, each of which is described below.

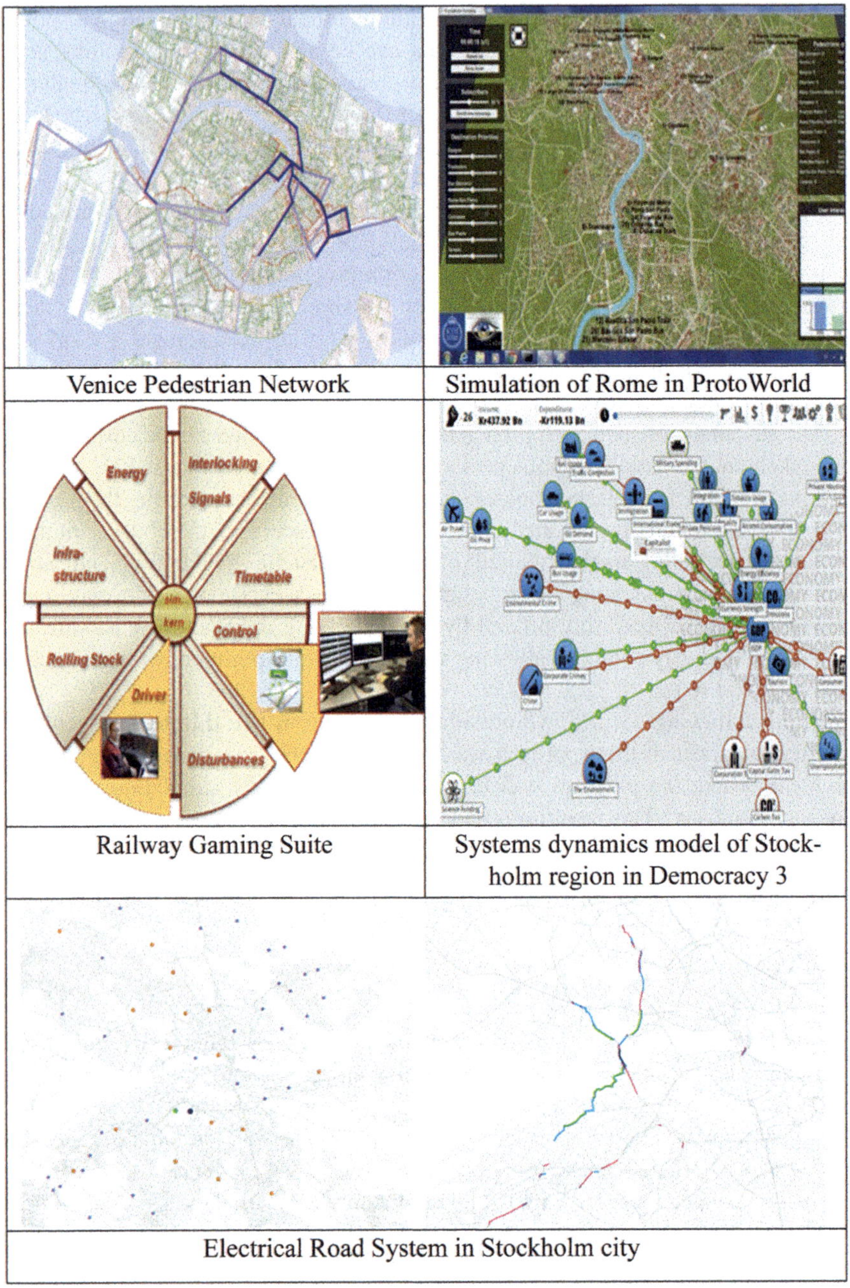

Fig. 8.2 Illustration of the different use cases

Venice Pedestrian Simulation: Replacing Data with Expertise

The historical city of Venice is situated on islands connected to each other through bridges forming a fascinating pedestrian network. This network is a free to move area that provides an environment of social interaction, economic activity, leisure, and cultural experience presenting a puzzling paradigm for modern cities [31]. Since Venice cannot expand due its geographical location and topology, it's capacity in terms of crowdedness has already been exceeded in many instances [47]. The resulting congestion has made the city less enjoyable for tourists and less comfortable for habitants. However, there is evidence that the city could manage the pedestrian network in ways that make flows of pedestrians spread across the whole network, by relying on providing information to city users and efficient use of public water transportation [40].

The challenge was to provide a simulation that responds to the complexity of the problem in hand and takes into account the lack of data in Venice. The extensive expert knowledge on Venice mobility has been used in a hybrid approach together with theory-driven models and available aggregated data. The simulation generalized in Flötteröd and Lämmel's [8] pedestrian model fits this problem well. The final simulation relied mainly on expert knowledge and highly aggregated low-dimensional data to simulate flows on the pedestrian network. A workshop with a multitude of stakeholders working with mobility in Venice assessed the simulation level of details and its participatory nature as a useful tool for operations and policy decision-making. The simulation provided travel times and levels of crowdedness of the city for different traffic scenarios, providing advice through an app for tourists to complete their journey while avoiding crowded parts.

Electrical Road Systems: Data Mining for Informing Design

Electrical road system (ERS) is a step in the direction of sustainable transport system. There are several competing technologies, which if deployed in a city can have a real impact in converting a large percentage of vehicles in the city – personal and heavy vehicles – to electric or hybrid. However, finding successful combinations of technologies in order to achieve this goal is complicated, as actors have their own optimized technology solutions, depending on their goals and interests, and there is no guarantee that summation of these technologies will lead the most suitable system.

The key component for all actors in this process is the location of charging installations, as this is a huge factor in the uptake of ERS. Big data from the movements of trucks all over Europe was mined to retrieve locations that have a potential for electrical road installations, along with metrics that characterized how these locations were relevant. The metrics were constructed based on expertise of stakeholders in ERS. Depending on these different metrics, the retrieved locations are subject to change. This gave an overview how the ERS would look for a city. This information was used to confront the experts in a design exercise to observe the changes in design [49].

The work in this use case, with the quantitative characterizations of results and qualitative validation from experts and actors, shows positive and important impact of using big data for design of future infrastructure systems. It also presents approaches to methodological and disciplinary boundaries [44].

Democracy: Gaming a Model of a Regional Policy Challenge

Democracy 3 is the latest version of the popular entertainment game, Democracy [19]. D3 is a political strategy game and simulates the desires, motivations, and loyalties of people in a country. The people are voters in the country, simulated through a neural network model and forming social groups, such as capitalists, socialists, environmentalists, and so on. The player takes on the role of a politically elected government and has to implement or change policies and keep the voters happy. Policies have effects, which are parameters that denote how the country is performing on various aspects. For example, increasing education subsidies will improve education as an effect. All policies are implemented by turning them on or by increasing their level in a slider. Policies and effects also influence how the voter groups feel about the government.

The core model of the game can be modified. A version of this model was created specific to the context of the Stockholm-Mälardalen region. The Stockholm region mod was created in a series of preparation workshops with policy makers from the region. In these workshops, the model as present in D3 was presented, and policy makers were asked to reflect on how the model can be changed to suit the local context. These changes were incorporated into the model to create the Stockholm model, and the Stockholm mod of Democracy was played in a workshop. Players were strategic planners from the region, and their objective was to act as the government of Sweden and reduce the CO2 emissions in the Stockholm region [37].

Railway Gaming Suite: Involving Operators in Design of the System

To address systemic innovations in the railway domain, one needs to be able to couple local expert knowledge to the design phase of the complex rail system. Over the period 2010–2014, ProRail (the Dutch railway administration) tested the use of gaming methods for enabling operator knowledge in design projects that affect train traffic control processes. The Railway Gaming Suite [22, 32] is a collection of analog and digital games built for this purpose. The general logic behind developing the suite was to meet a common perceived need by managers of engineering project to test the potential changes to a complex system before implementing changes in the real world. It therefore follows the logic of gaming for experimentation, by having operational experts participating as players. However, as discussed in Meijer [32] and van den Hoogen et al. [22], this logic often changes during the use and design phase of specific game instances for specific problems. Given the large uncertainties of the complex systems nature of railway engineering, especially when train operations get involved, the sessions change from pure testing of hypotheses to collecting evidence for hypothesized fundamental mechanisms and input to the design cycle of railway systems.

The suite consisted of board games, analog simulations of train traffic control with many (20+) humans in the loop, and digital gamified workplace simulators built on top of computer simulations. Extensive experience has been gained about the value of realism and fidelity in such environments. The concept of "Search Distance" [30] describes the difference between the current work setting and the simulated (future) setting and is of key importance to assess the mental models and situation awareness of the expert participants.

Smart City Dashboards in Rome: Interactive Simulation and Game for Shaping Policy Goals to Operational Realism

The city of Rome is attempting to build a dashboard where different kinds of big data analytics can be integrated and visualized to enable different agencies to manage its transportation and information services. The Mobility Agency of Rome needed a safe way to design and test different features of the dashboard. To enable such experimentation, a complex urban simulation was built using ProtoWorld, a framework to integrate, distribute, and play with models [35, 36]. The game built for the Mobility Agency integrated a vehicular simulation built using SUMO [27], a pedestrian simulation and a public transport simulation. The simulations run as separate component are aware of each other through integration mechanisms built within the ProtoWorld framework, which in turn is built within the Unity game engine. The results of the different simulations are animated and visualized in real time in a 3D city within the Unity engine, and traffic controllers can use different interfaces to control the live, virtual city.

The game was tested with controllers and managers from the Mobility Agency. Players could tweak certain parameters to calibrate the simulations according to their knowledge of how the city behaves. They could then manage this virtual city using interfaces and by sending information, closing links, increasing frequency of public transport, and so on. At the end, a log of their negotiated actions provided them with a strategy of how to handle that scenario. Experiencing such features enabled them to reflect on the features' usefulness or not, providing at the end of the session insights on requirements for the smart city dashboard.

Discussion

In the previous section, we described planning problems where models, simulations, and data and human expertise were applied, played with, and evaluated in a participatory setting by relevant stakeholders. The models ranged from simple (Venice) to complex (Rome), macro (Democracy) to micro (Rome, ProRail), supported by vast amounts of data (ERS) to very little (Venice). Each of these was evaluated in participatory gaming sessions where they interacted with the model system either during runtime (Rome, Democracy) to looking at static outputs (Venice, ERS) to a combination of the two approaches (Railway Gaming Suite). Applied in different contexts, the exercises yielded valuable insights on how planning can be achieved while expanding system and representational complexity in the planning process. The exercises also delivered insights on the form and function of such analytical tools. Despite the contrasting natures of the tools and problems, a pluralistic and participatory approach was vindicated, in that it delivered concrete and tangible insights into policy and design. Open, transparent, and flexible models are critical to planning in complex systems, as is the inclusion of human intuition and expertise in model development and model use. However, functional aspects of simulations such as validity and fidelity need to be redefined and addressed in this new approach to modeling. Below, we address some methodological dilemmas in detail, by taking into account the lessons learnt from the cases.

Dilemma 1: Quantitative and Qualitative Approaches

Quantitative approaches utilized in all use cases listed above speak to the modeling and characterizing phenomena in the complex systems. The quantitative information is very useful for comparing and perceiving the interactions within the system especially the ones that are well described and can be well defined.

The complex nature of the system and in some part the lack of tangible boundaries arise in which the expertise and knowledge of humans are inextricable from the operations of the system itself. The making of the quantitative information such as metrics in the case of ERS and the value of these metrics to the design process are needed to be validated from experts who had deep knowledge in aspects of ERS. The qualitative information was important to gather the impact of the quantitative information on the designs.

In case of Railway Gaming Suite, the responses of the operators during the exercise are irreplaceable and unquantifiable to assess the impact of the changes in system. Qualitative data was gathered from the operators to assess the impact of changes and limitations on the extent of the changes that can be handled by the operators/system.

In case of Venice pedestrian network, the lack and imprecision of data available in the city was compensated with an approach that combined an extensive local expert knowledge on the exact dimension of the problem combined with a model-based simulation that has as low dimensionality as possible. That is to avoid any suggestive over-precision and or incomprehension of simulations and to fit the simulation with the context of its use by policy makers.

There are some disciplinary imperatives that dictate the use of only one kind of data; and one is perceived to be more rigorous than the other. However, in the approaches and the use cases described above, the combination of quantitative and qualitative data was extremely useful in overcoming to a large extent, incompleteness and disciplinary boundaries.

Dilemma 2: Structural Uncertainty

Structural uncertainty presents unique challenges in the modeling of complex systems, where the system cannot be described adequately enough to model. Such uncertainty manifests itself in the form of epistemological uncertainties, concerning our conception of the system, methodological uncertainties concerning the appropriateness of methodologies, and tools and technical uncertainties concerning the appropriateness of data and its level of aggregation and detail. These affect model form, model completeness, and operation.

The game sessions in Rome with operators and managers provided insights on how this could be managed. The ProtoWorld framework and integrated simulations described only the physical and some technical components that could not change. For example, the road network, public transport network, population scales, and so on. Before interacting with the simulation, the players had to customize the integrated simulation by setting the values of a large set of parameters. For example, the players had to set tourist preference values for certain locations such as the Vatican. On a more fundamental level, players can rely on their expert knowledge and also change some assumptions in the model itself. For example, the pedestrian simulation exposes its fundamental state diagram through an interface to the player, which can be altered to change the decision-making logic of the modeled agent. Similar structures exist for other models. It allows players to endlessly tweak and change the structures of the simulation, capturing their knowledge within the model system, validating the simulation (to the extent of face validity), and allowing them to construct the simulated system according to their knowledge.

Overcoming some aspects of modeling uncertainty in this fashion proved valuable in the games in Rome, where players felt immediately comfortable with the model. Further, the process of negotiating the changes to the model among the players provided them with valuable insights on different strategies, requirements, and beliefs held about the city within and across organizational boundaries. The game proved to be an excellent vehicle for articulating and capturing these differences, generating further knowledge that could be used to improve the model.

Dilemma 3: Political Ambiguity

There is an inherent conflict between the technical-rational perspective espoused by engineering design and the political ambiguity in goals, values, and perspectives embraced by the communicative perspective of planning. Translating strategic goals into operational designs requires enormous coordination across institutional levels, with input from all levels going into the design [32]. The consequence of strategic decisions on the operational level is often unclear, and conversely the input of operational personnel is unasked for in strategic planning. However, the influence of big data technologies is slowly upending this trend, forcing operational levels to be strategic about their choices and forcing strategic levels to plan operational tasks.

In the ERS case, the data mining activity demonstrated potential locations of electrical road installations. When presented to engineers designing these systems at a very operational level, they could immediately reflect upon their engineering choices based on input at a very strategic level. The influence of this information relatively early in the design process prompted them to conceive alterations in technology. Conversely, based on this information and their expertise in the operating technology, they could immediately point out locations where it would be most feasible for installations, providing valuable insights on the strategic location choice.

Similarly in the ProRail case, redesign of infrastructure was necessary to achieve a strategic goal of performance improvement. Including the operators in this redesign exercise prompted different choices in the design, while also reconfiguring operational procedures to handle these changes.

Dilemma 4: Fidelity in Planning Tools

Fidelity as the degree to which a simulation is perceived to be physically similar to the part of the real world that it is meant to simulate [18]. The simulation should look, feel, and function like the reference system it is representing.

Despite lingering doubts about the necessity for such high fidelity, simulations tend to place large emphasis on achieving high fidelity, in both form and function. High fidelity is meant to support achieving stated goals, but while supporting planning and decision-making in complex systems, the goals might themselves be vague, be poorly articulated, or change during simulation development and application. Achieving high fidelity at very micro scales of technical components while technically possible doesn't always allow for validity or comprehensible analysis of the system.

The lessons drawn from the Democracy exercise suggest that it is more important to have a complete description of the system rather than a high fidelity but incomplete description of the system. An incomplete representation prompts cognitive dissonance among planners who are used to considering the system in its entirety.

Related systems should be included in the model, even if it is just as a simple variable. This exercise also suggests that planners are aware of modeling limitations and biases and would instead prefer models that are open in nature, with respect to transparency in assumptions and input data as well as open to customization of parameters and behaviors.

The lessons from the Rome games again suggest that a high fidelity description of the technical components is useful only to the extent that it provides face validity to the simulation. Once operators started to believe in the model, they tended to focus on the planning task, switching to strategic planning based largely on their expertise and intuition.

Conclusion

Health in a city involves contributions from many systems in the city. Each of these systems is usually functioning within their own borders and underpinned by different worldviews, practices, and bodies of theory. Health requires those systems to be harmonious and to complete each other pragmatically and epistemologically. It is therefore important to think of interfaces between systems and disciplines and hence on bridging different approaches. In a world where information, data, and expertise are increasingly fragmented, the strength of blended approaches presented in this paper is in identifying and bridging goals and knowledge of actors working in diverse systems but within the same city network. In methodological context of city planning, the onus is on using increasingly more computational intensive methods. Computation might increase the reach of science. But in the practice of implementing policies that have discernable effects on large populations, it is imperative to include relevant actors who understand the reality on ground and are able to mediate between complexity presented in model of models and reality. The outcomes of the cases presented have contributed to better designs to achieve desirable changes in the system. The future of these integrated approaches is in recursively transferring knowledge from empirical to computational methods to achieve comprehensive understanding and hence enrich the design process within complex systems.

References

1. Batty, M.: The new science of cities. MIT Press, Cambridge, MA (2013)
2. Batty, M.: Models again: their role in planning and prediction. Environ. Plan. B Plan. Des. **42**, 191–194 (2015). doi:10.1068/b4202ed
3. Cairncross, S., Hardoy, J.E., Satterthwaite, D.: The Urban Context. Earthscan Publications, London (1990)
4. Corburn, J.: Confronting the challenges in reconnecting urban planning and public health. Am. J. Public Health. **94**, 541–546 (2004). doi:10.2105/AJPH.94.4.541

5. Davidoff, P.: Advocacy and pluralism in planning. J. Am. Inst. Plann. **31**, 331–338 (1965). doi:10.1080/01944366508978187
6. Dearden, J., Wilson, A.: Using participatory computer simulation to explore the process of urban evolution. Trans. GIS. **15**, 273–289 (2011). doi:10.1111/j.1467-9671.2011.01257.x
7. Duke, R.D., Geurts, J.: Policy Games for Strategic Management. Rozenberg Publishers, Amsterdam (2004)
8. Flötteröd, G., Lämmel, G.: Bidirectional pedestrian fundamental diagram. Transp. Res. B Methodol. **71**, 194–212 (2015). doi:10.1016/j.trb.2014.11.001
9. Freudenberg, N.: Intersectoral approaches to health promotion in cities. In: Potvin, L., McQueen, D.V., Hall, M., de Salazar, L., Anderson, L.M., Hartz, Z.M.A. (eds.) Health Promotion Evaluation Practices in the Americas, pp. 191–219. Springer, New York (2008)
10. Freudenberg, N., Galea, S., Vlahov, D.: Cities and the Health of the Public. Vanderbilt University Press, Nashville (2006)
11. Galea, S., Vlahov, D.: Handbook of Urban Health: Populations, Methods, and Practice. Springer Science & Business Media, Heidelberg (2006)
12. Galea, S., Freudenberg, N., Vlahov, D.: Cities and population health. Soc. Sci. Med. **60**, 1017–1033 (2005). doi:10.1016/j.socscimed.2004.06.036
13. Geronimus, A.T.: To mitigate, resist, or undo: addressing structural influences on the health of urban populations. Am. J. Public Health. **90**, 867–872 (2000)
14. Goodspeed, R.: Smart cities: moving beyond urban cybernetics to tackle wicked problems. Camb. J. Reg. Econ. Soc. **8**, 79–92 (2015). doi:10.1093/cjres/rsu013
15. Greenblat, C.S., Duke, R.D.: Principles and Practices of Gaming-Simulation. Sage Publications, Beverly Hills (1981)
16. Gulati, R.: Tent poles, tribalism, and boundary spanning: the rigor-relevance debate in management research. Acad. Manag. J. **50**, 775–782 (2007)
17. Hajer, M.A.: Deliberative Policy Analysis: Understanding Governance in the Network Society. Cambridge University Press, Cambridge, CA (2003)
18. Hamstra, S.J., Brydges, R., Hatala, R., Zendejas, B., Cook, D.A.: Reconsidering Fidelity in simulation-based training. Acad. Med. **89**, 387–392 (2014)
19. Harris, C.: Democracy 3 [WWW Document]. Democr. 3. URL http://www.positech.co.uk/democracy3/index.php?src=positech. Accessed 30 Mar 2015
20. Healey, P.: Collaborative Planning: Shaping Places in Fragmented Societies. UBC Press, Vancouver (1997)
21. Healthy cities: Promoting health and equity – evidence for local policy and practice. Summary evaluation of Phase V of the WHO European healthy cities network [WWW Document]. URL http://www.euro.who.int/en/publications/abstracts/healthy-cities.-promoting-health-and-equity-evidence-for-local-policy-and-practice.-summary-evaluation-of-phase-v-of-the-who-european-healthy-cities-network (2015). Accessed 15 Feb 2017
22. van den Hoogen, J., Lo, J., Meijer, S.: Debriefing in gaming simulation for research: opening the black box of the von-trivial machine to assess validity and reliability. In: Proceedings of the 2014 Winter Simulation Conference, WSC'14. IEEE Press, Piscataway, pp. 3505–3516 (2014)
23. Hughes, J., Sadler, S.: Non-Plan: Essays on Freedom, Participation and Change in Modern Architecture and Urbanism. Routledge, London (2013)
24. Innes, J.E.: Planning theory's emerging paradigm: communicative action and interactive practice. J. Plan. Educ. Res. **14**, 183–189 (1995). doi:10.1177/0739456X9501400307
25. Innes, J.E.: Planning through consensus building: a new view of the comprehensive planning ideal. J. Am. Plan. Assoc. **62**, 460–472 (1996). doi:10.1080/01944369608975712
26. Jacobs, J.: The Death and Life of Great American Cities. Random House LLC, New York (1961)
27. Krajzewicz, D., Erdmann, J., Behrisch, M., Bieker, L.: Recent development and applications of SUMO – simulation of urban mobility. Int. J. Adv. Syst. Meas. **5**, 128–138 (2012)
28. Lee, D.B.: Requiem for large-scale models. J. Am. Inst. Plann. **39**, 163–178 (1973). doi:10.1080/01944367308977851

29. Leeuw, E. de Tsouros, A.D., Dyakova, M., Green, G., World Health Organization, Regional Office for Europe: Healthy cities, promoting health and equity–evidence for local policy and practice: summary evaluation of Phase V of the WHO European healthy cities network (2014)
30. Lo, J., van den Hoogen, J., Meijer, S.: Using Gaming Simulation Experiments to Test Railway Innovations: Implications for Validity, pp. 1766–1777. IEEE (2013). doi:10.1109/WSC.2013.6721557
31. Mamoli, M., Michieletto, P., Bazzani, A., Giorgini, B.: Venice as pedestrian city and tourist magnet: mass events and ordinary life. International forum on urbanism, pp. 1–10 (2012)
32. Meijer, S.: The power of sponges: comparing high-tech and low-tech gaming for innovation. Simul. Gaming. **46**, 512–535 (2015). doi:10.1177/1046878115594520
33. Miller, J.T.M.: Methodological issues for interdisciplinary research. In: Postgraduate english: a Journal and forum for postgraduates in english (2011)
34. O'Rourke, M., Crowley, S., Gonnerman, C.: On the nature of cross-disciplinary integration: a philosophical framework. Stud. Hist. Philos. Sci. Part C Stud. Hist. Philos. Biol. Biomed. Sci. **56**, 62–70 (2016). doi:10.1016/j.shpsc.2015.10.003
35. Raghothama, J., Meijer, S.: Gaming, urban planning and transportation design process. In: Geertman Jr., S., Ferreira, J., Goodspeed, R., Stillwell, J. (eds.) Planning Support Systems and Smart Cities, Lecture Notes in Geoinformation and Cartography, pp. 297–312. Springer International Publishing (2015a). doi:10.1007/978-3-319-18368-8_16
36. Raghothama, J., Meijer, S.: Distributed, integrated and interactive traffic simulations. In: Proceedings of the 2015 Winter Simulation Conference, WSC'15, pp. 1693–1704. IEEE Press, Piscataway (2015b)
37. Raghothama, J., Meijer, S.A.: What do policy makers talk about when talking about simulations? In: Proceedings of the 46th International Simulation and Gaming Association Conference. Presented at the 46th ISAGA Conference, Kyoto (2015c)
38. Rhoten, D.: Interdisciplinary research: trend or transition. Items Issues. **5**, 6–11 (2004)
39. Rittel, H.W.J., Webber, M.M.: Dilemmas in a general theory of planning. Policy. Sci. **4**, 155–169 (1973). doi:10.1007/BF01405730
40. Russo, A.P.: The "vicious circle" of tourism development in heritage cities. Ann. Tour. Res. **29**, 165–182 (2002). doi:10.1016/S0160-7383(01)00029-9
41. Salmon, J.W., Whiteis, D.G.: Improving public health care: lessons on governance from five cities. J. Health Care Poor Underserved. **3**, 285–304 (2010). doi:10.1353/hpu.2010.0314
42. Schoenwandt, W.: Planning in Crisis?: Theoretical Orientations for Architecture and Planning. Ashgate Publishing, Ltd (2012)
43. Sethi, D., Wood, S., Mitis, F., Bellis, M., Penhale, B., Marmolejo, I.I., Lowenstein, A., Manthorpe, G., Kärki, F.U., others: European Report on Preventing Elder Maltreatment. World Health Organization, Copenhagen (2011)
44. Shreenath, V.M., Meijer, S.: Spatial big data for designing large scale infrastructure: a case-study of electrical road systems. In: Proceedings of the 3rd IEEE/ACM International Conference on Big Data Computing, Applications and Technologies, BDCAT'16. ACM, New York, pp. 143–148 (2016). doi:10.1145/3006299.3006334
45. Smith, B.J., Tang, K.C., Nutbeam, D.: WHO health promotion glossary: new terms. Health Promot. Int. **21**, 340–345 (2006). doi:10.1093/heapro/dal033
46. Townsend, A.: Cities of data: examining the new urban science. Public Cult. **27**, 201–212 (2015). doi:10.1215/08992363-2841808
47. Van der Borg, J., Costa, P.: The management of tourism in cities of art. Tour. Rev. **48**, 2–10 (1993)
48. Vlahov, D., Freudenberg, N., Proietti, F., Ompad, D., Galea, S.: A conceptual framework for organizing determinants of urban health. In: Thematic Paper for KNUS Second Meeting, pp. 1–4 (2006)
49. Shreenath, Magal, V., Meijer, S.: Opening the search space for the design of a future transport system using 'Big Data'. In: Planning Support Science for Smarter Urban Futures, pp. 247–261. Springer International Publishing, Cham (2017)

Chapter 9
Aging in the Contemporary Urban Context: The Mortality Rates of Older Residents in Genoa, Italy

Mauro Palumbo and Stefano Poli

Abstract Understanding the need and the lines of the increasing numbers of older residents within the rapidly changing urban environment becomes a priority of urban architects, social gerontologists, city planners, and decision-makers. The aim of this study is to assess the possible association between the living conditions and the urban environments, where aging process is experienced. We propose also a model for secondary analysis of datasets provided by city registry offices that aims to realize a decision support system focused on detailed geo-mapping representation of the city of Genoa, Italy.

Keywords Demographic aging process • Data analysis • Decision support system • Geo-mapping

Introduction

The development of contemporary urbanization and the consequent growing complexity of social life have to be carefully reexamined considering the demographic aging process, increasingly interesting more and more Western as well as non-Western societies [19]. Understanding the lives of the increasing number of older residents within the rapidly changing context of urban environments becomes a priority issue for both urban architects and social gerontologists as well as city decision-makers [16, 25, 29].

While the article derives from several discussions between the authors, Mauro Palumbo is the author of the sections "Introduction" and "Conclusions", and Stefano Poli is the author of all the remaining sections.

M. Palumbo (✉) • S. Poli
Department of Education, University of Genoa, Genoa, Italy
e-mail: palumbo@unige.it; Stefano.Poli@unige.it

A. Karakitsiou et al. (eds.), *City Networks*, Springer Optimization and Its Applications 128, DOI 10.1007/978-3-319-65338-9_9

The changing demographics of aging must be observed in their composed heterogeneity as well as within the context of the challenges of urbanization [19]. Indeed, particularly when cities face environmental threats or experience major population decline, major risks can affect their most vulnerable residents, especially for older people.

At the same time, most of the urban administrations are involved in obtaining useful data on their residents. Collecting demographic statistics is typically one of the institutionally defined tasks of every municipality, and the capability to read and evaluate such large datasets becomes a strategic objective in order to improve the living conditions of residents and adequately intervene in case of difficulties or to promptly prevent threats [17].

This becomes particularly strategic when local administrations have to enhance or protect the quality of life of older residents. Knowing the specific context of the aging experience is important to understand when and where and how the typical breakdowns of the "formal" urban infrastructure may cause considerable damage to the "informal" networks and bonds that sustain older people in the local urban setting.

The aim of this contribution is to assess the possible implications of the urban settings for the conditions of older residents. To such purpose we will focus on the exam of the death rates of over 65 s, exploring the association of mortality of older residents with the local urban dwelling conditions. Undoubtedly, the mortality rates are affected by age-dependent effect, but environmental and socioeconomic factors, interesting the way the aging process is experienced, could play an important role in the risk of mortality, particularly among most vulnerable groups, like the older residents.

At the same time, we propose a model for a secondary analysis of datasets provided by city registry offices that aims to realize a decision support system focused on a deeply detailed geo-mapping representation of the town of Genoa, Italy. This urban location represents one of the demographically older urban settings in Europe that typically observes both the demographic processes of aging "from above" and "from below," showing a significant incidence of older residents combined with a long enduring reduction in newborns [22].

Aging in Contemporary Cities: A Brief Theoretical Review

Studying the relationship between the urban change and the conditions of older residents has become a priority because of the growing complexity of social life and the demographic aging process currently accompanying urbanization in Western society.

Considering the conditions of the growing elderly population, Phillipson [19] recalls three main dimensions of social exclusion suffered by older people in contemporary urban settings: the neighborhood change, the generic pressures operating in the urban environment, and the impact of globalization on the perception of places and local identities.

The first dimension typically interests those older people living in unpopular urban neighborhoods characterized by low housing demand and subsequent abandonment of housing by all but the poorest and least mobile residents. Both of these conditions often affect elderly profiles [5]. Still, social exclusion may also be observed in neighborhoods not threatened by socioeconomic decline but instead are undergoing gentrification processes. Indeed, such modifications often lead to social exclusion of older working-class inhabitants lacking the resources to match the new lifestyle conditions enacted by incoming upper-class residents [20, 28].

The second dimension refers to the environmental change in urban living where physical barriers, such as poor paving or street clutter, as well as structural problems, such as the lack of benches or green open spaces, can generate significant menace to the living conditions of older residents [10].

Lastly, we must consider that in the same place the impact of globalization on urban life is frequently accompanied by migrations and population displacement with consequent sociocultural modifications, while, on the other hand, the so-called aging in place process is experienced by older residents. So they live the paradox characterized by severely reduced mobility and strong attachment to places and local traditions [20], while these places suffer for great and negative modifications.

With respect to the first two dimensions, the neighborhood and the related environment can be a resource or a risk to health, particularly for older people who often suffer marginalization and limited mobility frequently associated with structurally inadequate living conditions and quite often socioeconomic deprivation.

Research has documented independent effects of neighborhood-level socioeconomic status on health after controlling for individual-level characteristics [21, 24]. These findings suggest that neighborhood-level associations with health are not due entirely to compositional effects of neighborhood residents (i.e., the demographic composition of residents, by age, gender, and ethnic factor, potentially producing differences in health conditions of residents) but that structural conditions of the neighborhood context itself may influence health. Excessive noise, inadequate lighting, and heavy traffic are related to a loss of physical function in elderly people [2]. Both individual perceptions of neighborhood disorders [27] and objective observations of neighborhood conditions [30] emerged as related to poorer health. In contrast, affluence and neighborhood social resources have been recognized as linked to better health [32]. Nevertheless, the specific mechanisms through which neighborhood conditions affect health are still under debate [1].

On such a basis, the hypothesis leading this study relies on both descriptive and explicative aims. On the one hand, our goal is to observe through statistical analysis how the mortality ratios of older people (here adopted as an antithetical proxy of good health conditions) could be associated with different dimensions related to demographic characteristics, contextual structural conditions, and issues related to isolation of the elderly population. On the other hand, we also aim to evaluate the different impacts of the aforementioned factors. This leads our study to provide a model for a decision support system focused on a deeply detailed geomapping representation of a metropolitan urban context and finalized to support local policymakers in evaluating the vulnerability risk for older residents in order to plan adequate urban interventions.

Methods: Assessing the Mortality Ratio Among Aging Population

Study Design and Variables

Our study was carried out in Genoa, a metropolitan context in Northern Italy containing a major aging demographic. Indeed, the aging index in Genoa was equal to 235.9 in 2013, significantly higher as compared to the Italian mean value of 152.7 and the European (EU28) mean value of 117.7 (https://open-data.europa.eu/en/data/publisher/estat).

This was a secondary data analysis combined with a geo-referencing model. The study was realized on big datasets provided by the official statistical registry of the city of Genoa, deepening the analysis to the most granular possible level of territorial observation. In order to perform a very detailed examination of the mortality ratios of older people and of potential associated factors, we assessed the census sections as units of analysis. The census section represents the smallest statistical territorial unit used for census purposes [11]. It is defined by a single territorial corpus delimited by a closed segmented line. Starting from census sections, it is possible to recreate, by sum, the bigger geographical and administrative units, such as municipalities, metropolitan areas, regions, and the entire national territory. Most urban census units approximately correspond to a city block, thus defining limited territorial aggregates and adequately estimating "contextual" or "neighborhood" effects [9].

All of the following indicators adopted for this study refer to several algorithms applied to the 3616 census sections composing the city of Genoa.

The dependent variable (i.e., the mortality of older residents) has been explored by means of the standardized mortality ratio (SMR). This indicator quantifies the increase or decrease in mortality of a study cohort with respect to the general population by expressing the ratio of observed deaths in the study group to expected deaths in the general population [3, p. 417]. It may be quoted as either a ratio or a percentage (multiplying by 100). If the SMR is quoted as a ratio and is equal to 1.0, then this means that the number of observed deaths equals that of expected cases. If the SMR is higher than 1.0, then there is a higher number of deaths than is expected. In our case, we used the over 65 SMR related to the older population. In the EU28, 82% of all deaths are deaths among over 65 s [7]. This represents a sort of "natural" mortality (if compared to the "preventable" mortality typically related to younger cohorts).

The SMR is a fluctuating indicator, year by year; thus, we adopted the average mean for the years 2011–2015 (represented at the urbanistic unit level in Fig. 9.1).

The average SMR in Genoa is generally higher than elsewhere because of the significantly higher aging index. Some neighborhoods describe greater ratios, particularly the western and northern suburban zones composed mainly by peripheral and less-advantaged working-class neighborhoods (almost in comparison with the eastern higher-status residential areas).

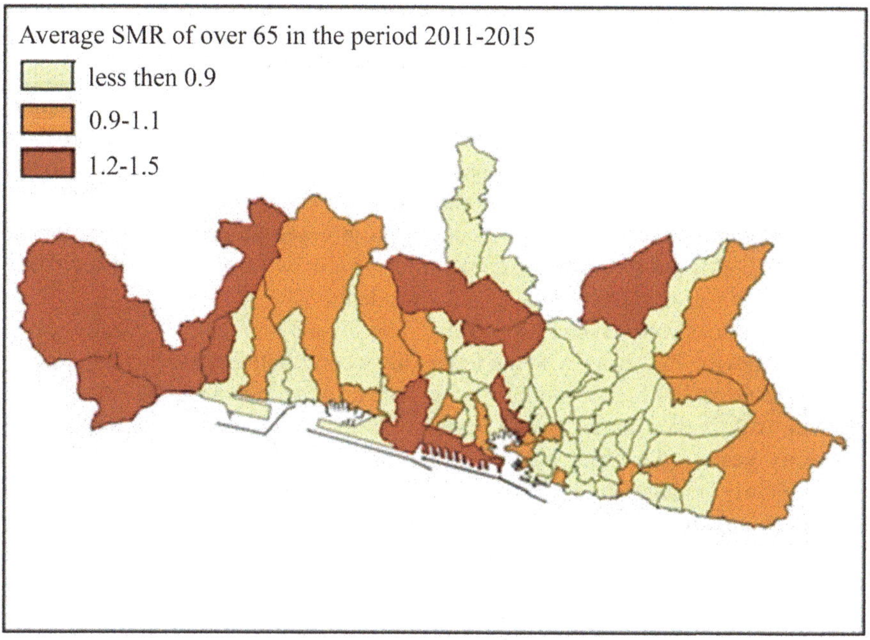

Fig. 9.1 Average SMR among over 65 residents for the period 2011–2015 in the city of Genoa represented at the urbanistic unit level (Source: Our elaborations on datasets provided by the statistical office of Genoa City)

Several factors can be potentially associated with average mortality of older residents. In our hypothesis, we explored the associations of the dependent variable with (a) the demographic dimension, considering the large incidence of over 65 s in the observed study context; (b) the structural environment, referring mainly to the wider socioeconomic conditions and the levels of architectural preservation of the urban setting; and (c) the isolation and the potential loss of autonomy among the older residents, considering the major factors of loneliness and care needs typically experienced by elderly population. Indeed, several studies showed how the mortality rates and life expectancy can be often associated with the aforesaid structural factors, particularly when they are combined with social deprivation [4, 26, 31].

In our hypothesis, the aforementioned dimensions can be considered as preliminary factors potentially contributing to the average SMR of older people for the period 2011–2015. Thus, we observed at the census section level all the independent factors adopting (where possible) their resulting averages for the period 2010–2014 (almost 1 year before the occurred death event).

Focusing on the demographic dimension, we first decided to adopt the standard OECD [15] definition of elderly population based on people aged 65 and over.[1]

The incidence of the elderly population has been calculated for the period of observation as the percentage share of over 65 residents on the total population, considering the average rate of over 65 s for the period 2010–2014 as a proxy of the demographic weight of older people in each census section. Such a variable is typically associated with natural mortality [8].

For the statistical analysis, we recorded the average rate of over 65 s for the period 2010–2014 as an ordinal variable, splitting it in two modalities between "less or equal to 50%" and "over 50%" of older residents.

Gender issues were taken into account. The female population typically shows higher longevity; however, by living longer, they contribute both to the major aging demographic and to higher mortality rates, because the increasing weight of older women, with higher mortality rate, leads to an increasing average rate [1]. Thus, for every year in the observed period, we calculated the percentage of females in the elderly population, and we considered for the analysis the average rate of women among over 65 s in the period 2010–2014 (recoding the female rate between "less or equal to 50%" and "over 50%" of older residents). We then shifted to the structural dimension in order to examine the contextual living environment for the over 65 s in each census section. One of the main factors potentially associated with the average mortality of older residents is the socioeconomic condition. The Social Disadvantage Index (SDI) [6, 14] defines a measure of potential socioeconomic difficulties in a given area (in our case, each census section) resulting from the weighted mean of deviations from specific indicators compared to the correspondent national average values observed in the population census in 2011. The considered indicators are the unemployment rate (UNEMP), the employment rate (OCC), the concentration rate of younger population (YOUNG), and the schooling rate. The SDI is calculated according to the following formula:

$$\text{SDI} = 0.40^{*}\,(\delta\ \text{UNEMP}) + 0.30^{*}\,(\delta\ \text{OCC}) + 0.15^{*}\,(\delta\ \text{YOUNG}) + 0.15^{*}\,(\delta\ \text{SCHOOL})$$

where, in details:

- δ UNEMP = UNEMP(i) – UNEMPNAT
  Calculating UNEMP(i) as the unemployment rate of each specific census section as given by the percent ratio between the over 15 population looking for a job and the labor forces of the same age class and UNEMPNAT as the national unemployment rate in 2011 (11.4%).

[1]This is mainly a statistical definition and not a proper sociological definition of older people. Indeed, aging factors are only partially related to chronological age [13]. Nevertheless, we performed separate analysis raising the chronological age and considering people aged 75 and over: this did not change the results, while the observed associations emerged even as reinforced.

- δ OCC = OCCNAT – OCC(i)
  Calculating OCC(i) as the employment rate of each specific census section as given by the percent ratio between the over 15 occupied population and the labor forces of the same age class and OCCNAT as the national employment rate in 2011 (45.0%).
- δ YOUTH = YOUTH(i) – YOUTHNAT
  Calculating YOUTH(i) as the concentration rate of younger population in each specific census section as given by the percent ratio between the under 25 population and the total population and YOUTHNAT as the national concentration rate of younger population in 2011 (24.0%).
- δ SCHOOL = SCHOOLNAT – SCHOOL(i)
  Calculating SCHOOL(i) as the schooling rate of each specific census section as given by the percent ratio between the population possessing at least a secondary school degree and the total over 25 population and SCHOOLNAT as the national schooling rate in 2011 (51.4%).
- The coefficients indicated in the SDI formula represent the official weights defined by the Ministry of Economic Development [14] for each one of the previous indicators: 0.40 for the variation of unemployment rate (δ UNEMP), 0.30 for the variation of the employment rate (δ OCC), 0.15 for the variation in concentration rate of younger population (δ YOUTH), and 0.15 for the variation of the schooling rate (δ SCHOOL).

A final value equal to "0" reflects equivalence to the national level, while a value greater than "0" defines a greater level of socioeconomic disadvantage in the census section compared to the average national level. For the successive bivariate analysis and the binary logistic regression model, we recorded SDI in two categories: "under or equivalent" (≤ 0) or "above" (>0) the national average. Also, the structural situations of the urban environment can produce effects on the living conditions and consequent mortality rates among older populations. The Building Decline Index (BDI 2011) [6] compares the state of maintenance of buildings of a specific urban area with the equivalent national mean level, according to the following formula:

$$\mathrm{BDI} = \left[\left(\mathrm{RBp} + \mathrm{RBm}\right) / \mathrm{RBT}\right] / 0{,}168$$

where RBp refers to the residential buildings in poor conditions in each census area, RBm refers to residential building in mediocre conditions, and RBT indicates the total residential buildings. The weighting coefficient reflects the national percentage of residential buildings in poor or mediocre conditions. A final value greater than "1" indicates the decline of residential buildings—i.e., the percentage of those in poor or mediocre conditions—and exceeds the national average (equal to 16.8% in 2011). For the descriptive analysis and the regression model, we recorded the BDI in two categories: "under or equivalent" (≤ 1) and "above" (>1) the national average.

The algorithms for both SDI and BDI[2] use data from the latest available census datasets; so, different from the other variables realized through arithmetical means for the 2010–2014 period, these two indicators refer only to 2011. Even so, considering the limited intervening changes in structural situation, the available data still represent a good proxy for the actual environmental conditions.

The last dimension refers to isolation factors. Among personal circumstances that are likely to increase vulnerability to loneliness in older age, besides lower social class and poverty, Jylhä and Saarenheimo [12] indicate living alone and living in an institution. Loneliness is measured by the share of the percentage of older people living alone of the total number of over 65 residents (recorded in two modalities: between "less or equal to 50%" and "over 50%" of older residents). This indicator reports the formal situation emerging from the municipal registry office, where the official record of "living alone" represents a simple and acceptable proxy of loneliness (for instance, even if living alone, other relatives can live nearby, reducing the effective isolation). Similarly, there are no data available at the census section level about the exact number of older people living in institutions. So we determined the exact number of active nursing homes operating mainly in elderly care sectors during 2010–2014. We also defined the presence of residential care units as a proxy of institutionalized assistance (recorded in a dichotomy between "1," corresponding to the "presence of almost one nursing home," and "0," equivalent to "no nursing home recorded") since the presence of even one nursing home in a census section could have a key effect on the local mortality rate because of the higher concentration of older people in worse health conditions.

Statistical Methods

The relationships between the average SMR among over 65 residents in the period 2011–2015 and the aforementioned independent factors were previously explored through bivariate analysis, verifying significances of p-values by means of Pearson's chi-square. Subsequently, we adopted a binary logistic regression model to verify the association of higher levels of average standardized mortality ratios among over 65 s (>1.5 vs. ≤ 1.5) with the various indicators in the census sections of the Genoa Municipality for the observed period. The two-sided Wald test was used to assess the statistical significance of each parameter in the model. Pearson's product-moment correlation coefficient or Spearman's rank correlation coefficient was calculated in order to estimate correlations between parameters. Once checked for multicollinearity, the examined variables were revealed to be mostly independent

[2]SDI and BDI are not the only possible indices for the observed dimensions; nevertheless, they have been adopted in this study because they are formally provided by the Italian official statistics to calculate deprivation dimensions at the census section level for those socioeconomically disadvantaged zones needing targeted and specific interventions (ZFU).

and not excessively correlated with each other (the only exception being the average incidence of over 65 s for the period 2010–2014, which was excluded from the regression model because it is implicitly correlated with most of the demographic variables). We adopted a stepwise backward-selection method (likelihood ratio test <0.2). Two-tailed probabilities were reported, and the p-value of 0.05 was used to define nominal statistical significance.

All statistical analyses were conducted by means of SPSS (version 23; SPSS Inc., Chicago, IL, USA) and were geospatially represented by a QGIS mapping system (version 2.18.1), providing a detailed chart of the whole town and describing the mortality ratio of older residents and its associations with the observed factors.

Results: Analyzing and Mapping the Mortality of Older Residents

Statistical Analysis

Considering results in average SMR among over 65 s in the period 2010–2014, we classified the 3616 census sections in three clusters (Table 9.1): (a) sections with a lower mortality among senior residents (2300 sections, equal to 63.6%), where the SMR remains below the mean of the municipality (<1); (b) sections with an average mortality among senior residents (816 sections, equal to 22.6%), where the SMR corresponds or slightly exceeds the mean of the municipality (≤1–≥1.5); and (c) sections showing higher mortality among over 65 s (500 sections, equal to 13.8%), where the SMR significantly exceeds the mean of the municipality (>1.5).

According to the results of bivariate analysis, those census sections showing higher SMR among over 65 s also show smaller average total populations. This can be partially related to less densely populated peripheral sections, but it can also hide a possible relationship of mortality with suburban isolation.

The average SMR among over 65 s increases, whereas in the previous years, (a) the average incidence of older people was higher (quite expected, due to the aging of population); (b) the average incidence of women among over 65 s exceeded 50% of the residents (which, due to female longevity, implies higher mortality rates among older women); (c) the average rate of over 65 s living alone surpassed the 50% of the older people residing in the census section; and (d) the presence of almost one nursing home was recorded. Remarkably, higher average SMR of over 65 s was associated with those census sections showing higher degrees of the Socioeconomic Disadvantage Index and the Building Decline Index (both exceeding the respective national means).

The analysis performed through the binary logistic regression model (Table 9.2) shows statistically significant associations of higher levels of average standardized mortality ratios among over 65 s (>1.5 vs. ≤1.5) with the observed factors.

Table 9.1 Characteristics of the census sections in Genoa City by average SMR among over 65 s ($n = 3616$)

	Average SMR over 65 s <1 2011–2015 (n = 2300)	Average SMR over 65 s ≤1–≥1.5 2011–2015 (n = 816)	Average SMR over 65 s >1.5 2011–2015 (n = 500)	*p*
Average total population 2010–2014	163.8	216.9	96.9	0.048
Average rate of over 65 s 2010–2014				
≤ 50%	2264 (99.4)	808 (99.0)	472 (94.4)	0.000
> 50%	36 (1.6)	8 (1.0)	28 (5.6)	
Average rate of women among over 65 s 2010–2014				
≤ 50%	2241 (97.4)	800 (98.0)	3503 (96.9)	0.000
> 50%	59 (2.6)	16 (2.0)	113 (3.1)	
Average rate of over 65 living alone among over 65 s 2010–2014				
≤ 50%	2166 (94.2)	765 (93.8)	3357 (92.8)	0.000
> 50%	134 (5.8)	51 (6.3)	259 (7.2)	
SDI 2011				
Under or equivalent to national average	1944 (84.5)	625 (76.6)	383 (76.6)	0.010
Above national average	356 (15.5)	191 (23.4)	117 (23.4)	
BDI 2011				
Under or equivalent national average	1752 (76.2)	568 (69.6)	311 (62.2)	0.013
Above national average	548 (23.8)	248 (30.4)	189 (36.8)	
Presence of nursing homes				
No	2297 (99.9)	801 (98.2)	462 (92.4)	0.000
Yes	3 (0.1)	15 (1.8)	38 (1.5)	

Data summarized as mean or numbers (%), expressing significance (*p*) with Eta squared and Pearson chi-squared tests, respectively

Table 9.2 Binary logistic regression model related to higher levels of average SMR among over 65 s (>1.5 vs. ≤1.5) in the census sections of Genoa City (period 2011–2015)

	N	OR[a]	(95% CI)[a]	p^b
Social disadvantage level (SDI 2011)				
Under or equivalent to national average (≤ 0)	2952	1.00		
Above national average (>0)	664	1.32	(1.04–1.67)	0.021
Building decline level (BDI 2011)				
Under or equivalent to national average (≤ 1)	2631	1.00		
Above national average (>1)	985	1.70	(1.38–2.09)	0.000
Average incidence of women among over 65 residents (2010–2014)				
≤ 50%	3503	1.00		
> 50%	113	2.49	(1.60–3.87)	0.000
Average incidence of people living alone among over 65 residents(2010–2014)				
≤ 50%	3357	1.00		
> 50%	259	2.71	(2.02–3.64)	0.000
Presence of nursing homes in the census section in period 2010–2014				
No	132	1.00		
Yes	312	13.72	(7.63–24.64)	0.000

[a]*OR* odds ratio, *CI* confidence interval; OR > 1 indicates higher association to higher levels of average standardized mortality ratios
[b]Two-sided Wald test

Significant associations of higher SMR levels of older residents emerged with (a) major deprivation and urban deterioration, particularly in those areas with above national average levels of the Social Disadvantage Index (odds ratio equal to 1.32 with a confidence interval equal to 1.04–1.67) and of the Building Decline Index (odds ratio equal to 1.70 with a confidence interval equal to 1.38–2.09); (b) major average incidence of older female population (odds ratio equal to 2.49 with a confidence interval equal to 1.60–3.87, where women exceed the 50% of the older population); and (c) major loneliness and vulnerability indicators, particularly in the case of higher average incidence of older people living alone (odds ratio equal to 2.71 with a confidence interval equal to 2.02–3.64, where older people living alone exceed the 50% of the over 65 s) and in the case of recording almost one nursing home for older people in the census section (odds ratio equal to 13.72 with a confidence interval equal to 7.63–24.64 in census section where almost one nursing home is recorded).

Mapping Mortality of the Elderly and Its Associated Factors: Evidence Among Different Neighborhood Conditions

Our results are confirmed when tested with a deeper empirical exploration using a detailed map of some specific quarters that, once again, showed the associations of mortality ratios of older people with the environmental conditions. The quarters of Genoa can be classified according to the economic criteria [18], mainly in peripheral, intermediate, and residential neighborhoods. We selected three specific neighborhoods with the aforementioned classifications in order to also show by data geo-referencing the emerged associations of SMR with the tested independent factors.

The first quarter is Cornigliano, a lower working-class neighborhood that was once an industrial pole of excellence. It is a typical example of contemporary postmodern metropolitan suburban area which is characterized by severe unemployment, deprivation, and difficulties in integrating the foreign immigrants attracted to the area by low housing rent. Interestingly, the major presence of foreign immigrants contributes to lower the average aging index of this area (in 2014, this was equal to 171.45 vs. a municipal average value of 242.7, according to the data provided by the statistical office of Genoa City). Here, older residents typically suffer both the diffused structural decline and the general loss of neighborhood identity. The average SMR among over 65 residents for the period 2011–2015 reaches 1.66, significantly exceeding the municipal value of 1.00. As shown in Fig. 9.2, many of the census units in this area encompass average mortality ratios among the over 65 residents. This is associated with widespread socioeconomic difficulties and structural decaying (as shown in the figure, local SDI and BDI frequently exceed the national threshold) in addition to the incidence of alone older people and the presence of several nursing homes hosting dependent elderly patients.

The area of Prè-Molo-Maddalena represents the old historical center of Genoa, an intermediate quarter of interest for both isolation issues and gentrification processes. Demographically, the local neighborhood has the lowest aging index of the municipality (in 2014, this was equal to 155.8 vs. 242.7, according to the data provided by the statistical office of Genoa City) because of the well-established incidence of foreign residents and of people living alone. On the one hand, the less qualified areas of the historical center have difficult living conditions that are due to social marginalization and severe building decline. On the other hand, in the last few decades, because of the high value of the historical and architectural heritage, a significant gentrification process is underway in certain areas, attracting both upper-class residents and city users. This has produced structural requalification but also the subsequent departure of lower-class residents who are unable to afford the increased cost of housing. When living in less qualified areas, older residents of the historical center often suffer from social marginalization and difficult structural conditions. As observed in Fig. 9.3, this is confirmed by higher levels in both the SMR among over 65 s (1.72 in the historical center vs. 1.00 at the municipal level) and the BDI (1.70 vs. the national average of 1.00). Typically, several of the older people who remain in the historical center live alone. In addition, the few nursing homes available are generally confined to the less qualified areas.

Lastly, we mapped the mortality among over 65 s and the associated factors in the residential quarter of the Foce, which is not only a typical downtown neighborhood but also the demographically older area of the town. In 2014 the aging index reached the remarkable value of 296.9 vs. 242.7 of Genoa (data provided by the statistical

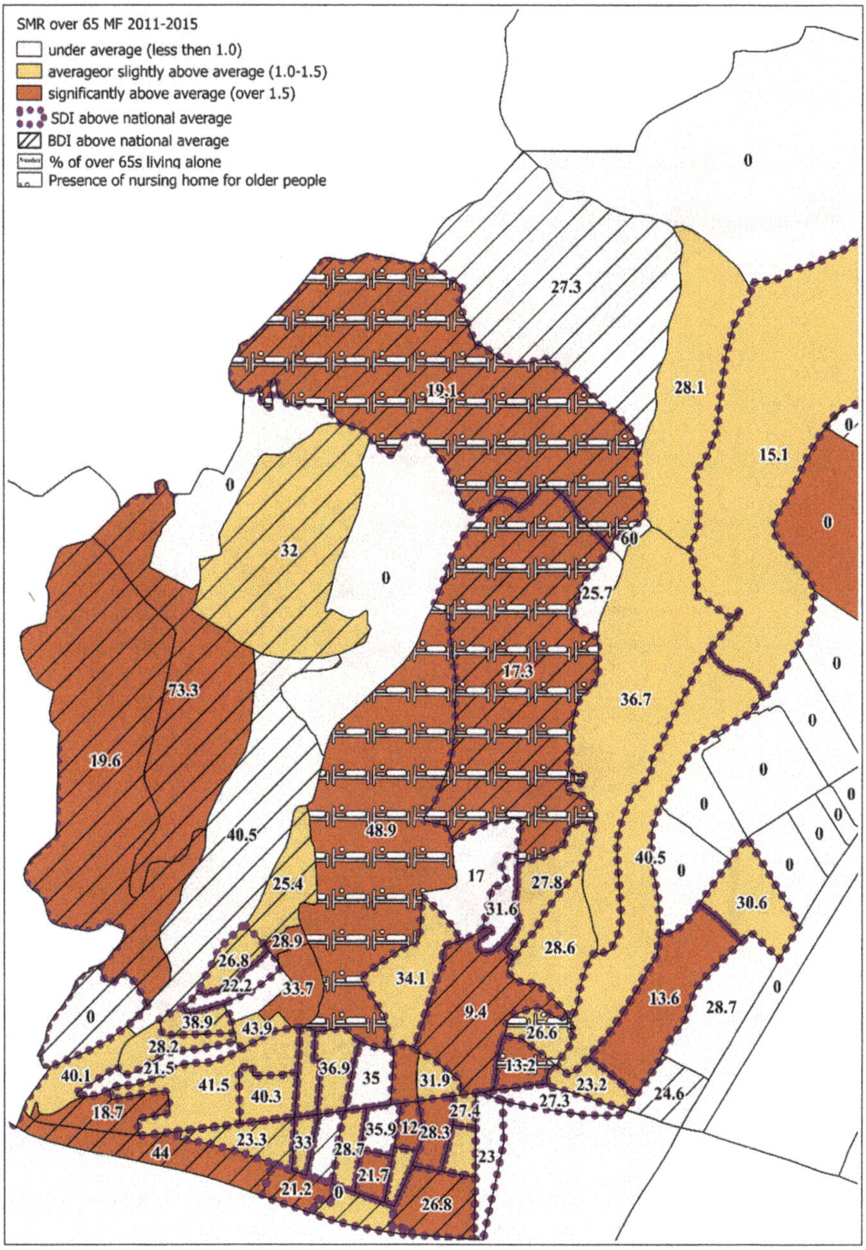

Fig. 9.2 Average SMR among over 65 residents for the period 2011–2015 in neighborhood of Cornigliano and associated factors (Source: Our elaborations on datasets provided by the statistical office of Genoa City)

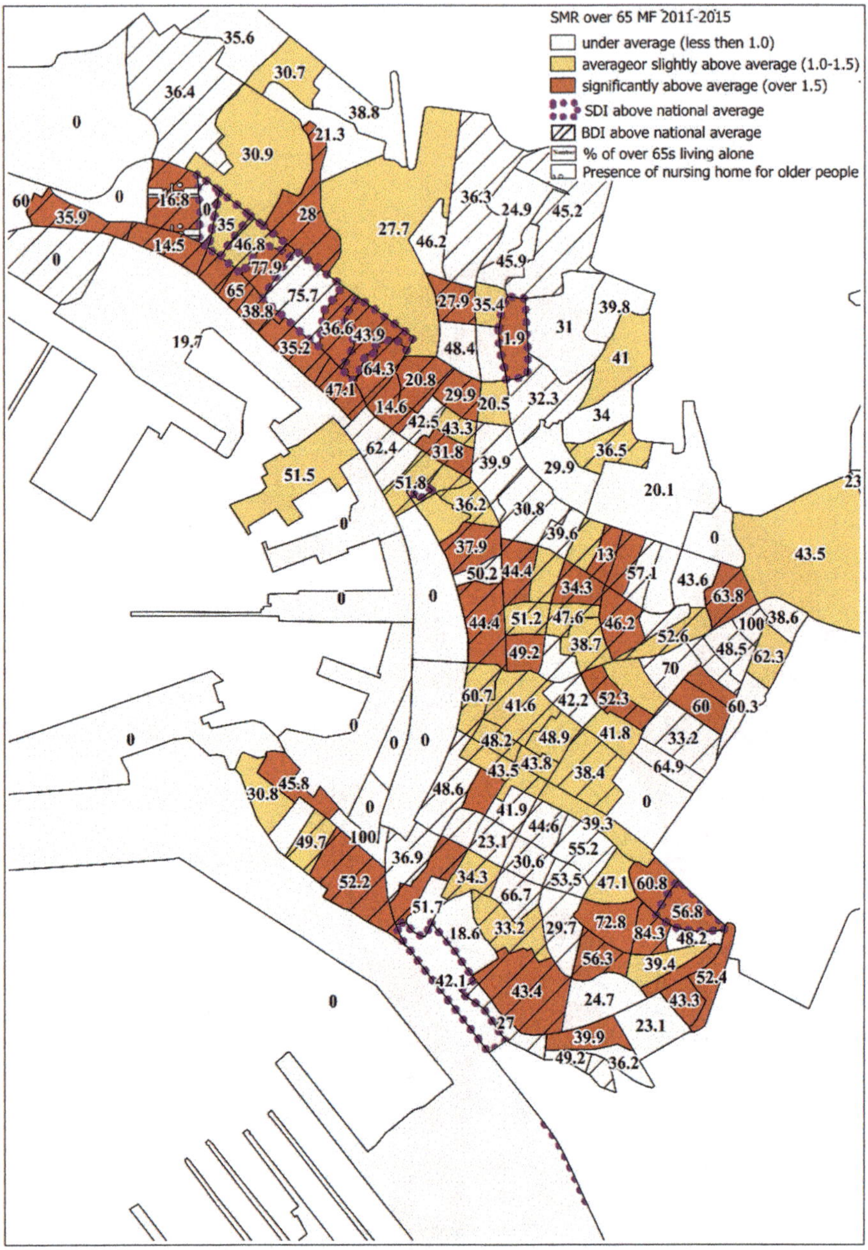

Fig. 9.3 Average SMR among over 65 residents for the period 2011–2015 in neighborhood of Prè-Molo-Maddalena and associated factors (Source: Our elaborations on datasets provided by the statistical office of Genoa City)

office of Genoa City). Interestingly, even if Foce has the highest aging index in all of the municipalities, the local average mortality rate among over 65 s is definitely restrained (1.37), especially if compared to the younger populations of the other two quarters previously described. As observed in Fig. 9.4, in this neighborhood, the elderly population, even if often living alone, benefits from better socioeconomic and structural conditions (also the mortality rates of the census sections showing the presence of almost one nursing home for older dependent patients do not show particularly significant mortality rates).

In sum, the detailed mapping of the average standard mortality rates of over 65 s showed significantly higher mortality in lower-class neighborhoods, particularly within those city blocks with the highest levels of social exclusion, economic deprivation, and environmental decay. The geo-referenced representation of results could provide a mapping alert system helpful for decision-makers to plan adequate interventions of social and urban requalification.

Discussing the Results: The Key Roles of Loneliness and Contextual Deprivation in Mortality Among Older People

Standard mortality rates of over 65 s were significantly higher within those city blocks showing the highest levels of social exclusion, economic deprivation, and environmental decay, with a major gender disadvantage for female older people and for older residents living in suburban areas. The geo-referenced representation of results provides a helpful tool for decision-makers in order to plan adequate interventions of social and urban requalification aimed at supporting living conditions of older residents.

Considering the observed associations of mortality ratios among older people with the different tested factors, we can draw the following conclusions.

Undoubtedly, the association of SMR with demographic factors, mainly the incidence of elderly people, particularly older women, among overall residents in each census section, is expected because of the relation of mortality to aging and female longevity. In contrast with the contemporary urban hypermobility in contemporary urban living, such results also reflect the prevalent residential immobility of those groups "aging in place," especially single older women.

This, on the one hand, stresses the importance of ensuring adequate living conditions. On the other hand, it underlines the higher vulnerability of elderly residents to the pressure of adapting to structural and social change in contemporary urban living. The latter consideration relates to the associations of the mortality of over 65 s with neighborhood structural conditions, particularly where older residents live in socioeconomically deprived or architecturally decaying contexts. Moreover, the aforementioned structural factors clearly combine with experiences of marginalization and social exclusion, particularly among elderly people living alone.

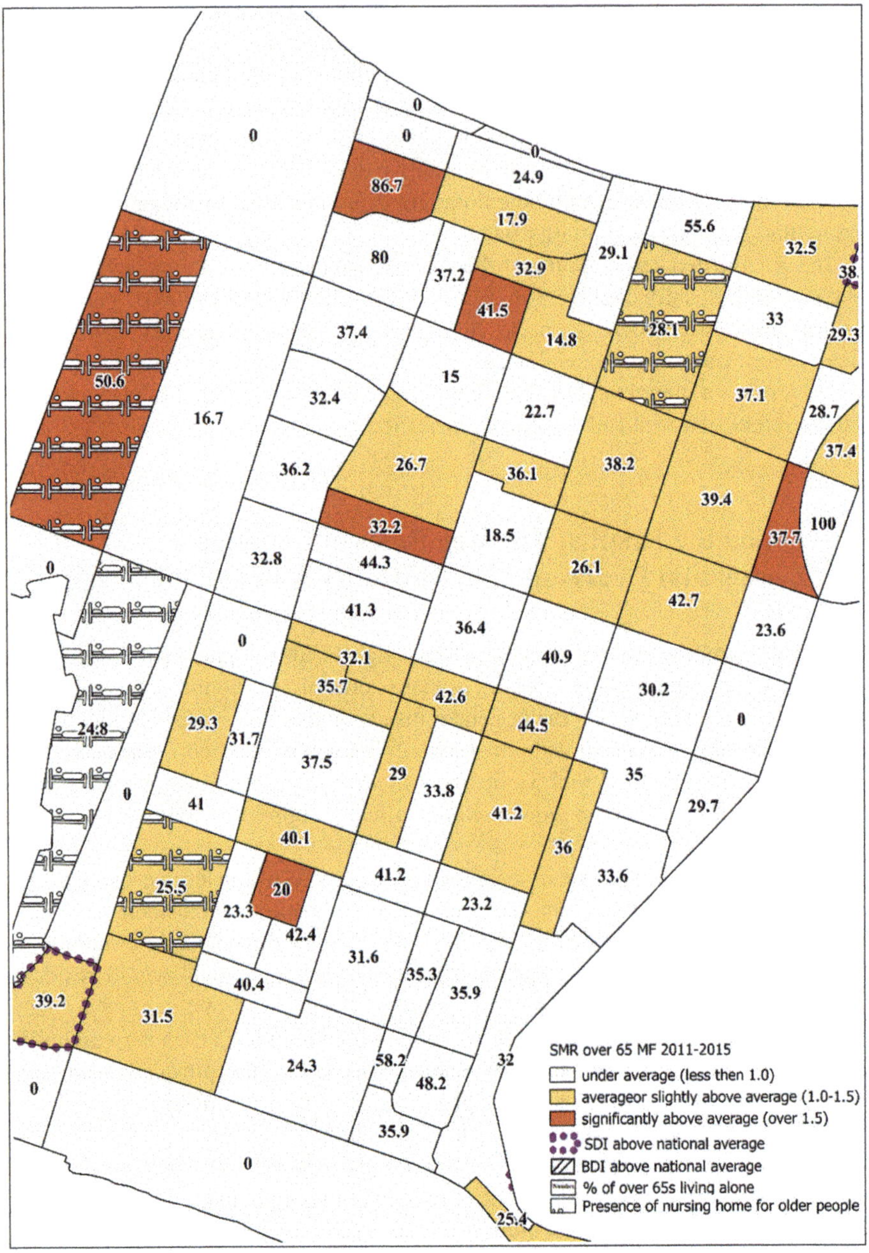

Fig. 9.4 Average SMR among over 65 residents for the period 2011–2015 in neighborhood of Foce and associated factors (Source: Our elaborations on datasets provided by the statistical office of Genoa City)

Our results show higher mortality rates of older people in a lower-class quarter with relatively lower aging index (Cornigliano) and proportionally lower mortality rates among over 65 s in the upper-class quarter with the highest aging index of the town (Foce). This underlines how the resulted neighborhood-level associations with mortality can be significantly related to structural factors and socioeconomic conditions observable in each quarter and not only to the mere demographic composition of residents in the area, giving evidence of how the deep structure of social inequalities can significantly influence the mortality risk of vulnerable groups like older people, and, consequently, the overall potential longevity of residents.

The main points of strength of our study are (a) the specific demographic setting, a metropolitan urban context that has historically demonstrated a strong aging process capable of anticipating the projected future growth of an older population at both the Italian and European level, and (b) the application and mapping of a large demographic dataset, mostly available at and regularly updated by any general registry office.

The main limits of the study are (a) the risk of ecological fallacy to be considered in interpretation of results, as the study was based on aggregated territorial data, and (b) the lack of update in some indicators, mainly the SDI and the BDI, which are dated 2011 (even though the local economic and urban requalification scenario of local census units remained quite unaltered).

We stress the fact that our results do not aim to demonstrate any etiological nexus between mortality rates of older residents and the observed neighborhood/environmental factors affecting living conditions, but due to the fact that we can rely only on simple aggregate territorial data, we just observed mere statistical associations between the dependent variable and the accounted factors.

On the one hand, our study, beyond implicit ecological limitations, demonstrates the importance of focusing specific attention on inclusion and exclusion processes affecting older people in the town environment. On the other hand, our study also demonstrates the opportunity of converting the already available and continuously updated datasets of city registry offices into useful planning tools, helping older residents to better experience later stages of life within the increasing complexity of the contemporary urban setting.

Conclusions

Our study, beyond implicit ecological limitations, demonstrates, on the one hand, the importance of focusing specific attention to inclusion and exclusion processes, interesting older people in town environment; on the other hand, it underlines the opportunity of converting the already available and continuously updated datasets of city registry offices in useful planning tools, helping older residents to better experience later stages of life within the increasing complexity of the contemporary urban setting.

The diffused demographic aging stresses the need to reconsider the urban setting to a necessarily proactive context ensuring well-being of more vulnerable groups, particularly older residents. Such reconsideration should be interpreted in a doubly “prosthetic” sense.

On the one hand, contemporary cities should promote the social change toward a new culture of aging, developing both more inclusive and participative processes to integrate older residents (especially those of the baby boomers' generation) and acquiring a larger consciousness of the deep heterogeneity of the aging process, in its staring phases as well in the very last part of life.

On the other hand, the urban settings must overpass the frequent structural and socioeconomic barriers, especially where frailty and disability in older life result from conditions of deprivation, poverty, and loneliness [24]. Indeed, the disparities in living and environmental conditions for older residents play a key role both in the effective development of successful aging paths and, on the contrary, in determining the old age as the unalterable crystallization of structural inequalities.

References

1. Alley, D., Crimmins, E.: Epidemiology of ageing. In: Dannefer, D., Phillipson, C. (eds.) The SAGE Handbook of Social Gerontology, pp. 75–95. SAGE, Los Angeles/London/New Delhi/Singapore/Washington DC (2013)
2. Balfour, J.L., Kaplan, G.A.: Neighborhood environment and loss of physical function in older adults: evidence from the Alameda County Study. Am. J. Epidemiol. **155**, 507–515 (2002)
3. Broeck, J., Brestoff, J.R., Kaulfuss, C.: Statistical estimation. In: Van den Broeck, J., Brestoff, J.R. (eds.) Epidemiology: Principles and Practical Guidelines, p. 417. Springer, Dordrecht (2013). doi:10.1007/978-94-007-5989-3_22
4. Cheshire, J.: Featured graphic. Lives on the line: mapping life expectancy along the London Tube network. Environ. Plan. **44**, 1525–1528 (2012). doi:10.1068/a45341
5. Dench, G., Gavron, K., Young, M.: The New East End. Profile Books, London (2006)
6. DPCM: Interventi per la riqualificazione sociale e culturale delle aree urbane degradate. Gazzetta Ufficiale, p. 249 (2015)
7. Eurostat: Causes of death statistics – people over 65. Retrieved from http://ec.europa.eu/eurostat/statistics-explained/index.php/Main_Page: http://ec.europa.eu/eurostat/statistics-explained/index.php?title=Causes_of_death_statistics_-_people_over_65&oldid=292547 (2016)
8. Ferrucci, L., Giallauria, F., Guralnik, J.M.: Epidemiology of aging. Radiol. Clin. N. Am. **46**(4), 643-v (2008.) doi:http://doi.org/10.1016/j.rcl.2008.07.005
9. Geronimus, T.A., Bound, J.: Use of census-based aggregate variables to proxy for socioeconomic group: evidence from national samples. Am. J. Epidemiol. **148**, 475–186 (1998)
10. Hunt, S., McKenna, S.: The impact of housing quality on mental and physical health. Housing Rev. **41**, 47–49 (1992)
11. Istat: Descrizione dei dati geografici e delle variabili censuarie per sezione di censimento Anni 1991, 2001, 2011. Istat, Roma (2015)
12. Jylhä, M., Saarenheimo, M.: Loneliness and ageing: comparative perspectives. In: Dannefer, D., Phillipson, C. (eds.) The SAGE Handbook of Social Gerontology, pp. 318–328. SAGE, Los Angeles/Llondon/New Delhi/Singapore/Washington DC (2013)
13. Larsson, K.: The social situation of older people. Int. J. Soc. Welf. **16**, 203–218 (2007)

14. Ministero dello Sviluppo Economico: Procedura tecnico-statistica per il calcolo dell'Indice di Disagio Socioeconomico (IDS). Ministero dello Sviluppo Economico, Roma (2008)
15. OECD: OECD Labour Force Statistics 2015. OECD Publishing, Paris (2016). doi:10.1787/oecd_lfs-2015-en
16. Ogg, J., Bonvalet, C.: Ageing in inner cities: the residential dilemmas of the baby boomer generation. Int. J. Ageing Later Life. **2**, 61–90 (2007)
17. Palumbo, M.: Conoscere per decidere: informazione statistica e programmazione regionale. ECIG, Genova (1991)
18. Palumbo, M., Poli, S., Torrigiani, C.: Genova: fra radicamento e trasformazioni in corso. In: Frudà, L. (ed.) La distanza sociale. Le città italiane tra spazio fisico e spazio socio-culturale, pp. 52–132. FrancoAngeli, Milan (2007.) doi:ISBN: 9788846496379
19. Phillipson, C.: Ageing and urban society: growing old in the 'Century of the City'. In: Dannefer, D., Phillipson, C. (eds.) The SAGE Handbook of Social Gerontology, pp. 597–606. SAGE, Los Angeles/London/New Delhi/Singapore/Washington DC (2013)
20. Phillipson, C., Bernard, M., Phillips, J., Ogg, J.: The Family and Community Life of Older People. Routledge, London (2001)
21. Pickett, K.E., Pearl, M.: Multilevel analysis of neighborhood socioeconomic context and health outcomes: a critical review. J. Epidemiol. Community Health. **55**, 111–122 (2001)
22. Poli, S.: Città vecchia, nuovi anziani. Invecchiamento e postmodernità in una periferia metropolitana. FrancoAngeli, Milan (2012)
23. Poli, S., Cella, A., Puntoni, M., Musacchio, C., Pomata, M., Torriglia, D., Molinari, B., Pandolfini, V., Torrigiani, C., Pilotto, A.: Frailty is associated with socioeconomic and lifestyle factors in community-dwelling older subjects. Aging Clin. Exp. Res. **29**, 721–728 (2016). doi:10.1007/s40520-016-0623-5
24. Robert, S.: Socioeconomic position and health: the independent contribution of socioeconomic status. Annu. Rev. Sociol. **25**, 489–516 (1999)
25. Rodwin, V., Gusmano, M. (eds.): Growing Older in World Cities. University Press, Vanderbilt Nashville (2006)
26. Romeri, E., Baker, A., Griffiths, C.: Mortality by deprivation and cause of death in England and Wales, 1999–2003. Health Stat. Q. **32 Winter 2006**, 19–34 (2006)
27. Ross, C.E., Mirowsky, J.: Neighborhood disadvantage, disorder, and health. J. Health Soc. Behav. **42**, 258–276 (2001)
28. Savage, M., Bagnall, G., Longhurst, B.: Globalization and Belonging. Sage, London (2005)
29. Scharf, T., Phillipson, C., Smith, A., Kingston, P.: Growing Older in Socially Deprived Areas. Help the Aged, London (2002)
30. Schootman, M., Andersen, E.M., Wolinsky, F.D.: Neighborhood conditions and risk of incident lower-body functional limitations among middle-aged African Americans. Am. J. Epidemiol. **163**, 450–458 (2005)
31. Sloggett, A., Joshi, H.: Higher mortality in deprived areas: community or personal disadvantage? BMJ. **309**, 1470–1474 (1994.) doi: https://doi.org/10.1136/bmj.309.6967.1470
32. Wen, M., Browning, C.R., Cagney, K.A.: Poverty, affluence, and income inequality: neighborhood economic structure and its implications for health. Soc. Sci. Med. **57**, 843–860 (2003)

Chapter 10
A Multicriteria Ranking of Thessaloniki's Public Hospitals Based on Their Infrastructure Adequacy

Georgios Chatzipoulidis, Georgios Aretoulis, and Glykeria Kalfakakou

Abstract The Healthy Cities project of the World Health Organization (WHO) is a global movement. It engages local governments in health development. There are many national Healthy Cities networks, and a large number of cities in Greece have already joined them. Meanwhile, especially in big cities like Thessaloniki, health infrastructures are very old and operationally and spatially suffocate. The current research thoroughly records the health infrastructure in Macedonia and Thrace. Data clearly highlights that the major "healthcare issues" are concentrated in Thessaloniki. Hospitals operate in old buildings with urgent need for reconstruction and a definite need for expansion. A methodology for identifying a hospitals' priority list is presented and applied taking into account the age of buildings, the building area/per bed, the field area/per bed, the coverage of beds, and the population being served. PROMETHEE methodology is implemented to rank healthcare units, on the basis of the urgency for improvement action. According to the analysis' results, the hospitals in Thessaloniki that require immediate attention, considering their infrastructure, are "Theagenio" Anticancer Hospital, Psychiatric Hospital, and "St. Paul" General Hospital. Various improvement actions were proposed including expansion of hospitals in adjacent land and buildings, transfer of specific clinics to new buildings within the city, relocation of certain healthcare activities or part of them to new hospitals on the outskirts of city, upgrading of primary healthcare, and reinforcement of neuralgic hospitals in other prefectures. Finally, a discussion about the appropriateness of each type of intervention per hospital is presented.

Keywords Healthy cities • Hospital infrastructure • Hospitals' prioritization • PROMETHEE methodology

---

G. Chatzipoulidis (✉) • G. Aretoulis • G. Kalfakakou
Department of Civil Engineering, Aristotle University of Thessaloniki, Thessaloniki, Greece
e-mail: ghtzip@civil.auth.gr; garet@civil.auth.gr; riak@civil.auth.gr

A. Karakitsiou et al. (eds.), *City Networks*, Springer Optimization and Its Applications 128, DOI 10.1007/978-3-319-65338-9_10

Introduction

The debate on the creation of networks of cities for sustainable development, environmental protection, and promotion of the health of their citizens was launched a few decades ago. The Healthy Cities project of the World Health Organization (WHO) is a global movement. It engages local governments in health development through a partnership-based planning and innovative projects. More than 100 cities are members of the WHO European Healthy Cities Network. Thirty national Healthy Cities' networks across the WHO European Region include more than 1400 cities and towns as members. The European Healthy Cities Network strategic goals provide the overarching umbrella for:

- Improving health for all and reducing health inequities
- Improving leadership and participatory governance for health [58]

Healthy Cities have been responsive to the great political and economic changes sweeping across Europe and transforming the global environment. The Healthy Cities movement promotes comprehensive and systematic policy and planning for health. City administrations are faced with major decisions in a highly complex and changing internal and external environment: ecological, public health and social demands, decentralization trends, economic development challenges and opportunities, metropolization, consumerism, social pressure, technological developments, and new democratic procedures and reforms. Above all they are needed to be open to new ideas and practices [36, 51].

The knowledge for the Healthy Cities approach became more extensive during the first three decades of network development. There are many instructions and related articles listed in guidance documents issued by the WHO Regional Office for Europe [51]. These guidelines include issues such as design for health development in cities, healthy urban planning, and instructions for community involvement [8, 56, 57].

Already cities in Greece have joined these networks and have initiated steps toward a more secure and healthy living for their citizens. In this direction, they carried out a series of projects and environmental interventions promoting healthier living conditions, creating organic parks, bicycle paths, protection of vulnerable groups, and prevention and health education programs [31].

At the same time though, in the Greek health sector, there are serious obstacles to promoting and maintaining the health of citizens especially in big cities where health infrastructures are very old and operationally and spatially suffocate. The lack of proper infrastructure and adequate staff also creates obstacles to cooperation of health structures. This, in turn, creates difficulties and barriers in the support and implementation of joint programs, for improving the health level of the population, in the context of networks such as Healthy Cities. Infrastructure deficiencies and limited support capabilities of the hospital facilities often have a negative impact on local programs, such as assistance at home, home care, but also informative health promotion programs, such as expectant parents' schools and information specific population groups. All these require support by specialized health staff, well-documented procedures, and infrastructure that the current health facilities do not sufficiently have. However, the ability to solve such issues goes beyond the jurisdiction of local authorities and requires the cooperation and support of main government agencies, in order to provide an overall design. The current research aims at applying a multicriteria-based approach for ranking the hospitals of the city of Thessaloniki on the basis of their needs for infrastructure improvement.

In the following sections, after the analytical presentation of the current situation and statement of the problem, the proposed methodology is presented aiming at the assessment and decision-making, regarding the planning and prioritization of interventions, focused on Thessaloniki's hospitals. A ranked list of the hospitals based on their improvement needs is produced, and then the type of appropriate interventions is discussed per each case. Finally, conclusions and further research are highlighted.

Thessaloniki and Health Infrastructure of Macedonia and Thrace Region

The city of Thessaloniki, located in northern Greece, is the second largest city in Greece, with almost over one million citizens, center of the administrative region of Central Macedonia and of the Decentralized Administration of Macedonia and Thrace, and historically one of Europe's oldest cities. Thessaloniki, with the corresponding municipalities, has started to move, with some tentative steps, in improving the environment and people's living conditions. Projects aimed to this direction include the regeneration of the coast front, creating bicycle paths, telematics in public transport, and the creation of reception facilities and hospitality of homeless and immigrants, among other actions.

But the city falls short in terms of qualitative and quantitative adequacy of health infrastructure. There is an attempt to approach this issue quantitatively at the 3rd and 4th Health Regions [1, 2], which include hospitals of Thessaloniki and cover all prefectures of Macedonia and Thrace. Hospitals of Thessaloniki cover, as far as healthcare is concerned, all these areas.

Greece could be divided in 13 administrative regions. Table 10.1 shows the distribution of population and hospital beds, as well as the ratio of beds per 1000 inhabitants in each administrative region of Greece. The 3rd and 4th Health Regions cover the population of Macedonia, Thrace, and Mount Athos. These areas include 2,773,740 inhabitants or a percentage equal to 25.65% of the Greek population. The number of beds per 1000 inhabitants for these regions is close to that of the average number of the whole country of Greece.

Table 10.1 Population and hospital beds

Region of	Region's population	Resident population density per square km	Beds per region	Inhabitants per bed	Beds per 1000 inhabitants
Eastern Macedonia and Thrace	608,182	42.96	1663	366	2.7
Central Macedonia	2,163,747	99.95	6826	317	3.2
Epirus	336,856	36.6	1634	206	4.9
Thessaly	732,762	52.2	1695	432	2.3
Central Greece	547,390		743	737	1.4
Ionian Islands	207,855	90.1	562	370	2.7
Western Greece	679,796	59.89	1780	382	2.6
Peloponnese	577,903	37.31	1746	331	30
Attica	3,827,624	1.00513	14,011	273	37
North Aegean	199,231	51.94	577	345	2.9
South Aegean	308,975	58.45	549	563	1.8
Crete	623,065	74.74	2289	272	3.7
Mount Athos	1,811	5.4			
Total	10,815,197		34,075	317	3.15

Greek National Statistical Authority [32]

Actual Condition of Hospital Buildings' Infrastructure

One of the critical parameters regarding the quality of healthcare is associated with the age of the hospitals' buildings. All the public hospital buildings of Macedonia and Thrace, operating in the 3rd and 4th Health Regions, were thoroughly recorded. Collected data included age of the buildings, the completeness, and the extent of the building infrastructure in relation to the beds offered by each hospital.

Two tables (namely, Tables 10.2 and 10.3) have illustrated all these data. There was a careful and organized effort to collect reliable data in order to objectively address the problem. Special attention was provided to verifying data and ensuring data reliability, by comparing findings from multiple alternative sources. In this context, Tables 10.1 and 10.2 recorded data originating from a significant number of sources. These tables depict the current situation and provide the essential data to identify the requirements to facilitate redesign of the National Health System (NHS), in the area of interest, namely, the 3rd and 4th Health Regions of Macedonia and Thrace. Regarding the age of the buildings, the hospitals were divided into two groups: those that were built before 1981 and those that were constructed later. The choice of the current reference year indicates that the buildings, constructed prior to this date, have not incorporated the first thermal insulation regulation of 1980 and the earthquake resistance regulation of 1984, which were adopted and implemented after the great earthquake of Thessaloniki. The importance of earthquake shielding

Table 10.2 Infrastructure of the 3rd Health Region

N	Hospital	Planned beds	Existing beds	Coverage of existing beds (%)	Building area (sqm)	Land area (sqm)	Building percentage older than 1981 (%)	Building area/ existing bed (sqm)	Land area/ existing bed (sqm)
1	General Hospital of Florina [22]	110	120	50	8500	14,975	35.3	70.83	136.14
2	General Hospital of Kastoria [25]	120	96	53	8172	50,725	45	85.13	422.71
3	Mamatsio General Hospital of Kozani [38]	210	178	57	14,000	16,000	44.3	78.65	76.19
4	"Bodosakio" General Hospital of Ptolemaida [9]	190	194	55	22,000	100,000	0.00	113.40	515.5
5	General Hospital of Grevena [24]	100	95	46	13,851	20,212	0.00	145.80	202.12
6	General Hospital of Veria [28]	210	176	59	15,302	62,000	65.80	86.94	352.27
7	General Hospital of Naousa [27]	120	120	41	8,014	9282	61.00	66.78	77.35
8	General Hospital of Edessa [21]	170	154	57	12,000	28,882	0.00	77.92	169.89
9	General Hospital of Gianitsa [23]	180	170	70	13,446	17,676	0.00	79.09	98.20
10	General Hospital of Katerini [26]	200	177	46	31,637	68,829	0.00	178.74	344.15
11	Psychiatric Hospital of Petra Olympos [43]	200	185	93	12,738	97,976	68.60	68.85	489.88
12	"Papageorgiou" General Hospital of Thessaloniki [40]	710	687	82	75,000	150,000	0.00	109.17	211.27

(continued)

Table 10.2 (continued)

N	Hospital	Planned beds	Existing beds	Coverage of existing beds (%)	Building area (sqm)	Land area (sqm)	Building percentage older than 1981 (%)	Building area/ existing bed (sqm)	Land area/ existing bed (sqm)
13	"Papanikolaou" General Hospital of Thessaloniki [41]	600	622	71	83,060	433,753	86.20	133.54	722.92
14	Psychiatric Hospital of Thessaloniki [44]	400	604	111	22,713	108,903	91.20	37.60	272.26
15	"G. Gennimatas" General Hospital of Thessaloniki [29]	280	272	52	13,432	10,385	69.10	49.38	37.09
16	"Saint Dimitrios" General Hospital of Thessaloniki [45]	175	147	58	11,921	19,765	82.10	81.10	112.94
Total of 3rd Health Region		3975	3997	72	365,786	1,209,363	40.5	91.5	302.5

Ministry of Health [39], Hellenic Regulatory Body of Nurses [53], 5th Health Region [3], Respective websites of each Hospital and Health Region, Archives of the 3rd and 4th Health Regions, and the Greek National Statistical Authority [32] for hospitals emerged from major seismic events which took place in Thessaloniki in 1978, the earthquake in Athens in 1999, and the earthquake in Lefkada in 2003 [13].

Table 10.2 indicates that the hospital buildings of the 3rd Health Region [1], at a rate of 40.5%, have been built before 1980. It is noted that for the city of Thessaloniki, where the seismic risk is particularly high, the percentage of old hospital buildings within the 3rd Health Region reaches approximately 54%.

The data presented in Table 10.2 includes the bed coverage. The mean percentage coverage of beds value indicates the mean number of times that a hospital bed accommodates a patient, per year. Average Coverage of existing beds (%) = Total Days of Hospitalization in a Year * 100/existing beds * 365 [35, 37].

Regarding the 4th Health Region of Macedonia and Thrace, it is evident in Table 10.3 that the hospital buildings, in a percentage of 36%, have been built before 1981, but the hospitals of Thessaloniki, belonging in the 4th Health Region, were built before 1981 at a rate of 72.4%. To summarize, 61.5% of operational hospital buildings in Thessaloniki are old, energy consuming, and unsafe (earthquake vulnerable). On the contrary, hospitals of the 3rd and 4th Health Regions' jurisdiction that are located outside the city of Thessaloniki have only a percentage of 15.5% old buildings.

Therefore, the health infrastructure and especially the hospital infrastructure in Thessaloniki are very old and inadequate. It is characteristic that some hospitals are still partly operating in unsuitable buildings that are over one century old. Typical examples include:

- The Hospital G. Genimatas: part of it was built to house the initial police academy. Since 1916 it has been operating as a military hospital and then for many years as the central hospital for refugees.
- The Ippokrateio Hospital: part of it consists of the building of the Jewish hospital Hirsh, which was built in 1904.
- The Saint Dimitrios Hospital: it was the first municipal hospital of the city. The construction works began in 1902–1903 [30, 55].

These buildings, besides the historical, cultural, and architectural value, respond with difficulty to modern needs and have many access restrictions that greatly limit their functionality. These in fact create obstacles to safe and functional integration of healthcare units, within a modern hospital. Finally, in most cases, the adjacent available land is very limited and is not sufficient for any further building expansion.

The following Table 10.4 summarizes the data from the previous two Tables (10.2 and 10.3) focusing on a health region-based view.

Improvement and Expanse of the Available Infrastructure

In recent decades, the health infrastructure has been renewed significantly within the health region of interest. Despite these interventions, few new hospitals were

Table 10.3 Infrastructure of the 4th Health Region

N	Hospital	Planned beds	Existing beds	Coverage of existing beds (%)	Building area (sqm)	Land area (sqm)	Building percentage older than 1981 (%)	Building area/existing bed (sqm)	Land area/existing bed (sqm)
1	"AHEPA," University General Hospital of Thessaloniki [4]	680	621	64	49,379	36,000	63.4	79.52	52.94
2	Skin Diseases Hospital of Thessaloniki [48]	10	10	70	3,165	3233	42.55	316.50	323.26
3	"Ippokrateio" General Hospital of Thessaloniki [33]	790	759	65	57,294	36.607	67.7	75.49	46.34
4	"Theagenio" Anticancer Hospital of Thessaloniki [50]	361	380	78	16,403	5434	100	43.17	15.05
5	"St. Paul" General Hospital of Thessaloniki [46]	250	207	90	8,557	18,640	100	41.34	74.56
6	General Hospital of Kilkis	180	205	57	21,239	69,111	69.32	103.60	383.95
7	General Hospital-Health Center of Goumenissa [17]	40	40	51	2,092	4478	89.29	52.31	111.95
8	General Hospital of Halkidiki [19]	180	146	61	16,714	47,603	0	114.48	264.46
9	General Hospital of Serres	400	365	59	34,283	110,000	0	93.93	275.00
10	General Hospital of Kavala [18]	420	392	61	37,504	60,338	0	95.67	143.66
11	General Hospital of Drama [16]	250	239	64	24,441	48,478	0	102.26	193.91
12	General Hospital of Xanthi [20]	240	222	70	28,580	100,545	0	128.74	418.94

(continued)

13	"Sismanoglio" General Hospital of Komotini, [47]	240	162	70	16,749	24,956	100	103.39	103.98
14	University General Hospital of Alexandroupolis [54]	590	500	71	87,800	200,000	0	175.60	338.98
15	General Hospital of Didymoteichon [15]	120	120	76	11,400	9544	0	95.00	79.54
Total of 4th Health Region		4751	4368	66.1	415,600	774,967	36	95.1	177.4

Ministry of Health [39], Hellenic Regulatory Body of Nurses [53], 5th Health Region [3], Respective websites of each Hospital and Health Region, Archives of the 3rd and 4th Health Regions, and the Greek National Statistical Authority [32]

Table 10.4 Hospital beds and their coverage in various Greek health regions

Health region	Planned beds	Existing beds	Coverage of existing beds (%)
Total of 1st Health Region	9830	9213	73.6
Total of 2nd Health Region	6233	6176	78.6
Total of 3rd Health Region	3975	3997	72
Total of 4th Health Region	4793	4461	66.1
Total of 5th Health Region	3000	2498	68.9
Total of 6th Health Region	5720	5412	68.5
Total of 7th Health Region	2335	2279	66.7
Total	35,886	34,036	71.7
Hospitals of Athens	14,298	14,011	77.4
Hospitals of Thessaloniki	4298	4402	75

Ministry of Health [39], Hellenic Regulatory Body of Nurses [53], 5th Health Region [3], Websites of the respective Hospitals and Health Regions

established in the limits of conurbation of Thessaloniki. This fact, combined with the tertiary nature of most hospitals located in this area and the subsequent increasing continuous flow of patients to the major urban centers to treat serious health issues, has created, nowadays, stifling operating conditions in most major hospitals in Thessaloniki. It is significant to note that in accordance with Tables 10.2 and 10.3, within the region of Thessaloniki, the average surface of the developed building infrastructure is 78.5 sqm/bed (No. 12–16 hospitals in Table 10.2 and No. 1–5 hospitals in Table 10.3), while in the respective hospitals outside Thessaloniki in the 3rd and 4th Health Regions (remaining hospitals of Tables 10.2 and 10.3), the surface increases to 106.5 sqm/bed. In addition, in old hospitals, where most of the clinics initially operated before 1981, the corresponding surface is 74.8 sqm/bed.

It is evident that the old hospitals, especially those located in large urban centers, suffocate in terms of adequacy of the building infrastructure. Hospitals with the smallest building surface per bed ratio include Psychiatric Hospital of Thessaloniki, "Saint Paul" General Hospital of Thessaloniki, "Theagenio" Anticancer Hospital of Thessaloniki, "G. Gennimatas" General Hospital of Thessaloniki, and General Hospital-Health Center of Goumenissa.

Another factor which indicates that hospitals of Thessaloniki suffocate in terms of space considers the available land. This factor has a direct impact on available parking spaces, the possibility of heliport, and the comfortable and rapid ambulance, patients, visitors, and supplier access. Tables 10.2 and 10.3 clearly state that most hospitals in Thessaloniki fall far short, compared to the provincial, on space availability for the abovementioned operations and facilities. Hospitals strangled by additional land availability, located far below the average of 177.4 sqm/planned bed in 4th Health Region or 302.5 sqm/planned bed in 3rd Health Region, include "Theagenio" Anticancer Hospital of Thessaloniki, "G. Gennimatas" General Hospital of Thessaloniki, "Ippokrateio" General Hospital of Thessaloniki, "AHEPA," University General Hospital of Thessaloniki, and "Saint Paul" General Hospital of Thessaloniki.

The overwhelming majority of Thessaloniki's hospitals are facing an acute problem of building infrastructure. From the above, it is evident that the hospitals "Theagenio," "G. Gennimatas," and "St. Paul," are facing a major problem in both available field area and buildings' infrastructure.

The consequences of this situation are the great difficulty of access, parking, the growth impossibility due to lack of space and land, and the increased danger for the patients. Hospitals are forced many times to operate out of the established safety procedures, protocols, and specifications, regarding the area of the various sections, putting a strain on hospital infections and hindering the safe movement of patients within or among buildings.

Conditions are so stressed in some of these hospitals that it becomes impossible to even consider of initiating renovation processes in the hospital vicinity, taking steps toward certification, or even establishing quality assurance systems. All these have a negative impact on hospitals' daily operations, image, and sustainable development.

Some solutions that could be considered include the expansion of these hospitals in adjacent land and buildings, transfer to new buildings elsewhere in the city of Thessaloniki, or relocate certain activities or part of them to a new hospital that will be created on the outskirts of city but at the same time maintaining a number of key functions in the old buildings located in the city center, for the easy and direct access of patients.

Some hospitals have adequate buildings, but their issues are mainly focusing on the inadequacy of the available land in the perimeter of the buildings. These hospitals are, namely, "Ippokrateio" and "AHEPA." In this case, solutions might be found in the creation of multistory parking places and the best arrangement of the available area for the convenience of patients. In any case, the design of sustainable development and improvement of Thessaloniki hospitals' infrastructure should take into account many factors and is definitely a sensitive, multicriteria problem.

Coverage of Existing Beds

Another very sensitive parameter that should be seriously considered is the coverage of existing hospital beds. The previously mentioned Tables 10.2, 10.3, and 10.4 present that the coverage of the existing beds is increased in Athens and Thessaloniki (77.4% and 75% correspondingly) than in the provincial hospitals (65.7%). The unbalanced development of health infrastructure in Greece, in recent decades and particularly in Macedonia and Thrace, is confirmed from these numbers, combined with the observed presence of old buildings.

Suggestions for Efficient Patients' Flow Management in Thessaloniki

From all the above evidence, it becomes clear that there is a need to redesign the health infrastructure and reallocate resources. Important interventions have to be implemented, targeting either the renewal of building and equipment facilities or in the redeployment, training, supplementation, and utilization of human resources to provide efficient services, with less cost and where necessary [42].

Special attention should be paid to hospitals of Thessaloniki. The task is challenging, especially for the old hospitals operating in the center of the city, due to lack of available space for the construction of new hospitals. A successful project requires careful planning and increased funding.

The above efforts to design a sustainable development for hospitals should yet take into account a number of additional parameters such as the type of hospital (university, general, specialized), population served, the main character of the services offered, the number of people hospitalized and coverage of beds, the number of visitors, etc.

Large hospitals in Thessaloniki are significantly burdened because they receive a very large number of patients from other regions, seeking the specialized services offered by the hospital or for better treatment of the health issue by more experienced scientists and better-organized sections than in the province. This patients' flow is considered to be particularly important but has not yet been assessed in actual numbers. The arrival of all these patients creates significant daily movement and accessibility problems in hospitals especially in the center of Thessaloniki and the surrounding area. All of the above highlight the imperative need of redesigning the hospital map of the agglomeration of Thessaloniki toward a sustainable development that takes into account all available parameters.

This planning and design are imperative to also take into account the need to reinforce certain neuralgic hospitals in other prefectures that will function as the intercepting centers for patients from the regional areas. Improvements to these hospitals will focus on skilled human resources and equipment which their exclusive presence in hospitals of Thessaloniki, attracts patients, and creates increased flows.

Hospitals that could play a similar role include the University Hospital of Alexandroupolis and prefectural hospitals in Macedonia and Thrace (such as Kavala, Serres, Kozani-Ptolemais, Katerini) which have relatively new infrastructure and equipment. Their appropriate reinforcement with additional equipment and specialized personnel could, in certain cases, decrease the significant flow of patients toward the central hospitals of Thessaloniki.

The attempted upgrading of primary healthcare will also help significantly to relieve the workload of the outpatient clinic and the emergency departments of hospitals. On the other hand, a restructuring and a holistic developmental planning of health infrastructures in the wider city of Thessaloniki are now imperative. This planning will contribute to the sustainable development of the city and the associated municipalities. When the health system's infrastructure itself is "sick," then it is not possible to effectively provide care for the health of citizens in the region.

Research Methodology

The aim of the current paper is to provide an approach for ranking Thessaloniki's hospitals based on the degree of interventions that they require, and at the same time, a corresponding discussion regarding the relevant appropriate responses and actions is presented. The methodology followed in the current research is depicted in Fig. 10.1:

The proposed approach is based on specific measurable and objective criteria, which will assist in making decisions for health planning and facilitate sustainable infrastructure development in healthcare. The criteria considered in the current approach include the following:

1. Age of buildings in relation to the safety regulations associated with the building design (percentage of hospital buildings constructed before 1980—obligatory application of earthquake resistance and thermal insulation regulations)
2. Ratio of building area per hospital bed
3. Ratio of available land area per bed
4. Each hospital's occupancy index (annual coverage of beds)
5. The estimated population being served by each hospital

The latter depends on the operation of the hospital as a university, special, or general Hospital. The presence of university clinics attracts patients from other

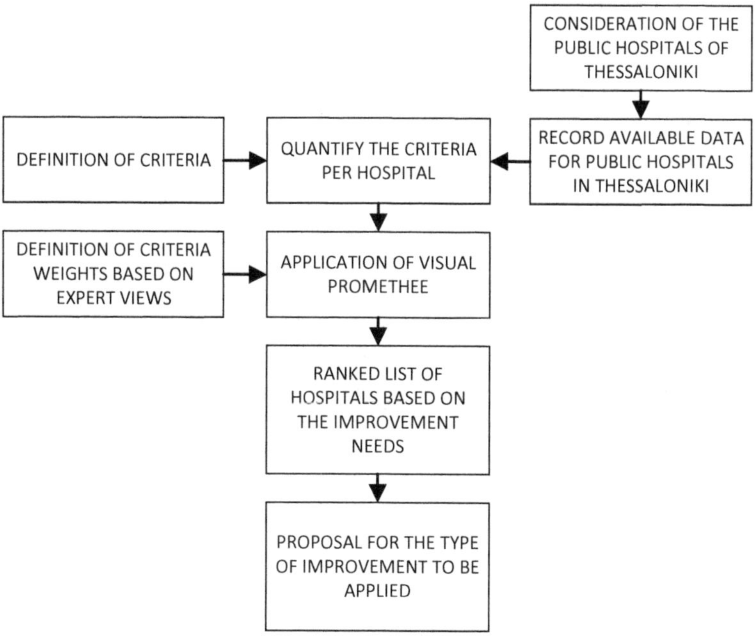

Fig. 10.1 Research methodology

prefectures that visit and allocate additional operational burden to the hospital. Ranking of hospitals and the creation of a priority list for improvement or replacement of infrastructure were realized with the multicriteria methodology PROMETHEE and the corresponding software Visual PROMETHEE.

The Multicriteria Methodology PROMETHEE

According to Tzeng and Huang [52] as cited in Antoniou et al. [6], the decision-making procedure is quite simple, when it is based on one single criterion, as the chosen alternative is simply, that which achieves the highest value of the single criterion. In the case of multiple criteria that must be considered when making a decision, things get complicated due to the need to compare the importance of each selection criterion.

The current research implements the Visual PROMETHEE application. The acronym PROMETHEE stands for "Preference Ranking Organization METHod for Enriched Evaluation" [34]. This method was proposed by Brans et al. (1984, [12]) and further extended by Brans and Mareschal [10], as a new outranking method for multi-attribute decision-making (MADM). According to this method, initially dual outranking relations are set up for the representation of the decision-maker's preferences with pairwise comparisons against each criterion [5].

As a result, PROMETHEE ranks the alternatives from best to worst based on the decision-maker's preferences. According to [34], as cited in Antoniou et al. [5], the method includes three main steps, focused on computation of:

- Preference degrees for every ordered pair of actions on each criterion
- Uni-criterion flows
- Global flows

Based on the global flows, a ranking of the actions will be obtained as well as a graphical representation of the decision problem. As a result the PROMETHEE methodology is based on the computation of preference degrees.

More specifically PROMETHEE II that was used in this paper is an outranking method of the European multicriteria analysis school and is a complete ranking; all the actions are ranked from the best to the worst. The outranking family of methods is based on pairwise comparisons; the decision-maker is assumed to naturally compare each of the actions with all of the rest on a one-to-one basis according to Brans and Mareschal [10] as cited in Aretoulis et al. [7]. The criteria and weights used are presented in Table 10.5:

At this point it should be noted that a number of assumptions have been made, based on the authors' experience and unpublished hospital data. These assumptions include the following:

- University hospitals serve the entire population of the administrative region to which they belong.

Table 10.5 Criteria and weighting factors

	Weighting factors	1	0.9	0.5	0.9	0.6
N	Hospital	Buildings constructed before 1980 (%)	Building area/per bed	Available surface/per bed	Coverage of beds (%)	Population being served (× 1000)
1	"AHEPA," University General Hospital of Thessaloniki	63.4	79.52	52.94	64	2163
2	"Ippokrateio" General Hospital of Thessaloniki	67.7	75.49	46.34	65	1514
3	"Theagenio" Anticancer Hospital of Thessaloniki	100	43.17	15.05	78	2771
4	"St. Paul" General Hospital of Thessaloniki	100	41.34	74.56	90	777
5	"Papageorgiou" General Hospital of Thessaloniki	0	109.17	211.27	82	1514
6	"G. Papanikolaou" General Hospital of Thessaloniki	86.20	133.54	722.92	71	1514
7	"G. Gennimatas" General Hospital of Thessaloniki	69.10	49.38	37.09	52	1514
8	"Saint Dimitrios" General Hospital of Thessaloniki	82.10	81.10	112.94	58	1514
9	Psychiatric Hospital of Thessaloniki	91.20	37.60	272.26	111	1940
10	Skin Diseases Hospital of Thessaloniki	42.55	316.50	323.26	70	1940

Data from Tables 10.2 and 10.3 and the [32]

- Hospitals which host a significant number of university clinics serve 70% of the population.
- The county hospitals serve 70% of the county population.
- The unique special hospitals along with university clinics serve the entire population of Macedonia and Thrace regions.
- The unique special hospitals with few or no university hospitals serve 70% of the population of Macedonia and Thrace regions.

Multicriteria Analysis' Results and Discussion

The results of this prioritization using the application Visual PROMETHEE by Gaia are presented in Table 10.6. The latter presents the net preference flow/global flow (Phi). The positive (Phi+) and negative preference flows (Phi-) are aggregated into the net preference flow, and as a consequence, the values of Phi range from −1 to +1, for each alternative (hospital). Hospitals are ranked according to the global flow. The hospitals with the smaller global flows (top of the list) are the ones in need of more urgent and extended attention and intervention.

According to the results of the ranking, there appears to be an urgent need for action, at least in the following hospitals: "Theagenio," Psychiatric Hospital, and "Saint Paul." Important interventions are also required in the case of the hospitals "Ippokrateio," "Genimatas," and "AHEPA," and limited interventions are required to take place in the remaining hospitals.

These interventions, of course, require particularly increased financial resources, which do not exist in Greece today. For these reasons, in recent years a debate has been launched on the inquiry for alternative ways of financing construction of new hospitals focusing especially on public-private partnerships (PPPs) [49]. It has been 10 years since the government considered, as a priority, to construct a new oncology

Table 10.6 Prioritizing hospitals of Thessaloniki based on their intervention needs

N	Hospital	Flow (Phi)
1	"Theagenio" Anticancer Hospital of Thessaloniki	−0.7151
2	Psychiatric Hospital of Thessaloniki	−0.6011
3	"St. Paul" General Hospital of Thessaloniki	−0.4473
4	"Ippokrateio" General Hospital of Thessaloniki	0.1168
5	"G. Gennimatas" General Hospital of Thessaloniki	0.1339
6	"AHEPA," University General Hospital of Thessaloniki	0.1339
7	"G. Papanikolaou" General Hospital of Thessaloniki	0.2479
8	"Saint Dimitrios" General Hospital of Thessaloniki	0.2934
9	"Papageorgiou" General Hospital of Thessaloniki	0.3504
10	Skin Diseases Hospital of Thessaloniki	0.4872

hospital for the accommodation of "Theagenio" and the creation of a new pediatric hospital that does not exist in the city. The implementation process was based on the application of PPPs [14].

According to the present research results, "Theagenio" is ranked, indeed, first in the list of hospitals that require intervention, confirming the state's decision. The current hospital doesn't have the required space. One of the options considered was complete relocation of the hospital; the second was to partly relocate the hospital to another area and finally expand the existing hospital by using adjacent property of the "Euclides" Technical School, which is an empty public building that is no longer operational.

Furthermore, the decision to establish a stand-alone pediatric hospital was taken in light of the increased demand and the relocation perspectives of specialized pediatric departments that operate mainly in hospitals "Ippokrateio," "AHEPA," and "Papageorgiou." Unfortunately, for various reasons, the implementation of this design has not yet been possible. Instead, a new hospital wing for AHEPA was implemented that will upgrade the operating rooms and the intensive care unit, with funding originating from the European Union. Nevertheless, according to the ranking of Table 10.6, there is an urgent need for immediate improvements in hospitals Theagenio, Psychiatric, and St. Paul and then follow Ippokrateio, Genimatas, AHEPA, and Papanicolaou.

The existence of adequate field of land facilitates interventions at the Psychiatric Hospital and the Papanicolaou. In the case of the other large hospitals (Ippokrateio and AHEPA), relief practices are implemented, which include the establishment of a pediatric hospital, strengthening of refurbishment, and better use of existing facilities.

Regarding the St. Paul General Hospital, the creation of new buildings within walking distance appears as the only viable solution, due to the relatively small existing field area and the high occupancy. As far as Genimatas Hospital is concerned, new facilities are also needed. But until this becomes possible, consideration should be given to a better utilization of existing facilities due to the relatively low occupancy and the possibility offered by the interface between the neighboring hospital Saint Dimitrios. Small corrective actions could even be made in Papanicolaou, Papageorgiou, and Skin Diseases hospitals.

The order of interventions may be modified, depending on the priorities set by the political leadership and the importance of each criterion. Table 10.6 gives a general insight into future planning and prioritization of upgrading the infrastructure of Thessaloniki's hospitals.

In any case, there is the opportunity of a more analytical approach to the issue, prioritizing interventions and examining different solutions, oriented, for example, to renovation of an existing establishment or to the possibility of expansions, reallocation of clinics, or moving staff between hospitals.

Conclusion and Further Research

The goal of the current study was to firstly highlight the need for improvements and interventions to the hospitals of Thessaloniki. This was achieved through the presentation of reliable data regarding hospitals' infrastructure and the number of patients that they treat. Keeping in mind the limited funds available for such projects, the study applies a multicriteria-based approach for ranking the existing public hospitals on the basis of their needs as far as infrastructure and operational resources adequacy is concerned. The results defined a priority list for the hospitals. This information is valuable for the central government agencies to prioritize their corresponding actions. The current research proposes a road map for improving Thessaloniki's public hospitals.

The municipal authorities can play an important role in planning for the reorganization of the health infrastructure of Thessaloniki in cooperation with the state. The most important of these is finding suitable premises and land to be allocated for the construction of new hospitals or expansion of existing ones. Additional important consideration is even the plan to connect the new hospitals to the social fabric. The design will include a minimum of road and transport infrastructure that will serve the new hospitals and any other action that could facilitate their functional integration into city life, in the light of the WHO guidelines for healthy cities.

The results of the proposed actions will produce significant improvements in the everyday life and in the quality of the healthcare received by patients. This is also true for the hospital personnel, which will be dealing with more manageable patient flows. The study makes one more contribution toward the proposal of specific actions to provide effective solutions. These suggestions include:

- Expansion of hospitals in adjacent land and buildings
- Transfer of specific clinics to new buildings elsewhere in Thessaloniki
- Relocation of certain healthcare activities or part of them to a new hospital that will be created on the outskirts of city
- Upgrading of primary healthcare, which will relieve the workload of the outpatient clinic and the emergency departments of hospitals
- Reinforcing of certain neuralgic hospitals in other prefectures that will function as the intercepting centers for patients from the regional areas

As part of the future work a number of issues, which could be further examined include:

- The consideration of more criteria for the decision-making process
- Application of different methodologies for the multicriteria decision-making

At the same time, it is very interesting to extend the proposed methodology to focus on ranking possible available interventions and improvements per hospital. Finally, the cost and time for implementation of each intervention could be considered in the process of decision-making.

References

1. 3rd Health Region Administration of Macedonia.: http://www.3ype.gr/ Accessed 5 Feb 2015
2. 4th Health Region Administration of Macedonia & Thrace.: http://www.4ype.gr/ Accessed 2 Feb 2015
3. 5th Health Region. PivotTable – Hospital beds 2011 http://civil.dypethessaly.gr/hospitals (2012). Accessed 17 Jul 2013
4. "AHEPA", University General Hospital of Thessaloniki: http://www.ahepahosp.gr/ Accessed 2 Feb 2015
5. Antoniou, F., Aretoulis, G.N., Konstantinidis, D., Papathanasiou, J.: Choosing the most appropriate contract type for compensating major highway project contractors. J. Comput. Optim. Econ. Finance. **6**(2), 77–95 (2014)
6. Antoniou, F., Konstantinidis, D., Aretoulis, G.N.: Application of the multi attribute utility theory for the choice of project procurement system for highway projects. Int. J. Manag. Decis. Mak. **15**(2), 83–112 (2016)
7. Aretoulis, G.N., Triantafyllidis, C.H., Papathanasiou, J.B., Anagnostopoulos, I.K.: Selection of the most competent project designer based on multi-criteria and cluster analysis. Int. J. Data Anal. Tech. Strateg. **7**(2), 172–186 (2015)
8. Barton, H., Tsourou, C.: Healthy urban planning – a WHO guide to planning for people. E&FN Spon, London (2000)
9. "Bodosakio" General Hospital of Ptolemaida: www.mpodosakeio.gr Accessed 5 Feb 2015
10. Brans, J.P., Mareschal, B.: The PROMCALC and GAIA decision support system for MCDA. Decis. Support. Syst. **12**(4–5), 297–310 (1994)
11. Brans, J., Vincke, P.: A preference ranking organization method: The PROMETHEE method for multiple criteria decision making. Manag. Sci. **31**, 647–656 (1985)
12. Brans, J., Mareschal, B., Vincke, P.: How to select and how to rank projects: the PROMETHEE method. Eur. J. Oper. Res. **24**, 228–338 (1986)
13. Chatzipoulidis, G.: Natural disasters and Hospital. The case of the earthquake: preventive risk reduction action. Health Rev. (Athens). **20**(120), 15–21 (2006)
14. Chatzipoulidis, G., Aretoulis, G., Kalfakakou, G.: PPP's application in Greek health infrastructure and personnel's response. Int. J. Healthcare Deliv. Reform Initiat. **3**(4), 1–15 (2011)
15. General Hospital of Didymoteichon.: http://www.did-hosp.gr/ Accessed 2 Feb 2015
16. General Hospital of Drama.: www.dramahospital.gr Accessed 2 Feb 2015
17. General Hospital of Halkidiki.: http://www.hospitalchalkidiki.gr/ Accessed 3 Feb 2015
18. General Hospital of Kavala. http://kavalahospital.gr/ Accessed 2 Feb 2015
19. General Hospital of Serres.: http://www.hospser.gr/ Accessed 2 Feb 2015
20. General Hospital of Xanthi.: http://www.hosp-xanthi.gr/ Accessed 2 Feb 2015
21. General Hospital of Edessa.: www.gnedessas.eu Accessed 5 Feb 2015
22. General Hospital of Florina.: http://www.florinahosp.gr/index.htm Accessed 5 Feb 2015
23. General Hospital of Gianitsa.: www.gng.gr Accessed 5 Feb 2015
24. General Hospital of Grevena.: www.nosgrevenon.gr Accessed 5 Feb 2015
25. General Hospital of Kastoria.: www.kastoriahospital.gr Accessed 5 Feb 2015
26. General Hospital of Katerini.: www.gnkaterini.gr Accessed 5 Feb 2015
27. General Hospital of Naousa.: http://www.gnnaousas.gr/ Accessed 5 Feb 2015
28. General Hospital of Veria.: http://www.verhospi.gr/ Accessed 5 Feb 2015
29. "G. Gennimatas" General Hospital of Thessaloniki.: http://www.gennimatas-thess.gr Accessed 5 Feb 2015
30. Gerolimpos, A., Kolonas, B.: "A Cosmopolitan Urbanism", Thessaloniki, 1850–1918. The "City of the Jews" and the Awakening of the Balkans, pp. 171–189. Hecate, Athens (1994)
31. Greek Intermunicipal Network of Healthy Cities: Report for Health actions 2014–2016. http://www.eddyppy.gr/apologismos-draseon-kep-ygeias-2014-2016 (2016). Accessed 4 Jan 2017

32. Greek National Statistical Authority: Press release: announcement of census results – Housing 2011 for the permanent population of the country http://www.statistics.gr/portal/page/portal/ESYE/BUCKET/General/resident_population_census2011.xls (2012). Accessed 24 Nov 2015
33. 'Ippokrateio' General Hospital of Thessaloniki.: http://www.ippokratio.gr/ Accessed 3 Feb 2015
34. Ishizaka, A., Nemery, P.: Multi-Criteria Decision Analysis Methods and Software. John Wiley & Sons, Ltd, Chichester (2013)
35. Kalogeropoulou, M., Mpourdoukoutas, P.: Health Services, vol. A. Kleidarithmos, Athens (2007)
36. Lawrence, R., Fudge, C.: Healthy cities in a global and regional context. Health Promot. Int. **24**, i11–i18 (2009)
37. Liaropoulos, L.: Health Systems and Services Organization, vol. A. Beta, Athens (2007)
38. "Mamatsio" General Hospital of Kozani: www.mamatsio.gr Accessed 5 Feb 2015
39. Ministry of Health: Planned number of beds, September 2013 http://www.moh.gov.gr/articles/esynet/...2013/2191-9mhno-leit-2013 (2013). Accessed 24 Nov 2015
40. "Papageorgiou" General Hospital of Thessaloniki: www.papageorgiou-hospital.gr Accessed 6 Feb 2015
41. "G. Papanikolaou" General Hospital of Thessaloniki: www.gpapanikolaou.gr Accessed 6 Feb 2015
42. Pérez, S.V., Ignacio Echániz Salgado, J.I.E., Bartolomé, J.B.: Implementation of a new public health structure in Madrid. In: Our Cities, Our Future: Policies and Action Plans for Health & Sustainable Development. Edited: Charles Price & Agis Tsouros.WHO Healthy Cities Project Office, Copenhagen (1996)
43. Psychiatric Hospital of Petra Olympos.: http://www.psynpo.gr/ Accessed 5 Feb 2015
44. Psychiatric Hospital of Thessaloniki.: http://www.psychothes.gr/ Accessed 5 Feb 2015
45. "Saint Dimitrios" General Hospital of Thessaloniki: http://www.oagiosdimitrios.gr/ Accessed 6 Feb 2015
46. "Saint Paul" General Hospital of Thessaloniki: http://www.agpavlos.gr/ Accessed 3 Feb 2015
47. "Sismanoglio" General Hospital of Komotini: http://www.komotini-hospital.gr/ Accessed 2 Feb 2015
48. Skin Diseases Hospital of Thessaloniki.: www.nadn.gr Accessed 2 Feb 2015
49. Special Secretariat for PPPs: Public – Private Partnerships. Health projects implemented. http://www.sdit.mnec.gr/el/sectors/sector/sector0002.html (2012). Accessed 03 Jul 2014
50. "Theagenio", Anticancer Hospital of Thessaloniki: http://ww2.theagenio.gr/ Accessed 2 Feb 2015
51. Tsouros A.: City leadership for health and sustainable development: The World Health Organization European Healthy Cities Network. Health Promotion International, 24(S1), Oxford University Press (2009)
52. Tzeng, G.H., Huang, J.J.: Multiple Attribute Decision Making. Methods and Applications. Taylor and Francis Group, Boca Raton (2011)
53. Hellenic Regulatory Body of Nurses: Hospital Organizations http://enne.gr/organismoi_nosokomeiwn (2013). Accessed 24 Nov 2015
54. University General Hospital of Alexandroupolis.: www.pgna.gr Accessed 2 Feb 2015
55. Velonaki, V.S., Kalokairinou-Anagnostopoulou, A.: Historical evolution of Health System and Nursing in Greece. Nursing. **45**(4), 491–499 (2006)
56. WHO Regional Office for Europe: City Health Planning: The Framework. WHO Regional Office for Europe, Copenhagen (document EUR/ICP/HCIT 94 01/ MT06/7) (1996)
57. WHO Regional Office for Europe: Community Participation in Local Health & Sustainable Development (European sustainable development & health series: 4). WHO Regional Office for Europe, Copenhagen (2002)
58. WHO Regional Office for Europe: Healthy Cities http://www.euro.who.int/en/health-topics/environment-and-health/urban-health/activities/healthy-cities (2017). Accessed 12 Jan 2017

Chapter 11
Sustainable Urban Living and Social Capital: Some Evidence from Crisis-Hit Greece

Irene Daskalopoulou

Abstract Developed societies have for long been engaged in finding ways to increase, deepen, and strengthen their social capital levels as a means to sustain institutional quality and their overall growth potential. However, this is a multifaceted and complex task, especially when linked to urban sustainability. Analysis of social capital as a spatial phenomenon is rather limited, despite the commonly held view that locally embedded associations, partnerships, and initiatives can foster community empowerment and regeneration by building and sustaining well-functioning communities. To that extent, social capital is inexorably linked to the territorial capital that regions and localities might mobilize toward addressing the wider social, economic, environmental, and developmental challenges that they face. Within this context, the present study analyzes the civic engagement pattern of urban residents in Greece and tests for the effect of soft social capital constructs such as social trust, social altruism, equality, tolerance, and humanitarianism, upon this pattern. Analysis is based on microlevel data drawn from the European Social Value surveys round 4 (2008) and round 5 (2010). Results show that, while controlling for the sociodemographic and economic profile of respondents, the onset of the economic crisis in the country has negatively affected the social capital scores of both urban and nonurban residents, with urban residents showing higher levels of civic engagement compared to nonurban residents.

Keywords Sustainable living • Urban living • Social capital • Civic engagement • Social trust • Greece

---

I. Daskalopoulou (✉)
Department of Economics, University of Peloponnese, Thesi Sechi (Proin 4o Pedio Volis), Tripoli, 22100, Greece
e-mail: daskal@uop.gr

A. Karakitsiou et al. (eds.), *City Networks*, Springer Optimization and Its Applications 128, DOI 10.1007/978-3-319-65338-9_11

Introduction

Sustainable urban living has been at the core of research for many decades. The increasing focus on cities and, in particular, on urban sustainability follows the dramatic urbanization trends that the human society experiences from the beginning of the twentieth century onward [57]. With the number of population living in cities and towns growing from less than a tenth to about half of the world's population within the span of the twentieth century, the concern over the problems associated with massive urbanization worldwide is anticipated [27]. According to the UN, the pace of urbanization will accelerate, and by 2025 urban population is projected to reach 60% of the world's population [63]. While this unprecedented appeal of cities and of urban living draws from greater social and, in particular, economic opportunities that are usually met in an urban context, increasing urbanization does not come without disadvantages [27]. Social, economic, and environmental problems have come to gradually change the conditions of urban living [30, 44, 63]. Summarizing the importance of these challenges, Street [57] states that "If cities are to continue to function effectively as nodal points for people and their activities, their whole economic development needs to be put on a more sustainable basis."

Within this context, finding ways to promote sustainable urban living gained increased interest at both the research and policy levels [2, 3, 27]. The main aim of research is to provide vital information on how to promote sustainable cities defined as "cities where socio-economic interests are brought together in harmony (co-evolution) with environmental and energy concerns in order to ensure continuity in change" [44: 4]. A common realization of research findings suggests that the goal of achieving urban sustainability goes beyond the issues of designing local, regional, or national policies to address the challenges that cities and towns might face. At the policy level therefore, urban sustainability requires a shift in the way in which urban living conditions are formed not only in environmental terms but also at social and economic terms. In particular, addressing issues such as the need to reverse urban decay and fuel economic regeneration requires that a more integrated approach is applied to the matter [30, 57]. As such, the development of relevant strategies in the field is acknowledged as a very complex issue with many studies providing different views over the appropriate processes, activities, and policies for their formation and implementation [28, 30, 44]. The one factor that is found common to all these approaches is "the need for increased participation and wider consultation in the strategy formulation and policy making process" [57: 142], if sustainable urban living is to become a reality.

Thus, the role of urban citizens, their actions, associations, etc. in promoting sustainable cities has for long been acknowledged. Research on urban social movements has traditionally aimed at enhancing our knowledge and information with regard to promoting sustainable urban living conditions. Focusing at issues such as the urban ways of life, housing, the role and forms of kinship, neighboring and friendship, the role of voluntary associations, and urban social problems such as crime and the spatial structure [48: 102], this strand of research emphasized actions and outcomes (i.e., the effects of actions) in terms of urban renaissance [13, 48]. Research evidence suggest that these issues are quite complex to address. In particular, the usually observed gap between citizens' actions and anticipated effects suggests that we need to use more rich concepts such as the social capital concept, in order to more fully understand the sociopolitical and economic processes taking place at the urban level, i.e., at the dynamic evolution context of cities and towns [23, 24, 35, 42, 48, 55, 58].

Given the above, the present study sets out to explore the civic engagement pattern of urban residents in Greece and tests for the effect of soft social capital constructs upon this pattern. This way the link between the various forms of social capital is sought to be analyzed within the context of the specificities owing to urbanity and the onset of the economic crisis in Greece. At the empirical level, analysis is based on individual level (micro) data drawn from the European Social Value surveys round 4 (2008) and round 5 (2010). Using these data we measure the social capital levels of individuals by measuring six main constructs, namely, civic engagement (i.e., manifested forms of social capital, the so-called hard social capital element) and social trust, social altruism, equality, tolerance, and humanitarianism (the so-called soft social capital elements) [10]. Following this operational definition of social capital, we then analyze the effect of soft social capital elements upon the civic engagement scores of individuals. Overall, empirical results indicate that urban residents show higher levels of civic engagement scores compared to nonurban residents. In addition, while controlling for the sociodemographic and economic profile of respondents, evidence is provided that the onset of the economic crisis in the country has negatively affected the social capital scores of both urban and nonurban residents.

The rest of the chapter is organized as follows: part 2 is devoted to a brief discussion about social capital as a civil culture quality and the relationship between social capital and territorial capital in an urban context. Part 3 is devoted to the presentation of the empirical model about Greece and of the data and variables used in the study. Part 4 presents the results of the empirical analysis and Part 5 concludes the study.

Theoretical Context

Social Capital and Civic Culture

The early definitions of the concept of social capital drew mainly from an anthropological and sociological background and tended to focus on the individual (see indicatively, [6, 7]). Bourdieu and Wacquant [7], for example, define social capital as "... the sum of the resources, actual or virtual, that accrue to an individual or a group by virtue of possessing a durable network of more or less institutionalized relationships of mutual acquaintance and recognition." Later works have come to place the concept on the sphere of *relations* rather than the individual sphere (see [17, 53]) and argue that social capital is the product that networks offer to individuals, i.e., social capital is the links within and between networks and groups [50, 62]. As such, Coleman [17] defines social capital as "... an attribute of the social structure in which a person is embedded ... [and thus it is] ... not the private property of any of the persons who benefit from it," whereas Putnam [53] suggests that social capital relates to "... features of social organization, such as trust, norms, and networks, that can improve the efficiency of society by facilitating coordinated actions." The use and outcomes of social capital generation frameworks depend largely on the wider socioeconomic, cultural, and political structures of a society [5, 26, 29, 36, 45], and thus social capital can take up many forms following a country's specific sociocultural and associational relations [5, 53, 62].

In the context of the present study, social capital is used in order to denote a civic culture set of characteristics or, else, the qualities of a civic culture [22, 50]. Following Edward's [21: 137] theorization, we might understand the hard and soft social capital elements of a civic culture, such as the ones analyzed here, as a schema wherein "People [...] coordinate their behavior and govern the compact through just rules. The maintenance of the cooperation rests on maintaining the legitimacy of the rules." This is an inclusive and informative theorization of the relationship between civic engagement and social trust, social altruism, equality, tolerance, and humanitarianism or else between the hard and soft social capital qualities of a society. Accordingly, the empirical results can be interpreted into a civic culture framework that is meaningful about the importance of the changes observed in these societal characteristics and their potential effects upon sustainable urban living conditions.

Research regarding the ways in which social capital contributes to building strong civil communities and its relation to territorial capital is so far limited [8, 14, 34, 43]. This gap in the literature might be attributed, among other factors, to the fact that research on urban social movements has grown largely in isolation from other writings on social movements [48], and thus the social capital literature was only recently brought into the discussion of addressing urban problems and challenges. Following Castells' [13] original definition of the concept of urban social movements, emphasis has been placed almost entirely on analyzing the urban ways of life and urban social problems. As such, research on urban social movements has grown to exclude from its analysis the core themes of power and conflict as basic parameters and determinants of the observed urban social movements [48]. Thus, important urban politics issues, such as the role of the local government and the existence and bargaining power of pressure groups and parties, were left out of the analysis. The restrictive definition of Castells [13] was later on abandoned in favor of a generic one linking urban social movements to any or all citizen actions irrespective of their actual (or potential) effects. Nevertheless, it is important to note that the isolated manner in which research on urban social movements has grown has had both positive and negative results [48] in terms of the ways in which the urban challenges are studied.[1] To that extent, Mayer [43] argues that the social capital concept has been appropriated by the urban movements' strand of research mainly due to its popularity as a concept that can bring about social cohesion in modern capitalist economies. As such a number of distorting effects in the use of the concept have occurred[2] causing urban movements not to be analyzed as a political action form which she considers their main characteristic.

Social Capital in an Urban Context

As mentioned earlier, research on the dynamic evolution of the social capital phenomenon reveals that different types of societies are "born" and survive as a result of the different types of social capital which they possess [5, 53, 62]. Analysis of social capital as a spatial phenomenon has been primarily grounded on community empowerment and regeneration research under the commonly held view that locally embedded associations, partnerships, and other types of formal and informal initiatives can build and sustain well-functioning communities [14, 34, 42, 54]. To that extent, social capital is inexorably linked to, and an integral part of, the territorial capital that regions and localities might mobilize toward addressing their wider social, economic, and developmental challenges [11, 20, 25, 31, 38, 40, 59, 61]. In addition, social capital is the basis of urban movements demanding grassroots empowerment and citizen participation as a more democratic mode of governance [8, 12, 14, 34, 43, 49, 51, 52].

As Kearns [34] suggests social capital is linked to neighborhood renewal goals such as community empowerment and to urban policy goals such as community cohesion. As such, social capital might contribute to the objective of "joined-up policy" and the broader aim of "democratic renewal." This is true since urban movements demanding increased grassroots participation and the democratization

[1] The positive results are (a) a focus on studying effects, (b) an interest in the strategies and tactics employed in the local political process, and (c) a focus on the political context in which urban movements developed and in particular the specific sociopolitical conditions that sustained them (Melucci, 1989; [23, 24, 48, 58]. On the other hand, the negative results of the isolated manner in which the urban social movements literature has grown were (a) the evolution of a strand of research that was largely cut off from general social movement theory; (b) an empirical lacunae, e.g., with regard to the process of people mobilization; and (c) a separation was established between studies of voluntary associations and their interaction with authorities and studies on urban movements [35, 42, 55].

[2] These distorting effects relate to (a) a focus on the relation of the concept with increasing economic competitiveness and achieving social cohesion while leaving out of the discussion its effect on making and demands that are in conflict with the ruling elites; (b) a focus on the concept as an entirely positive concept and a purposeful ignorance of issues such as patron-client relationships, crime, and corruption; and (c) a focus on interpersonal relations that diverts attention from the equally important wider restructuring processes occurring at the political economy level and the state [43].

of urban politics have gradually succeeded in placing grassroots empowerment and citizen participation at the core of the official urban development agenda [8, 14, 43]). In this sense, grassroots participation and local activism are perceived as part of a new mode of governance that has emerged in, and in favor of, neglected and disadvantaged areas and communities in order to successfully address their "exclusion" and inequality or segregation problems [43].

Social capital as embedded in horizontal networks and reciprocity is the mechanism via which local participation might drive communities toward economic growth and democratic institutions [43]. Civic engagement along with processes of state transformation [49] actually frames the contemporary reconfigurations of local state-society relations and thus significantly impacts upon the development of initiatives and sectors (e.g., the voluntary sector development) that are conducive to building strong and well-functioning civil societies [43]. Many governments fund area-based initiatives as a means to create linking social capital that can play a positive role in establishing trust and relations between civil society and a municipality [1], albeit their impact and long-term effectiveness is yet to be evidenced [1, 37]. The key characteristics of social capital as embedded in all these processes are seen to be produced along the lines of a "social capital policy-making continuum" ranging from engagement to participation and from such engaged participation to common decision-making outcomes [14, 34, 42, 49].

Empirical Analysis: Model and Data

Urban Social Capital in Greece: A Proposed Model

Following the general trends observed in the developed countries of the west, urbanization in Greece has also been dramatic (Table 11.1) causing important problems and challenges related to the development of the urban nexus in the country (see indicatively, [41]). On the other hand, Greece is characterized by low social capital levels [4, 16, 18, 19, 32, 33, 39, 47] and declining social cohesion levels [46, 60]. The civil society has grown to respond to the challenges of the economic crisis [56], albeit the sustainability of such practices and initiatives needs to be further sustained.Within this context it is important to analyze the civic participation scores of urban residents in the country and in particular to analyze the potential effect of soft social capital elements upon the observed civic participation scores. This way information might be provided on the drivers of urban residents' civic mobilization. Thus, the following empirical relationship is identified and tested here:

$$CE_{it} = \alpha_0 + \beta \boldsymbol{D}_{\boldsymbol{it}} + \gamma U_{it} + \delta T_i + \varepsilon\, \boldsymbol{sc}_{\boldsymbol{it}} + u_{it} \tag{11.1}$$

Table 11.1 Change of population, population density, and urbanization trends (1951–2011)

	1951	1961	1971	1981	1991	2001	2011
Greece total	7,632,801	8,388,553	8,768,372	9,739,589	10,259,900	10,934,097	10,815,197
Change	287,941	755,752	379,819	971,217	520,311	710,705	−118,900
% change		9.5	4.4	10.6	5.2	6.7	−1.01
Population density	57.8	63.6	66.5	73.8	77.8	83.1	81.96
Urban population (%)	37.7	43.3	53.2	58.1	58.8	74.8	n.a.
Suburban population (%)	14.8	12.9	11.6	11.6	12.8	13.8	n.a.
Rural population (%)	47.5	43.8	35.2	30.3	28.4	24.8	n.a.
% of Attica in urban and suburban population	38.8	43.7	49.2	49.7	47.9	47.1	n.a.
% of Attica in total population	20.4	24.5	31.9	34.6	34.3	34.3	35.4

Source: ELSTAT, Population Census

where CE_{it} stands for the civic engagement score of person i in time t; $\boldsymbol{D}$ stands for a vector of control variables referring to individual level socioeconomic and demographic characteristics; U is an urban dummy variable accounting for urban vs nonurban respondents; T is a time trend variable accounting for the two periods of the analysis; $\boldsymbol{sc}$ is a vector of the soft social capital constructs, namely, social trust, social altruism, equality, tolerance, and humanitarianism; α, β, γ, δ, and ε are the coefficients to be estimated; and u is an error term. Pooled regression analysis is performed in order to estimate equation. Since the available data refer to random samples of individuals for two different time periods, panel analysis techniques cannot be used [64]. Thus, pooled regression analysis is performed in order to estimate Eq. (11.1). The potential structural breaks in our model, caused by time effects and by belonging to the urban respondents group, are tested via the coefficients of the two dummy variables, T and U [15, 64]. In addition, regression results have been tested for multicollinearity using the collinearity diagnostics of tolerance and variance inflation factor (VIF).

In relation to the discussion presented here, we expect the urban residents group of people to be positively associated with higher levels of civic engagement, i.e., we expect $f_i'(U) > 0$ to hold. Further, given the negative shock that the country experienced during the 2008–2010 period, we expect the change analyzed via the time trend variable to be negative, i.e., we expect $f_i' < 0$ to hold. Finally, as regards the effect of social capital constructs upon civic engagement, we expect positive signs, i.e., we expect $f_i'(\boldsymbol{sc}) > 0$ to hold, denoting a positive effect of soft social capital constructs upon civic participation scores. The effect of the sociodemographic and economic variables is expected to be positive.

Data and Variables

Following Brehm and Rahn [9] and Brewer [10], we measure people's level of social capital via the measurement of six closely related constructs referring to social trust, social altruism, equality, tolerance, humanitarianism, and civic participation. In addition, we identify civic participation as our dependent variable denoting hard behavioral evidence of social capital and test for the effect of soft social capital elements (i.e., social trust, social altruism, equality, tolerance, and humanitarianism) upon it. The civic participation variable is approximated by the sum of 12 items including interest in politics and political behavior (voted, contacted politician, interest in politics, feel closer to a political party, member of political party, worked in political party) and other social activities including work in another organization, worn or displayed badge, signed petition, taken part in lawful public demonstration, boycotted certain products, and being an active member of a union. The independent variables of the model can be categorized into three sets. The first set includes the sociodemographic and economic characteristics of the respondents; the second set includes two dummy variables controlling for possible structural break in the model as a result of time (2008–2010) and as a result of belonging to the urban group of respondents, as opposed to nonurban respondents; and the third set refers to the soft social capital attributes. In particular, the first set of independent variables includes eight variables controlling for age, gender, education, marital status, children, the type of employment, and the level of household income. The second set includes a time trend dummy variable, controlling for changes owing to the period of the analysis, and an urban resident dummy variable, controlling for changes owing to belonging to the group of respondents that resigns in an urban place. The third set includes five social capital variables that are composed as follows:

1. Social trust, approximated by the sum of seven items including generalized trust and trust in important institutions such as the country's parliament, the legal system, the police, the politicians, the political parties, and the European Parliament
2. Social altruism, approximated by the sum of two items including participation in social activities and general perceptions of helpfulness characterizing people
3. Equality, approximated by the sum of three items including general perceptions of people's fairness, perceptions on the importance that people are treated equally and have equal opportunities and perceptions over the importance to understand different people
4. Tolerance, approximated by the sum of three items referring to effects of immigrants on the country's economy, cultural life, and living conditions
5. Humanitarianism, approximated by the sum of two items referring to perceptions over the importance to care for others' well-being and time devoted to friends and people close to them

For an analytical description of how these constructs are measured in the case of Greece, see Daskalopoulou [18, 19].

Data for the construction of the dependent and independent variables of the analysis are drawn from the European Social Value (ESV) surveys rounds 4 and 5, referring to the 2008 and 2010 waves. Table 11.2 presents the definitions and basic descriptive statistics of all the variables used in the analysis.

Results

Overall, a sample of $N = 4787$ observations have been used in the study, consisting of a subsample of $N = 2072$ observations for 2008 (ESV r4) and a subsample of $N = 2715$ observations for 2010 (ESV r5). In line with the main goal of the analysis, we have identified urban and nonurban residents in our sample (and in the subsamples) and performed a two-sample t-test for comparing two means in order to test whether the observed difference in the means of the social capital variables of interest is statistically significant or not. This analysis is performed twice for the 2008 and 2010 periods. In all cases, standard statistical significance levels are used

Table 11.2 Definitions and descriptive statistics of used variables

Variable	Definition and measurement	Mean	St. dev.
Dependent variable			
Civic engagement	The sum of the 12 items used to proxy civic participation	3.511	1.830
Independent variables			
Sociodemographic and economic variables			
Age	Respondents' age in years	46.489	17.972
Education	Respondents' education in years	11.321	4.132
Gender	Dummy variable, 1 = male	0.445	0.497
Married	Dummy variable, 1 = respondent is married	3.767	6.315
Household size	Number of persons in the household	2.694	1.283
Child	Dummy variable, 1 = child lives at home	0.396	0.489
Income	Respondents' household income category[a]	3.894	2.438
Public servant	Dummy variable, 1 = respondent is a public servant	0.179	0.383
Social capital variables			
Social trust	The sum of the 7 items used to proxy social trust	22.648	12.318
Social altruism	The sum of the 2 items used to proxy social altruism	4.996	2.419
Equality	The sum of the 3 items used to proxy equality	4.450	2.368
Tolerance	The sum of the 3 items used to proxy tolerance	9.521	6.352
Humanitarianism	The sum of the 2 items used to proxy humanitarianism	0.428	0.878
Structural break (control) variables			
Urban	Dummy variable, 1 = respondent is urban resident	0.601	0.489
Time	Dummy variable, 1 = wave 5	0.567	0.495

Source: Author's calculations
Notes: Variables are defined and coded as in Appendix A4 (Variables lists) of ESS4–2008 ed.4.0. The same variables have been used from the ESS5–2010 wave. Descriptive statistics are based on non-missing observations
[a]The variable of total household income is recorded into ten monthly income categories ranging from less than €500 to more than €3001

in order to decide on whether or not to reject the null hypothesis of equality of means. For high t-stats we reject the null hypothesis of equality of means, and thus, the observed differences are statistically significant.

Table 11.3 summarizes the results of the analysis regarding the differences in the level of social capital held by urban and nonurban residents. As the subsamples used for the tests are quite large, the corresponding degrees of freedom are also quite large, and thus they are not reported here. Of the total sample of $N = 2072$ observations for 2008, a subsample of 1353 urban residents has been identified based on their description of their place of residence. Results show that, in 2008, urban residents score higher in all constructs of social capital. The difference in these mean scores is statistically significant in the case of the civic engagement, the social altruism, and the equality constructs. Of the total sample of $N = 2715$ observations for 2010, a subsample of 1522 urban residents has been identified again based on their description of their place of residence. Results show that, by 2010, a decline is observed in the mean scores of all social capital constructs yet

Table 11.3 Mean values of social capital elements for urban and nonurban residents

	2008			2010		
	Urban ($N = 1353$)	Nonurban ($N = 719$)	T-test	Urban ($N = 1522$)	Nonurban ($N = 1193$)	T-test
Civic engagement	3.891	3.655	2.781[a]	3.309	3.253	0.800
Social trust	26.620	25.925	1.207	20.423	18.877	3.326[a]
Social altruism	5.379	4.768	5.621[a]	5.054	4.620	4.519[a]
Equality	4.536	3.969	5.203[a]	4.646	4.396	2.704[a]
Tolerance	10.169	10.087	0.267	9.800	8.076	7.190[a]
Humanitarianism	0.492	0.523	−0.688	0.351	0.395	−1.422

Source: Author's calculations

Notes: All estimations are based on non-missing observations. The two sample equality of means t-test are reported. For high t-stats we reject the null hypothesis of equality of means. Thus, the observed differences are statistically significant

[a]indicate significance at the 0.5% and above

the difference between urban and nonurban residents still holds. At the onset of the economic crisis, the two groups present statistically significant differences in terms of the social trust, social altruism, equality, and tolerance constructs, again with urban residents presenting higher scores.

Table 11.4 reports the results of estimating Eq. 11.1. The first three columns of the table summarize the regression results of predicting civic engagement scores as the outcome of soft social capital constructs while controlling for urbanity and time as well as for the sociodemographic and economic characteristics of the sample (model I). As shown, civic engagement scores are affected by soft social capital constructs and in particular by social trust and social altruism which exercise positive effects. Of the sociodemographic and economic variables, the positive and statistically significant effects of the age and education variables are reported. The change during the 2008–2010 period, as shown by the time trend variable, is negative and statistically significant, while urbanity is also found to exert a positive and statistically significant effect upon civic engagement scores. To further analyze the finding regarding urbanity, we split the sample into two subgroups referring to only urban residents (model II) and to nonurban residents (model III). Results of model II show that the civic engagement scores of urban residents are positively affected by social trust and social altruism but negatively affected by equality scores. Of the sociodemographic and economic variables, the positive and statistically significant effects of age, marital status, and employment status (being a public servant) are reported. The effect of time is again negative and statistically significant. For nonurban residents results show that only age and education have a statistically significant effect, while the effect of the time control variable is again negative and statically significant (model III). For nonurban residents none of the soft social capital variables have been found to exert a statistically significant effect upon civic engagement scores.

Table 11.4 Civic engagement in an urban context

	Pooled regression I			Pooled urban II			Pooled nonurban III		
	β	S.E.	Sig.	β	S.E.	Sig.	β	S.E.	Sig.
Sociodemographic and economic variables									
Constant	−0.451	0.404	0.265	−0.231	0.570	0.685	−0.688	0.590	0.245
Age	0.178	0.078	0.002	0.124	0.109	0.097	0.270	0.115	0.003
Education	0.125	0.069	0.028	0.075	0.098	0.293	0.174	0.098	0.052
Gender	0.057	0.046	0.198	0.047	0.063	0.417	0.050	0.069	0.495
Married	0.029	0.058	0.596	0.128	0.080	0.086	−0.110	0.086	0.191
Household size	0.001	0.068	0.986	−0.112	0.089	0.188	0.174	0.111	0.109
Child	0.063	0.062	0.285	0.048	0.084	0.531	0.060	0.093	0.535
Income	0.060	0.011	0.251	0.071	0.015	0.292	0.073	0.018	0.382
Public servant	0.051	0.062	0.272	0.101	0.080	0.099	−0.031	0.105	0.686
Social capital variables									
Social trust	0.089	0.037	0.059	0.140	0.055	0.025	0.027	0.048	0.717
Social altruism	0.115	0.053	0.024	0.157	0.080	0.024	0.072	0.071	0.357
Equality	−0.080	0.048	0.121	−0.144	0.069	0.039	−0.042	0.066	0.596
Tolerance	−0.031	0.031	0.478	0.030	0.043	0.597	−0.114	0.045	0.104
Humanitarianism	−0.065	0.049	0.140	−0.030	0.065	0.592	−0.099	0.076	0.171
Structural break control variables									
Time	−0.168	0.050	0.000	−0.176	0.068	0.006	−0.131	0.073	0.092
Urban	0.086	0.049	0.064						
Summary statistics									
N	4787			2875			1908		
Adj R^2	0.122			0.136			0.079		
F-test (prob)	5.565 (0.000)			4.273 (0.000)			2.224 (0.008)		

Source: Author's calculations
Notes: All estimations are based on non-missing observations. Dependent variable is civic engagement scores. Regression results are reported
*β*s refer to the estimated coefficients, *S.E.* reports standard errors and *Sig.* reports significance level

Conclusion

In the context of the present study, it is acknowledged that social capital is inexorably linked to territorial capital which constitutes a form of capital that regions and localities might mobilize toward addressing the wider social, economic, environmental, and developmental challenges that they face. To that extent, we analyze the civic engagement pattern of urban residents in Greece, which is also referred to as hard social capital, and test for the effect of soft social capital constructs such as social trust, social altruism, equality, tolerance, and humanitarianism, upon this pattern. Analysis is based on microlevel data drawn from the European Social Value surveys round 4 (2008) and round 5 (2010). The changes in the country's stock of urban social capital occurring as a result of the outburst of the economic crisis are also analyzed.

Results indicate that the country experienced a statistically significant decline in its social capital level during the 2008–2010 period, while the two subgroups of urban citizens and nonurban citizens show statistically significant differences in terms of their social capital levels. These findings hold while controlling for the sociodemographic and economic profile of the two subgroups of respondents. Furthermore, the civic engagement scores of urban residents are significantly affected by soft social constructs, with social trust and social altruism exercising positive effects and equality scores exercising negative effects. The positive and statistically significant effects of the age, marital status, and employment status variables are also reported.

The present findings suggest that urban residents in the country might be carriers of a distinct profile in terms of their stock of social capital. This is important in terms of motivation and the latent (unobserved) drivers of public actions that might occur at the urban landscape. The higher social and political engagement scores of urban residents qualify them as an interest group that might actively support its assertions. This might be particularly important given the harsh socioeconomic situation currently characterizing the urban centers in Greece. Indeed, the continuing economic crisis in the country and exogenous factors such as increased illegal migration trends have placed extra burden upon the country, harshening even further the existing social and economic challenges that major cities and towns face. Thus, the design and successful implementation of policy measures to reverse this negative situation and promote sustainable living conditions and social cohesion in the country might be much dependent upon finding ways to strengthen local community's ties and initiatives at the urban centers that carry the bulk of the current socioeconomic and developmental challenges. Finally, it should be noted that while the pervasiveness of the changes reported here is a subject for further research, we might argue that a challenge is ahead in terms of how the country might manage to empower the social capital levels of different groups of citizens, by reducing participatory inequalities and enforcing social cohesion and stronger civil societies.

References

1. Agger, A., Jensen, J.O.: Area-based initiatives—and their work in bonding, bridging and linking social capital. Eur. Plan. Stud. **23**(10), 2045–2061 (2015)
2. Banister, D., & Button, K. J. (Eds.). (1993). Transport, the Environment and Sustainable Development. E & FN Spon: London
3. Barton, H., Bruder, N.: A Guide to Local Environmental Auditing. Routledge, Routledge (2014)
4. Bitros, G.C.: From riches to rags or what went wrong in Greece. J. Econ. Soc. Meas. **38**(1), 5–39 (2013)
5. Bjørnskov, C.: The multiple facets of social capital. Eur. J. Polit. Econ. **22**(1), 22–40 (2006)
6. Bourdieu, P.: The forms of capital. In: Richardson, J. (ed.) The Handbook of Theory and Research for the Sociology of Education, pp. 241–258. Greenwood Press, New York (1986)
7. Bourdieu, P., Wacquant, L.: An Invitation to Reflexive Sociology. Polity Press, Cambridge (1992)

8. Boyte, H.C.: The Backyard Revolution: Understanding the New Citizen Movement. Temple University Press, Philadelphia, PA (1980)
9. Brehm, J., Rahn, W.: Individual –level evidence for the causes and consequences of social capital. Am. J. Polit. Sci. **41**, 999–1023 (1997)
10. Brewer, G.A.: Building social capital: civic attitudes and behavior of public servants. J. Public Adm. Res. Theory. **13**(1), 5–26 (2003)
11. Brueckner, J.K., Largey, A.G.: Social interaction and urban sprawl. J. Urban Econ. **64**(1), 18–34 (2008)
12. Carpenter, J.P., Daniere, A.G., Takahashi, L.M.: Cooperation, trust, and social capital in Southeast Asian urban slums. J. Econ. Behav. Organ. **55**(4), 533–551 (2004)
13. Castells, M.: The Urban Question. Edward Arnold, London (1977)
14. Castells, M.: The City and the Grassroots: A Cross-Cultural Theory of Urban Social Movements, vol. 7. Univ of California Press, Berkeley (1983)
15. Chow, G.C.: Tests of equality between sets of coefficients in two linear regressions. Econometrica. **28**(3), 591–605 (1960)
16. Christoforou, A.: On the determinants of social capital in Greece compared to countries of the European Union, Nota di Lavoro, Fondazione Eni Enrico Mattei, No. 68.2005 (2005)
17. Coleman, J.S.: Foundations of Social Theory. Cambridge, MA, Belknap (1990)
18. Daskalopoulou, I.: Rent seeking or corruption? An analysis of income satisfaction and perceptions of institutions in Greece. Soc. Sci. J. **53**(4), 477–485 (2016a)
19. Daskalopoulou, I.: Social capital at the outburst of the economic crisis in Greece, 2008–2010. Int. J. Bus. Econ. Sci. Appl. Res. **9**(2), 79–86 (2016b)
20. DiPasquale, D., Glaeser, E.L.: Incentives and social capital: are homeowners better citizens? J. Urban Econ. **45**(2), 354–384 (1999)
21. Edwards, L.M.: Ideational social capital and the civic culture: extricating Putnam's legacy from the social capital debates. Soc. Epistemol. **23**(2), 125–144 (2009)
22. Edwards, B., Foley, M.W.: Social capital and the political economy of our discontent. Am. Behav. Sci. **40**(5), 669–678 (1997)
23. Eyerman, R., Jamison, A.: Social Movements: A Cognitive Approach. Ardent Media, Cambridge (1991)
24. Fainstein, S.S., Fainstein, N.I.: Economic restructuring and the rise of urban social movements. Urban Aff. Rev. **21**(2), 187–206 (1985)
25. Forrest, R., Kearns, A.: Social cohesion, social capital and the neighborhood. Urban Stud. **38**(12), 2125–2143 (2001)
26. Fukuyama, F.: Trust. Basic Books, New York (1995)
27. Gilbert, R., Stevenson, D., Girardet, H., Stren, R.: Making Cities Work. The Role of Local Authorities in the Urban Environment. Earthscan, London (1996)
28. Grubb, M.: The Earth Summit Agreements: A Guide and Assessment; An Analysis of the Rio'92 UN Conference on Environment and Development. Earthscan and the Energy and Environmental Programme of the Royal Institute of International Affairs, London (1993)
29. Hardin, R.: Conceptions and explanations of trust. In: Cook, K.S. (ed.) Trust in Society, pp. 3–39. Russell Sage Foundation, New York (2001)
30. Haughton, G., Hunter, C.: Sustainable Cities. Routledge, London (2004)
31. Hofferth, S.L., Iceland, J.: Social capital in rural and urban communities. Rural. Sociol. **63**, 574–598 (1998)
32. Jones, N., Malesios, C., Iosifides, T., Sophoulis, C.M.: Social capital in Greece: measurement and comparative perspectives. South Eur. Soc. Pol. **13**(2), 175–193 (2008)
33. Kazakos, P.: Social capital and collective action: Implications for economic structures, performance and reform. Case study Greece. Science and Society. J. Polit. Moral Theory. **16**, 107–138 (2006.) [in Greek]
34. Kearns, A.: Social capital, regeneration and urban policy. In: Imrie, R., Raco, M. (eds.) Urban Renaissance? New Labour, Community and Urban Policy, pp. 37–60. The Policy Press, Bristol (2003)

35. Klandermans, B.: New social movements and resource mobilization: The European and the American approach revisited. Polit Individ. **1**(2), 89–111 (1991)
36. Knack, S., Keefer, P.: Does social capital have an economic payoff? A cross-country investigation. Q. J. Econ. **112**(4), 1251–1288 (1997)
37. Lawless, P.: Area-based urban interventions: rationale and outcomes: the New Deal for Communities programme in England. Urban Stud. **43**(11), 1991–2011 (2006)
38. Leana, C.R., Pil, F.K.: Social capital and organizational performance: evidence from urban public schools. Organ. Sci. **17**(3), 353–366 (2006)
39. Lyberaki, A., Paraskevopoulos, C.J.: Social capital measurement in Greece. Paper at the OECD-ONS International Conference on Social Capital Measurement, London, Sept 25–27, http://www.oecd.org/greece/2381649.pdf (2002)
40. Lyons, M., Snoxell, S.: Creating urban social capital: some evidence from informal traders in Nairobi. Urban Stud. **42**(7), 1077–1097 (2005)
41. Maloutas: (Εδ.) οι Πολεις, Κοινωνικοας και Οικονομικος Ατλας της Ελλαδας, Τόμος 1, Οι Πόλεις, β' εκδοση. Athens – Volos: National Centre for Social Research and University of Thessaly Publications. (in Greek) (2002)
42. Mayer, M.: Social movements as political actors in European cities. In: Bagnasco, A., Le Galès, P. (eds.) Cities in Contemporary Europe. Cambridge University Press, Cambridge (1999)
43. Mayer, M.: The onward sweep of social capital: causes and consequences for understanding cities, communities and urban movements. Int. J. Urban Reg. Res. **27**(1), 110–132 (2003)
44. Nijkamp, P., Perrels, A.: Sustainable Cities in Europe. A Comparative Analysis of Urban Energy-environmental Policies. Earthscan, London (1994)
45. Norris, P.: Making Democracies Work: Social Capital and Civic Engagement in 47 Societies, No. rwp01–036. Harvard University, Cambridge, MA (2001)
46. Pantazidou, M.: Treading new ground: a changing moment for citizen action in Greece. Dev. Pract. **23**(5–6), 755–770 (2013)
47. Petrou, A., Daskalopoulou, I.: Is Greece a rent-seeking society? A research on the relationship between entrepreneurship, trust and institutions. Soc. Indic. Res. **119**, 1495–1515 (2014)
48. Pickvance, C.: From urban social movements to urban movements: a review and introduction to a symposium on urban movements. Int. J. Urban Reg. Res. **27**(1), 102–109 (2003)
49. Pierre, J. (ed.): Debating governance: Authority, steering, and democracy. OUP, Oxford (2000)
50. Portes, A.: Social capital: its origins and applications in modern sociology. Annu. Rev. Sociol. **24**, 1–24 (1998)
51. Portney, K.E., Berry, J.M.: Mobilizing minority communities social capital and participation in urban neighborhoods. Am. Behav. Sci. **40**(5), 632–644 (1997)
52. Purdue, D.: Neighbourhood governance: leadership, trust and social capital. Urban Stud. **38**(12), 2211–2224 (2001)
53. Putnam, R.D.: The prosperous community: social capital and public life. Am. Prospect. **13**, 35–42 (1993)
54. Sampson, R.J., Morenoff, J.D., Earls, F.: Beyond social capital: SPATIAL dynamics of collective efficacy for children. Am. Sociol. Rev. **64**(5), 633–660 (1999)
55. Saunders, P.: Urban Politics: A Sociological Interpretation, vol. 9. Routledge, London (2006)
56. Sotiropoulos, D.A., Bourikos, D.: Economic crisis, social solidarity and the voluntary sector in Greece. J. Power Polit. Gov. **2**(2), 33–53 (2014)
57. Street, P.: Scenario workshops: a participatory approach to sustainable urban living? Futures. **29**(2), 139–158 (1997)
58. Tarrow, S.G.: Struggling to Reform: Social Movements and Policy Change During Cycles of Protest (No. 15). Center fore International Studies, Cornell University, Ithaca, NY (1983)
59. Tóth, B.I.: Territorial capital: theory, empirics and critical remarks. Eur. Plan. Stud. **23**(7), 1327–1344 (2015)
60. Voulgarelli-Christidou, C.: Dangers of an urban crisis within the European Union: fueling xenophobia and undermining democracy. Perspect Global Dev. Technol. **15**(1–2), 236–245 (2016)

61. Wacquant, L.J.: Negative social capital: State breakdown and social destitution in America's urban core. Neth. J. Hous. Built Environ. **13**(1), 25–40 (1998)
62. Westlund, H.: Social Capital in the Knowledge Economy. Theory and Empirics. Springer – Verlag Berlin Heidelberg, New York (2006)
63. Wilken, E.: Urbanization spreading. In: Brown, L.R., Lenssen, N., Kane, H. (eds.) Vital Signs: The Trends that are Shaping our Future, 1995–1996, pp. 100–101. Worldwatch Institute, Washington (1995)
64. Wooldridge, J.M.: Introduction to Econometrics: A Modern Approach, 4th edn. South-Western, Mason (2009)

Chapter 12
Power Distribution Networks Planning Optimization in Smart Cities

Virgil Dumbrava, Theodor Miclescu, and George Cristian Lazaroiu

Abstract For meeting the demands imposed by the transition from the current tate networks to smart grids and smart cities, the planning of distribution networks will operate. We propose a method of optimizing urban power distribution networks using the idealized networks using the idealized concept.

Keywords Power distribution network • Idealized network • Optimization of the number of transition networks

Introduction

Considering the impetuous development of new technologies integrated in power distribution networks, for meeting the demands imposed by the transition from the current state of networks to smart grids and smart cities, the planning of distribution networks must consider the new conditions in which the networks will operate [7–9, 12].

In the scientific literature, the definition of smart grids and smart cities concepts are intensively discussed, of great importance mainly in the urban distribution networks, specific to cities [10, 11].

In accordance with the European Smart Cities Framework [13, 17], the electrical networks must evolve to smart grids, integrating as much as possible clean and renewable generation sources; electric vehicles, with storage capacity; as well as elements of information technology capable of providing technical support for safe and economic operation of networks and transfer and data management [4–6, 14, 19, 20].

The urban electrical distribution networks are mainly operated in a not looped configuration; also most of them are looped. Thus, the design of weakly meshed networks is discussed in [1, 2]. Considering that now we are assisting to the appearance of new residential areas, clusters of buildings that for eco-friendly

V. Dumbrava (✉) • T. Miclescu • G.C. Lazaroiu
University POLITEHNICA of Bucharest, Bucharest, Romania
e-mail: v_dumbrava@yahoo.com; clazaroiu@yahoo.com

A. Karakitsiou et al. (eds.), *City Networks*, Springer Optimization and Its Applications 128, DOI 10.1007/978-3-319-65338-9_12

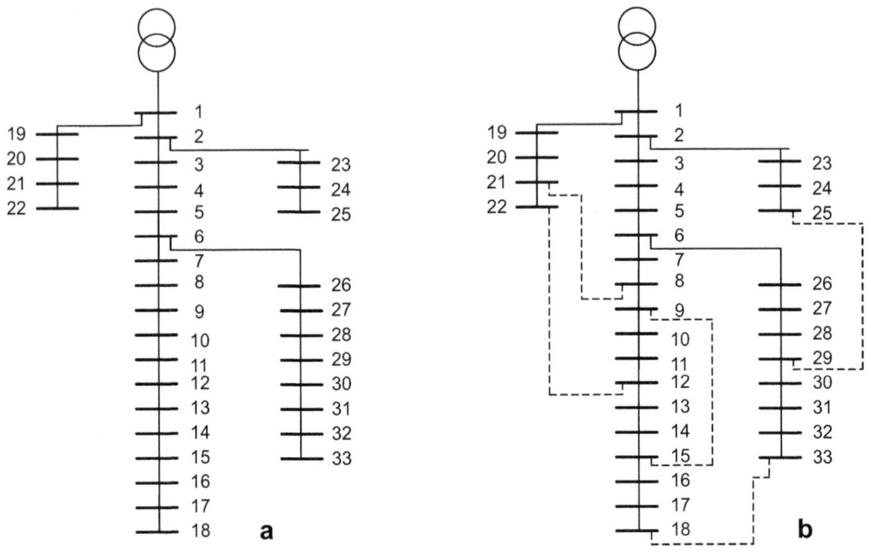

Fig. 12.1 Test system of 33-bus distribution network (**a**) radial, (**b**) weakly meshed

objectives located in green areas, where the density of consumers is not very high, the design of radial distribution networks is justified [3, 15, 16, 18] (Fig. 12.1).

The chapter describes a method of optimizing urban networks using the idealized network concept. This concept involves a layer structure of distribution networks and fulfilling common principles.

The optimum number of medium-voltage/low-voltage (MV/LV) transformer stations, supply points, and high-voltage/medium-voltage (HV/MV) transformer stations is obtained for an urban network whose average power consumption and the distance between two consumers are known.

Given the power demands of known future consumers and their location, the optimal solution is studied for:

- The location and capacity of MV/LV transformer substations as well as of the low-voltage network supplying consumers
- The location of the medium-voltage supply points and their connections to the MV/LV transformer stations
- The location and capacity of the HV/MV transformer stations and the network connecting them with the supply points

All these results, location, number, and capacity of the nodes and the configuration of the network corresponding to the three electrical distribution levels must represent an optimal structure.

The Idealized Network

Finding a solution to this problem represents a challenging task. By knowing the optimum capacity and location for transformer stations, supply points, and MV/LV transformer stations, the problem becomes simpler, reducing to the determination of optimal configuration for the three networks. A method of idealization allowing to solve the problem formulated in the previously presented way is proposed.

The idealized network is represented by a grid whose consumers have the same capacity and are placed in the vertices of regular, simple (squares or equilateral triangles). The side length of the simple figure forming the network is called the pitch of the network. The characteristics of the idealized network are therefore the demand at each node and the pitch of the network. To pass from the real network to the idealized network, the total demand of the real network is divided by the number of consumers, thus obtaining the average power of a consumption center:

- Considering the consumers placed in the nodes of a rectangular grid within a square having the same area as the real network, the pitch of the network (the distance between two nodes of the idealized network) can be determined with

$$L_1 = \sqrt{S}/\left(\sqrt{N} - 1\right) \tag{12.1}$$

where L_l is the first-order network pitch (the base network), S the surface area corresponding to the actual network, and N the number of consumer centers.

The idealized network is used in the same way for determining the optimal structure for each of the three superimposed networks.

The optimum LV network is determined, for example, in the case where a transformer supplies 1, 2, 3, or more consumers. The resulting network is made up of identical components composed of a transformer and the optimal network around it.

The optimum network of supply points is determined for the situation where a supply point is connected with 1, 2, 3, or more transforming stations. The capacity of the transformer station and the distance between two stations were obtained successively in the previous stage. The same is done for obtaining the optimum network, when a transformer station supplies several consumption nodes.

In each of these three successive steps, the characteristics of the idealized network are the results of the previous step: the demand in the node (consumer, transformer point, supply point) and the distance between two nodes.

In each step, the optimal structure of the elementary network, for a source which supplies 1, 2, 3, or several nodes, remains invariable.

All the possibilities of realization of the idealized network structure are found in a three-level tree presented in Fig. 12.2.

In Fig. 12.2, n_c represents the number of consumers supplied by a transformer point (first-order network), n_{pT} represents the number of transformer points fed by a supply point (2nd order network), and n_{pA} is the number of feed points linked to a transformer station (3rd order network).

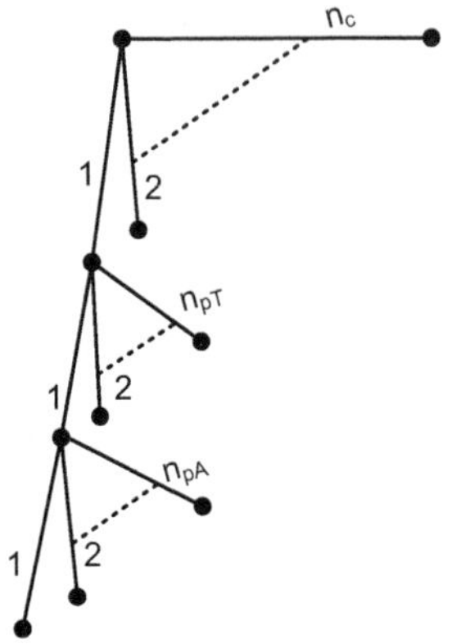

Fig. 12.2 Three-level tree

In the case where at most 20 branches are connected to a transformer point, at most ten branches are connected to a supply point, and at most 20 branches are connected to a transformer station, up to 4000 variants must be analyzed.

Adopting an optimization criterion, we determine the optimal solution of the idealized network corresponding to the real network. This is expressed by the optimum number of consumers supplied from a supply point and by the optimum number of supply points connected to a transformer station.

Using the results afore obtained and knowing the number of consumers of the real network, the optimal number of transformer points, supply points, and transformer stations for the real network can be determined.

The location of transformer stations, supply points, and transformer stations is established through a simulation process.

If the consumers were uniformly distributed over the considered surface, the simulation process would also be uniform. In reality, the distribution of consumers is variable, and this fact must be reflected in the process of random simulation. The simulation is realized by dividing the real surface into equal areas, to which probabilities weighted by the number of consumers of the respective area are matched (the Monte Carlo method).

Once a location variant is established, the associated supply zone is determined so that the source consumption point distance is minimum (Fig. 12.3).

By establishing the supply area, it results the capacity of the source (transformer point, supply point, transformer station). Then is determined the partial tree of minimum cost. If to this tree it corresponds a value of the optimization criterion smaller than that of the radial tree, the network layout corresponding to the partial tree of minimum cost is adopted as optimal solution (for the partial tree of the minimum cost, the values of the currents are not previously known). The process is repeated identically for the other two networks.

By repeating the simulation process, the value of the optimization criterion corresponding to the optimal solution can be reduced.

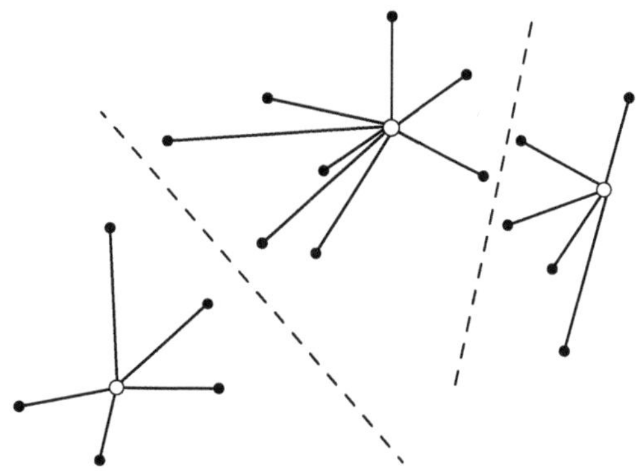

Fig. 12.3 Distribution of consumers on power sources

Mathematical Model and Optimization Criterion

The criterion according to which the optimal configuration of the distribution network is studied will be that of annual minimum calculation costs:

$$\min Z_{calc} = \min (C + p_n \cdot I_v) \tag{12.2}$$

where C is the annual production cost, p_n the value fixed for the relative coefficient of economic efficiency of investments, and I_v investment in the network.

The annual costs C are composed of operating costs, repair, and maintenance costs, C_L, plus the cost of electrical energy losses, C_p:

$$C = C_L + C_p \tag{12.3}$$

The operating costs, the repair, and maintenance costs can be expressed as follows:

$$C_L = k_1 \cdot I_v = k_1 \cdot L \cdot i_L = k_2 \cdot L \tag{12.4}$$

where k_1 is the rating for operation, maintenance, and repair and calculated as a percentage of investments, L the length of the line, i_L the line specific investment of a certain nominal voltage, and k_2 coefficient accounting for the exploitation, maintenance expenses, as well as investment.

The cost of electrical energy losses can be calculated using the formula

$$C_p = \Delta P \cdot c_p \cdot \tau = 3 \cdot R \cdot I^2 \cdot c_p \cdot \tau = 3 \cdot \rho \cdot L/s \cdot I^2 c_p \cdot \tau = 3 \cdot \rho \cdot L \cdot j \cdot I \cdot c_p \cdot \tau, \tag{12.5}$$

where ρ represents the electrical resistivity, L the length of the line, j the density of the electric current, I the intensity of the electric current, c_p the specific cost of electrical energy losses, The compute time of the power losses.

In the case where the whole network is made of the same material, with lines having the same current density, the cost of losses can be expressed by

$$C_p = k_3 \cdot L \cdot I. \tag{12.6}$$

The investment in lines (depending on the cross-section of the active conductors and the voltage level used) is

$$I_v = (k_4 + k_5 \cdot s + k_6 \cdot U) \cdot L = \left(k_{4,6} + k_5 \cdot \frac{I}{j}\right) \cdot L = (k_{4,6} + k\prime_5 \cdot I) \cdot L. \tag{12.7}$$

Hence,

$$\begin{aligned} Z_{calc} &= k_2 \cdot L + k_3 \cdot L \cdot I + p_n \cdot (k_{4,6} + k_5 \cdot I) \cdot L \\ &= (k_2 + p_n \cdot k_{4,6}) \cdot L + (k_3 + p_n \cdot k\prime_5) \cdot L \cdot I = A \cdot L + B \cdot L \cdot I \end{aligned} \tag{12.8}$$

Consequently, for determining the total costs necessary for the comparison of the power distribution network variants, apart from the constants A and B, the electrical current I absorbed by the consumers is required and the pitch length L of the consumer network.

Case Studies

Two basic types of networks have been studied which ensure a uniform distribution for the location of consumers: a network where consumers are placed in the vertices of a square (rectangular network), and a network where consumers are placed in the vertices of an equilateral triangle (triangular network). The results obtained for the two types of network are presented in the following.

The Rectangular Idealized Network

A statistical method was used to calculate the pitch of the higher-order idealized network, for example, a second-order network (MV network), starting from the first-order network (LV network).

The method consists in filling a surface, as large as possible, with network elements identical and arbitrarily located, such that all nodes of the basis network are contained in the elementary networks. Then the side of the mean square whose vertices are occupied by nodes of the network of higher order is calculated, with expression (12.9)

$$L_{n+1} = \frac{N_n \cdot L_n}{\sqrt{N_{n+1}} - 1} \tag{12.9}$$

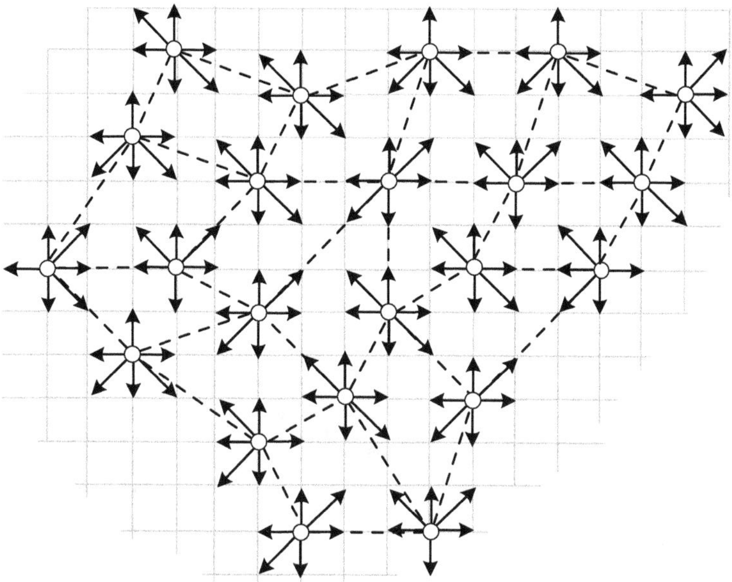

Fig. 12.4 Elementary rectangular network with seven consumers per power source

where N_n represents the number of nodes for the elementary network, on the calculation surface; L_n is the pitch length of the starting network; N_{n+1} is the number of nodes of the higher-order network; and L_{n+1} is the average pitch length of the higher-order network.

The method is illustrated in Fig. 12.4, in the case where the elementary network supplies seven consumers. The average pitch length of the higher-order network tends to a determinable value as the surface increases. In fact, the area of the surface around a node of the starting network is equal to $L_n{}^2$.

The area of the surface all around a node of the higher-order network, supplying N nodes of the basis network, is therefore

$$L_{n+1}^2 = N_n \cdot L_n^2 \tag{12.10}$$

From (12.10) is obtained

$$L_{n+1} = \sqrt{N_n} \cdot L_n \tag{12.11}$$

Table 12.1 and Fig. 12.5 show the value for L_{n+1} obtained by the statistical method, $L_{n+1,STAT}$ and the value for L_{n+1} when the surface grows indefinitely, $L_{n+1,LIM}$.

For calculating the optimal structure, the value of the annual computed costs that corresponds to the elementary rectangular network, $Z_{R,CALC}$, must be known. This

Table 12.1 Values of the average step of the higher-order network L_{n+1} for the two hypotheses

N_n	$L_{n+1,STAT}$	$L_{n+1,LIM}$	N_n	$L_{n+1,STAT}$	$L_{n+1,LIM}$
1	1.000	1.000	8	3.039	2.828
2	1.430	1.414	9	3.000	3.000
3	1.796	1.732	10	3.366	3.162
4	2.000	2.000	11	3.532	3.317
5	2.236	2.236	12	3.568	3.464
6	2.644	2.449	13	3.606	3.606
7	2.838	2.646	14	3.742	3.742

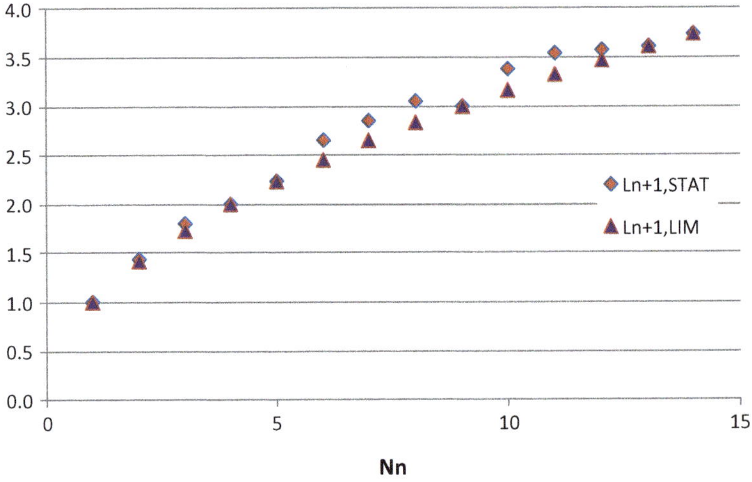

Fig. 12.5 Variation of L_{n+1} in relation to N_n for the two hypotheses

value depends on the number of nodes of the successive higher-order network. For example, for $N_n = 4$, these costs have the expression

$$Z_{R,\ CALC} = 2.828 \cdot A \cdot L + 2.828 \cdot B \cdot L \cdot I \tag{12.12}$$

where L is the pitch length of the starting network and I is the electrical current intensity absorbed by the consumer located in the nodes of the idealized network.

For $N_n \geq 6$ there are different modes for power supplying the consumers of the elementary networks, each supplying mode having different values for $Z_{R,\text{CALC}}$.

The optimum realization of the elementary network depends on the particular values of the measures of A, B, and I. Thus, for example, for $N_n = 17$, we can realize networks similar to those in Fig. 12.6.

For the two supplying modes, the calculations of cost values are given by

$$Z_{R,\ CALC,\ a} = 22.6 \cdot A \cdot L + 26.6 \cdot B \cdot L \cdot I \tag{12.13}$$

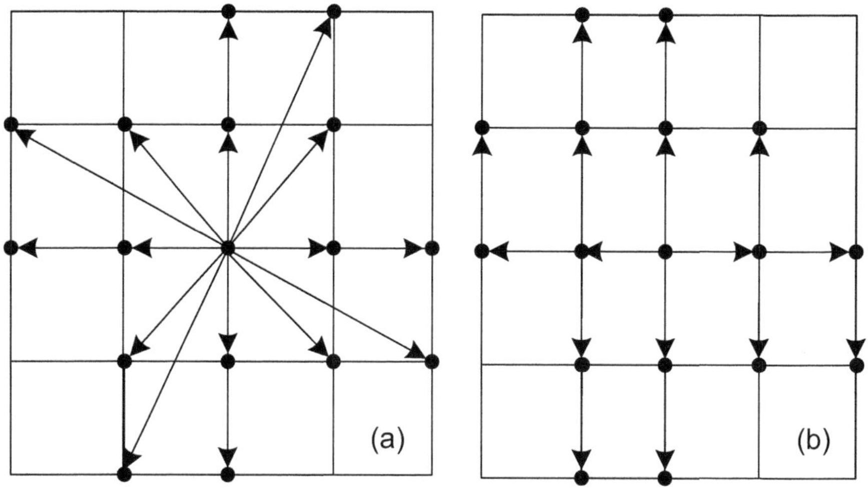

Fig. 12.6 Elementary rectangular networks for $N_n = 17$

Table 12.2 Values of annual computed costs for the elementary network shown in Fig. 12.6

N_n	$Z_{R,\ CALC,\ a}$	$Z_{R,\ CALC,\ b}$	$I_{CRIT} \cdot B/A$
6	$5.41\ AL + 5.41\ BLI$	$5\ AL + 6\ BLI$	0.707
7	$6.82\ AL + 6.82\ BLI$	$6\ AL + 8\ BLI$	0.707
8	$8.24\ AL + 8.24\ BLI$	$7\ AL + 10\ BLI$	0.707
9	$9.64\ AL + 9.64\ BLI$	$8\ AL + 12\ BLI$	0.707
10	$10.64\ AL + 11.64\ BLI$	$9\ AL + 14\ BLI$	0.707
11	$11.66\ AL + 13.66\ BLI$	$10\ AL + 16\ BLI$	0.707
12	$12.66\ AL + 15.66\ BLI$	$11\ AL + 18\ BLI$	0.707
13	$13.66\ AL + 17.66\ BLI$	$12\ AL + 20\ BLI$	0.707
14	$15.89\ AL + 19.89\ BLI$	$13\ AL + 23\ BLI$	0.931
15	$18.13\ AL + 22.13\ BLI$	$14\ AL + 26\ BLI$	1.067
16	$20.36\ AL + 24.36\ BLI$	$15\ AL + 29\ BLI$	1.158
17	$22.60\ AL + 26.60\ BLI$	$16\ AL + 32\ BLI$	1.223
18	$24.84\ AL + 28.84\ BLI$	$17\ AL + 35\ BLI$	1.272
19	$27.07\ AL + 31.07\ BLI$	$18\ AL + 38\ BLI$	1.310
20	$29.31\ AL + 33.31\ BLI$	$19\ AL + 41\ BLI$	1.340

$$Z_{R,\ CALC,\ b} = 16 \cdot A \cdot L + 32 \cdot B \cdot L \cdot I \tag{12.14}$$

If $I > 1.223\ A/B$, configuration (b) is more economical than (a) and conversely in the opposite case.

Table 12.2 reports the values for $Z_{R,\ CALC,\ a}$, $Z_{R,\ CALC,\ b}$, and I_{CRIT} for values of N_n between 6 and 20.

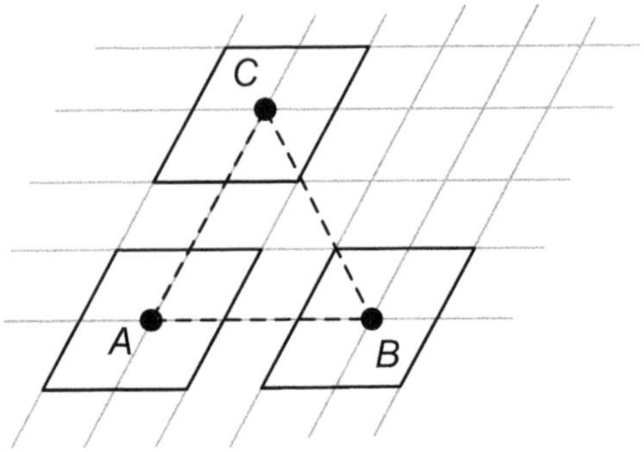

Fig. 12.7 Elementary triangular network for $N_n = 6$

Triangular Idealized Network

For calculating the pitch of the higher-order network, a deductive method and a statistic method are chosen. The deductive method for determining the pitch for the network with sources consists in building the layouts of consumer supply from the sources, obtaining a triangular network of sources, more are less regularly, function of the supplied consumers. Taking $N_n = 6$, the case occurring in Fig. 12.7 is obtained.

The area of the triangle *ABC* from Fig. 12.7 is $S = L_{n+1}^2\sqrt{3}/4$, where L_{n+1} is the edge of the equilateral triangle *ABC*. It can be expressed as function of the area of equilateral triangle of side L_n as

$$\begin{aligned} S = \frac{L_{n+1}^2\sqrt{3}}{4} &= \left(\frac{N}{\sqrt{N_n}} - 1\right)^2 \cdot \frac{L_{n+1}^2\sqrt{3}}{4} + \left(\frac{2 \cdot N_n}{\sqrt{N_n}} - 1\right)^2 \cdot \frac{L_n\sqrt{3}}{4} \\ &= \underbrace{\left(\sqrt{N_n - 1}\right)^2 \cdot \frac{L_{n+1}^2\sqrt{3}}{4}}_{A} + \underbrace{\left(2 \cdot \sqrt{N_n}-1\right)^2 \cdot \frac{L_n\sqrt{3}}{4}}_{B} = \frac{N_n \cdot L_n^2\sqrt{3}}{4} \end{aligned} \tag{12.15}$$

Therefore,

$$\frac{L_{n+1}^2\sqrt{3}}{4} = \frac{N_n \cdot L_n^2\sqrt{3}}{4} \tag{12.16}$$

where

$$L_{n+1} = L_n \cdot \sqrt{N} \tag{12.17}$$

Therefore, the pitch of the network of sources depends on the pitch of the network of consumers supplied from a source. In (12.15), the terms *A* and *B* represent *A* the part of the area of the triangle *ABC* contained within the contours of the supply scheme for N consumers from a sources and *B* the part of the area of the triangle *ABC* formed by the area contained between the contours of the supply scheme.

The principle of the statistical method is as follows: all the possible schemes for supplying 1, 2 … N_n consumers are realized (when using a type of diagram, all

Table 12.3 Values of the average step of the higher-order network $L_n + 1$ for the two hypotheses and for the triangular network

N_n	$L_{n+1,LIM}$	$L_{n+1,STAT}$	Er (%)	N_n	$L_{n+1,LIM}$	$L_{n+1,STAT}$	Er (%)
1	1.00	1.00	0	11	3.31	3.34	0.90
2	1.41	1.45	2.83	12	3.46	3.52	1.73
3	1.73	1.68	02.89	13	3.60	3.60	0
4	2.00	2.00	0.00	14	3.74	3.75	0.26
5	2.23	2.34	4.93	15	3.87	3.98	2.84
6	2.45	2.54	3.67	16	4.00	4.00	0.00
7	2.64	2.64	0.00	17	4.12	4.21	2.18
8	2.82	2.91	3.19	18	4.24	4.34	2.35
9	3.00	3.00	0	19	4.35	4.35	0
10	3.16	3.23	2.21	20	4.47	4.52	1.11

consumers must be supplied). Then, the sources are arbitrarily combined to form an array of triangles, and the average length of the side of the equilateral triangle is determined as the arithmetic mean of lengths of all sides of triangles found for a given number of consumers. This equilateral triangle constitutes the cell of the network of sources.

For $N_n = 20$, Table 12.3 reports the results obtained using the two methods described above.

From the aforementioned, the criterion for selecting the optimal variant of the supply scheme is the criterion of the minimum annual calculation costs.

$$\min Z_{\mathrm{CALC}} = \min\left(A \cdot I + B \cdot L \cdot I\right) \tag{12.18}$$

The variants of the schemes have been sorted according to the criterion mentioned, for $N_n \leq 20$. Starting from $N_n = 8$, the optimum variant of the diagram is obtained as function of the particular values for A, B, and I.

In Table 12.4, some variants of supply schemes with the values of the objective function are reported.

The correlation between the MV distribution network and the LV electrical network is illustrated in Fig. 12.8.

Conclusions

Using the idealized network concept, the values of the pitch of the superimposed networks and of the annual cost for the elementary networks of these superimposed networks have been calculated as a function of the number of the nodes of the elementary network. The optimal structure of the elementary network depends on the number of nodes of the elementary network and the relation of the two coefficients depending on the fixed part of the operating costs, the part of the investments

Table 12.4 Supply scheme variant and value of objective function

$N_n = 5$	Supply scheme variant	$Z_{R,\ CALC}$
5	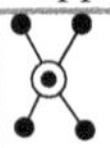	$4L_n \cdot (A + B \cdot I)$
12	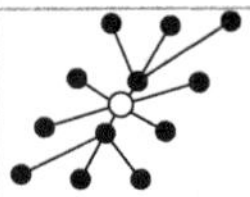	$11.37L_nA + 14.37L_nBI$
	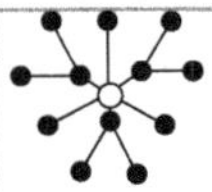	$11.19L_nA + 14.66L_nBI$
	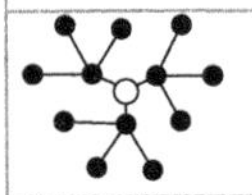	$10.73L_nA + 15.92L_nBI$
	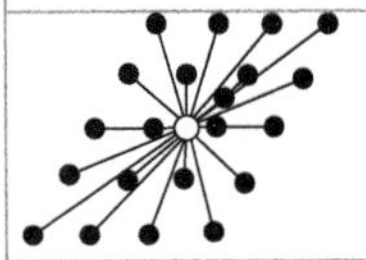	$30.75L_nA + 31.25L_nBI$
	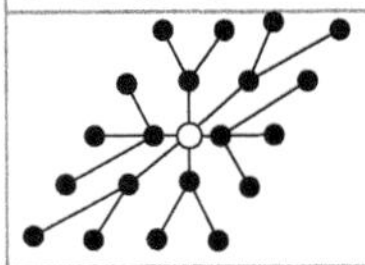	$22.30L_nA + 34.06L_nBI$
20	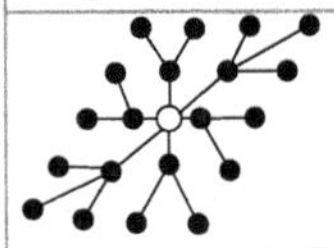	$20.84L_nA + 34.24L_nBI$
	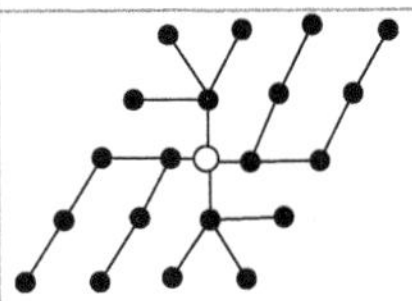	$18.73L_nA + 36.92L_nBI$
	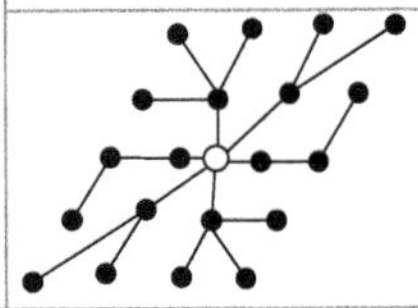	$20.19L_nA + 36.39L_nBI$

independent of the section, the cost of electrical energy, the material used for the conductor, the used electrical current density, and the part of investments variable with the cross-section. This structure does not depend on the pitch value of the starting network.

From the results obtained, the proposed method for the optimization of urban networks within the future smart cities can be applied.

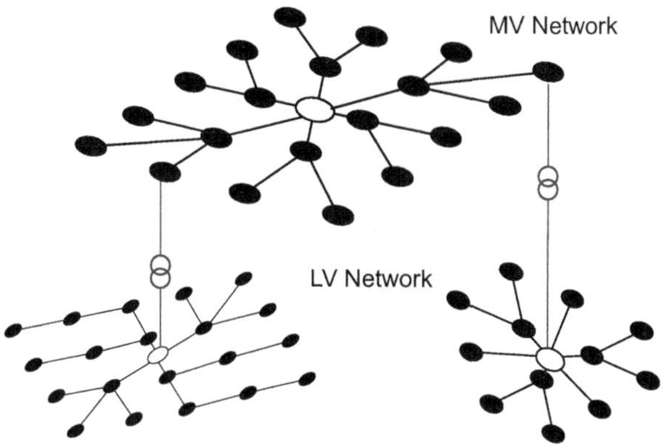

Fig. 12.8 Correlation between the MV and LV networks

References

1. Alvarez-Herault, M.C., N'Doye, N., Gandioli, C., et al.: Meshed distribution network vs reinforcement to increase the distributed generation connection. Sustain. Energy Grids Netw. **1**, 20–27 (2015)
2. Celli, G., Pilo, F., Pisano, G., et al.: Meshed vs. radial MV distribution network in presence of large amount of DG. In: power systems conference and exposition, New York, 10–13 Oct. 2004 (2004)
3. Dumbrava, V., Bazacliu, G., Nicoara. B., et al.: Load profiling for the electricity market. In: 43rd international universities power engineering conference, Padova, 1–4 Sept. 2008 (2008)
4. Dumbrava, V., Lazaroiu, G.C., Roscia, M., et al.: Expansion planning and reliability evaluation of distribution networks by heuristic algorithms. In: 10th international conference on environment and electrical engineering, Rome, 8–11 May 2011 (2011)
5. Georgilakis, P.S., Hatziargyriou, N.D.: Optimal distributed generation placement in power distribution networks: models, methods, and future research. IEEE Trans. Power. Syst. **28**(3), 3420–3428 (2013)
6. Georgilakis, P.S., Hatziargyriou, N.D.: A review of power distribution planning in the modern power systems era: models, methods and future research. Electr. Power Syst. Res. **121**, 89–100 (2015)
7. Haffner, S., Pereira, L.F.A., Pereira, L.A., et al.: Multistage model for distribution expansion planning with distributed generation – Part I: problem formulation. IEEE Trans. Power Deliver. **23**, 915–923 (2008)
8. Haffner, S., Pereira, L.F.A., Pereira, L.A., et al.: Multistage model for distribution expansion planning with distributed generation – Part II: numerical results. IEEE Trans. Power Deliver. **23**, 924–929 (2008)
9. Humayd, S.B., Bhattacharya, K.: Distribution system planning to accommodate distributed energy resources and PEVs. Electr. Power Syst. Res. **45**, 1–11 (2017)
10. Lazaroiu, G.C., Dumbrava, V., Costoiu, M., et al.: Energy-informatic-centric smart campus. In: 2016 IEEE 16th international conference on environment and electrical engineering, Florence, 7–10 June 2016 (2015)
11. Lazaroiu, G.C., Dumbrava, V., Costoiu, M. et al.: Smart Campus-an energy integrated approach. In: 4th international conference on renewable energy research and applications, Palermo, 22–25 Nov 2015 (2015)
12. Nahman, J., Peric, D.: Path-set based optimal planning of new urban distribution networks. Int. J. Electr. Power. **85**, 42–49 (2017)

13. Setis–European Initiative on Smart Cities.: https://setis.ec.europa.eu/set-plan-implementation/technology-roadmaps/european-initiative-smart-cities (2017). Accessed 20 Mar 2017
14. Sharma, I., Canizares, C., Bhattacharya, K.: Smart charging of PEVs penetrating into residential distribution systems. IEEE Trans. Smart Grid. **5**(3), 1196–1209 (2014)
15. Soedibyo, A., Ashari, M., Syahputra, R.: Power loss reduction strategy of distribution network with distributed generator integration. In: IEEE information technology, computer and electrical engineering, Semarang, 8 Nov 2014 (2014)
16. Syahputra, R.: Fuzzy multi-objective approach for the improvement of distribution network efficiency by considering DG. Int. J. Comput. Sci. Inf. Technol. **4**(2), 57–68 (2012)
17. The European Innovation Partnership on Smart Cities and Communities.: http://ec.europa.eu/eip/smartcities/ (2017). Accessed 20 Mar 2017
18. Willis, H.L.: Power Distribution Planning Reference Book. CRC Press, New York (2010)
19. Zeng, B., Zhang, J., Yang, X., et al.: Integrated planning for transition to low-carbon distribution system with renewable energy generation and demand response. IEEE Trans. Power Syst. **29**(3), 1153–1165 (2014)
20. Ziari, I., Ledwich, G., Ghosh, A., et al.: Optimal distribution network reinforcement considering load growth, line loss, and reliability. IEEE Trans. Power Syst. **28**(2), 587–597 (2013)

Chapter 13
Sustainable Urban Water Management

Antigoni Zafirakou

Abstract Sustainability in urban water management is of utmost importance. The distribution of water should depend not only on the availability of the valuable natural resources but also on its efficient use. Water quality plays an equally significant role since if different water uses are taken into account could provide a more sustainable water distribution. We establish that water quantity and quality should be jointly studied and managed in order to acquire a beneficial sustainable water environment for urban areas.

Keywords Water management • Water sustainability

Introduction

> "Urbanization brings opportunities for more efficient water management and improved access to drinking water and sanitation. At the same time, problems are often magnified in cities, and are currently outpacing our ability to devise solutions."
>
> Ban Ki-moon, UN Secretary General, 2014

The exploding urban population growth creates unprecedented challenges, among which provision for water and sanitation for millions of people. Even though water supply networks are as ancient as human civilizations, water management is as urgent as ever, since water is not distributed evenly on Earth and among countries. Climate change is always an additional source of uncertainty in water quantity prediction and water supply planning. Therefore, sustainability in urban water management is prominent to arid or remote regions, as well as to big industrial cities. The distribution of water should depend not only on the availability of this valuable natural resource but also on its efficient use. On the other hand, water quality plays an equally significant role, as different water uses set different requirement standards, which, if taken into account, would provide a more sustainable distribution of water on Earth. As it will be developed in this

A. Zafirakou (✉)
Civil Engineering Department, Faculty of Engineering, Aristotle University of Thessaloniki, 54124, Thessaloniki, Greece
e-mail: azafir@civil.auth.gr

A. Karakitsiou et al. (eds.), *City Networks*, Springer Optimization and Its Applications 128, DOI 10.1007/978-3-319-65338-9_13

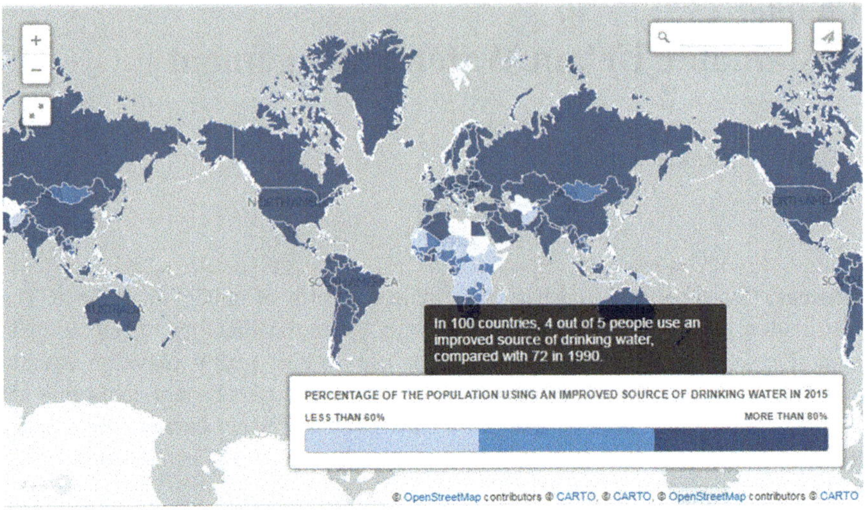

Fig. 13.1 Percentage of population using an improved source of drinking water in 2015 [1]

chapter, water quantity and quality should be jointly studied and managed in order to acquire a beneficial sustainable water environment for urban areas. Coping with the growing needs of water and sanitation services within cities is one of the most crucial issues of this century.

Half of humanity now lives in cities, and by the middle of the twenty-first century, it is projected that it will approach 60%. Urban growth is more eminent in the developing world, where cities are augmented by an average of five million residents every month. Global water demand is expected to increase, as countries develop and populations grow. The growing population creates problems, such as water deficiency and deterioration of water quality, which at the same time invokes challenges to the scientific and engineering community to develop new methods to better manage the water resources. Sustainable, efficient, and equitable management of water in cities has never been as important as it is now.

Sustainability in cities cannot be met without ensuring reliable access to safe drinking water and sanitation. The latest published percentages of the population using an improved source of drinking water and improved sanitation facilities in 2015 are shown in Figs. 13.1 and 13.2 [1]. However, less people are gaining access to drinking water and sanitation needs than the corresponding global population growth, as Fig. 13.3 clearly shows, based on data collected for approximately two decades, between 1990 and 2008 [2].

In most developed countries, access to safe water and sanitation is taken for granted. However, in the middle of the nineteenth century, big cities were hit by infectious diseases, due to misfortunate neglect of sanitation measures and legislations. Child death rates were as high as they are now in much of sub-Saharan Africa (see Figs. 13.1 and 13.2). The significant progress that was acquired was

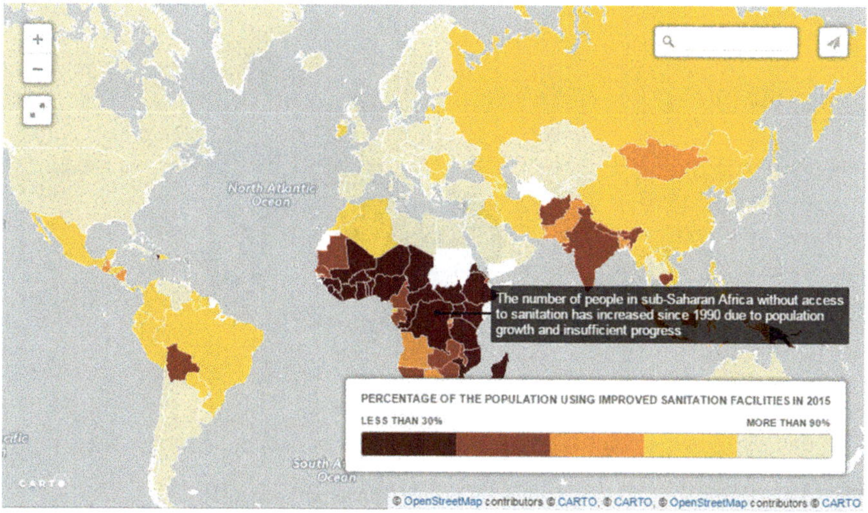

Fig. 13.2 Percentage of population using improved sanitation facilities in 2015 [1]

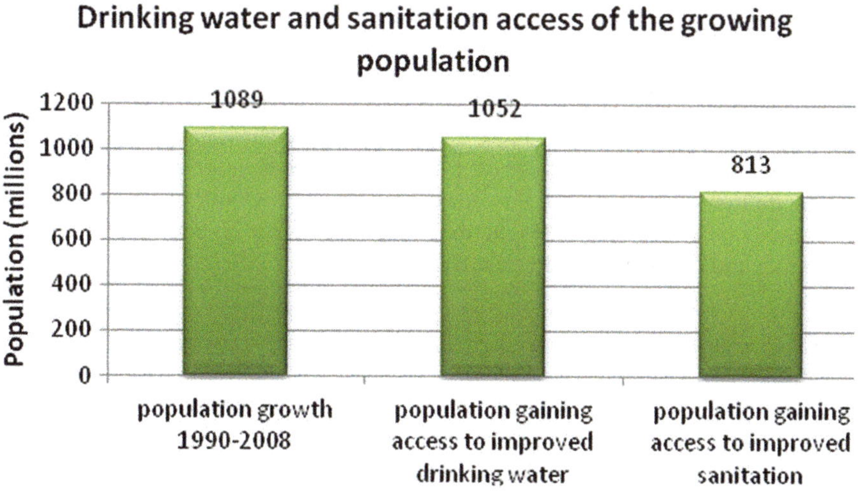

Fig. 13.3 Population gaining access to improved drinking water and sanitation vs population growth, 1990–2008, on a global scale

due to extensive reforms in water and sanitation. In fact, in a 2007 poll by the British Medical Journal, clean water and sanitation access was designated the most important medical advancement since 1840. On March 2015 Springer published a book on *Water Policy in Canada* which dedicated one chapter on health issues related to inefficient water management in developed countries, pinpointing two major waterborne disease outbreaks in the USA and Canada within a period of 11 years [3]. However, the developing countries are facing health-related problems

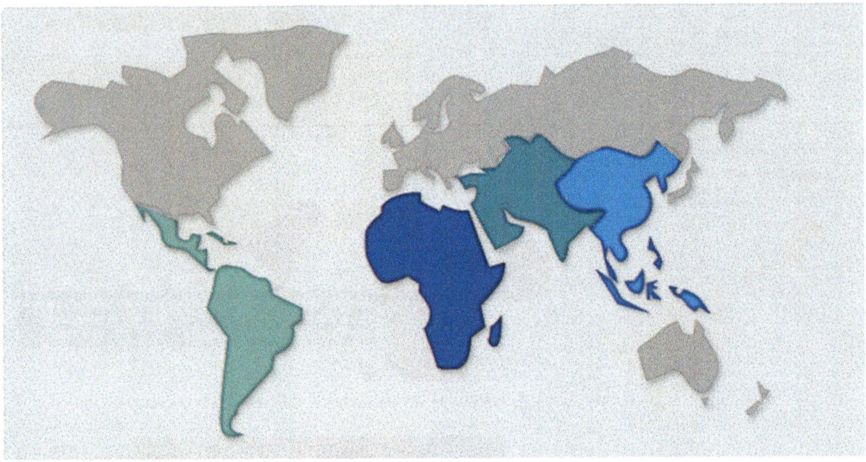

Fig. 13.4 The water crisis leaves millions of people without safe water [5]

continuously due to poor water quality. According to the World Health Organization and UNICEF [4], nowadays:

- In low- and middle-income countries, 1/3 of all healthcare facilities lack a safe water source.
- Water-related diseases affect more than 1.5 billion people every year; one million of them die each year due to water, sanitation, and hygiene-related diseases.
- 160 million children suffer from chronic malnutrition, poor quality water, and lack of sanitation. Diarrhea is the third leading cause of children's death. Every 90 s a child dies from a water-related disease.

In Fig. 13.4, the world is divided into regions that show the populations that lack access to water in descending order [5]. The Southern hemisphere, with the only exception of Australia, has still a long road ahead to reach sustainability in terms of water supply. Moreover, these regions are facing more rapid population growth, water scarcity, higher temperatures, and economic deficiencies (poverty).

The populations that lack access to water worldwide, in descending order, are:

- 332 million people in Africa
- 155 million people in South, West, and Central Asia
- 131 million people in Southeast, East Asia, and Oceania (except Japan and Australia)
- 32 million people in Latin America and the Caribbean
- 13 million people in developed countries (USA, Canada, Europe, Russia, Australia)

These global aggregates also pinpoint large inequalities between nations and regions, rich and poor, between rural and urban populations, as well as disadvantaged groups and the general population. It is also documented that:

- More than 1.7 billion people live in river basins where water use exceeds natural recharge, leading to the dehydration of rivers, depletion of groundwater, and the degradation of ecosystems.
- It is projected that by 2025, two thirds of the world's population could be living in water-stressed countries, if water consumption continues in the same ratio.

The water crisis is the number one global risk, based on impact to society (as a measure of devastation), as announced by the World Economic Forum in January 2015 [5, 6]. The outcome document adopted at Rio + 20 "The Future We Want" [7] states that "Water is at the core of sustainable development, as it is closely linked to a number of key global challenges." Water is vital for reducing the global burden of diseases and toward improving health, welfare, and productivity of populations. It is also central to the production and preservation of a whole of benefits and services for people. Water is also at the heart of adaptation to climate change, serving as the crucial link between the climate system, human society, and the environment.

The Millennium Development Goals (MDGs), agreed in 2000, set the goal to decrease the percentage of people without sustainable access to safe drinking water and basic sanitation to half, between 1990 and 2015. According to UNDESA [8]:

- From 1990 to 2015, 2.1 billion people progressively gained access to a latrine, flush toilet, or other improved sanitation facilities, helping to comprise a total of 68% of the global population.
- Nevertheless, 2.4 billion people (1 out of 3) still lack access to a toilet [5, 9].
- In 2015, however, 663 million people (1 out of 10) lack access to safe water and still withdraw water from unimproved sources [4, 5].

The United Nations declared the years 2005–2015 as the "Water for Life" decade [8]. Its goal was to promote efforts to fulfill international commitments on the globe by 2015. They tried to establish further cooperation between governments and other stakeholders, between nations and diverse communities, and between economic interests and the needs of ecosystems and the poor, in order to achieve the water goals of the Millennium Declaration, the Johannesburg Plan of Implementation of the World Summit for Sustainable Development and Agenda 21. In July 2010, the General Assembly adopted a resolution, which "recognized the right to safe and clean drinking water and sanitation as a human right that is essential for the full enjoyment of life and all human rights [10]." In September 2015, the General Assembly finally agreed on a water goal to "ensure the availability and sustainable management of water and sanitation for all." This reflects that water and sanitation has become a key priority for UN member states. Water should no longer be considered a "free resource", but rather a valuable resource for life.

The MDG target for sanitation was a pressing challenge. To reach the requirements of the right to access to safe drinking water requires real improvements for several billions of people. There are many smart ideas to be implemented by individuals and governments to resolve world's water crises, by securing and/or transforming water use.

Building a sustainable water supply system consists of a viable infrastructure for the collection, transfer, treatment, storage, and distribution of water for public and private use, in residential, commercial, industrial, and irrigational areas. Public use refers, but is not limited, to schools, hospitals, and other public buildings, but also firefighting, gardening, and street cleaning. Of all municipal needs, provision of potable water is undoubtedly the most important. Citizens depend on water for drinking, cooking, washing, and other domestic everyday needs. However, urban water supply systems must also cover quantitatively all other public, commercial, and industrial activities.

Public or private water supply companies worldwide set as their primary goal to provide to their consumers best quality water, for both urban and non-urban use. Back in the 1990s in the USA, however, it was raised as an issue whether that approach was the designated one. The possibility of running two parallel pipes of different quality water, to different destinations, was under serious consideration. Two decades later, small steps are made toward this direction, as it is very hard to apply it to big developed cities. It's rather appropriate to design new cities or extensions to old ones with a more sophisticated water supply system.

In 2004 the term water footprint (WF) was introduced by Prof. Arjen Y. Hoekstra and Dr. Ashok K. Chapagain at UNESCO-IHE [11], to assess how much water is used per person. Since then numerous publications focused on quantifying and mapping national WF, as a measurement of water use, under-use, or overuse [12–14]. It can also serve as an indicator of a nation's socioeconomic development. WF's increase indicates allegedly an escalation in living standards, as it is closely related to water consumed by an individual for his/her personal needs. Water footprint of an individual, community, or business is defined as "the total volume of freshwater that is used to produce the goods and services consumed by the individual or community or produced by the business." This is due to the fact that people use water for personal needs (drinking, cooking, and washing), but even more for producing things (such as food, paper, cotton clothes, etc.) [15]. The water footprint (WF) is an indicator of water use that looks at both direct and indirect water use of a consumer or producer [16]. Or else, the WF of national consumption is estimated as the sum of the direct WF of consumers and the indirect WF components of agricultural and industrial water use:

$$\text{WFcons} = \text{WFcons, dir} + \text{WFcons, ind (agricultural)} + \text{WFcons, ind (industrial)}$$

The water footprint of national consumption for countries (both developing and developed) with a population larger than five million, for the decade 1996–2005, distributed to agricultural, industrial, and domestic use [17], illustrates a wide range of 500–3500 m^3/yr/cap, which depicts the disparity in water use among nations. The extreme values can possibly be related to insufficient and/or unreliable collected data on consumption and water productivity in those (mainly developing) countries. The world average consumption is 1385 m^3/yr./cap. Water consumption for agricultural use is leading the global WF, by 92%, whereas that for industrial products and domestic use account only for 4.7% and 3.8%, respectively. This

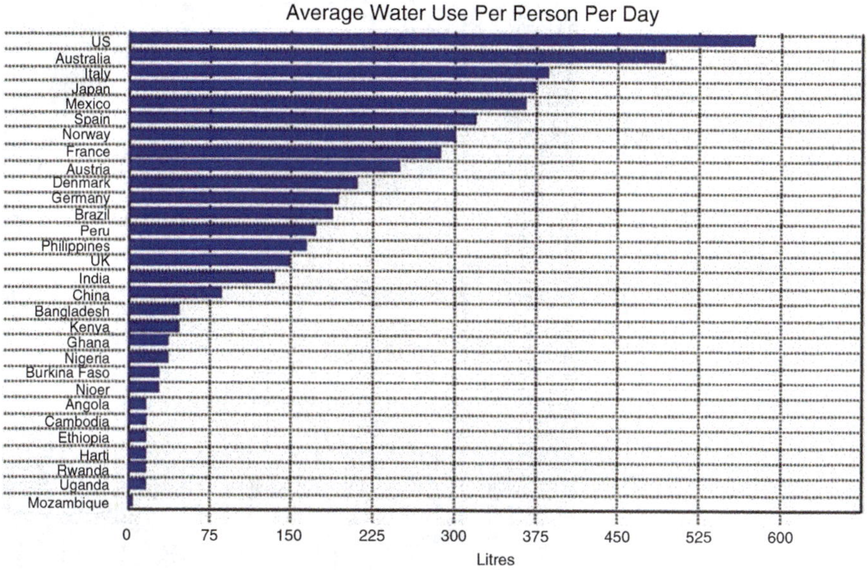

Fig. 13.5 Average daily water consumption per person for various countries [18]

ranking of countries also shows that industrialized countries have WFs per capita in the range of 1250–2850 m^3/y. The UK, with a WF of 1258 m^3/y, is at the low end of this range, whereas the USA, with a footprint of 2842 m^3/y, is at the high end. Surprisingly enough Greece is close to the high end, with approximately 2300 m^3/yr./cap, next to Canada and Australia.

In accordance with the aforementioned WF, the average daily water consumption per person, for different countries of the world, is given in the following more comprehensive chart [18]. Similarly, it is denoted that a person in the USA is consuming more than 550 L/d, whereas in the UK a citizen is consuming approximately 150 L/d (Fig. 13.5).

Conventional methods of estimating national water use take into account only water withdrawals. By introducing WF, as a measure of humans' appropriation of freshwater resources, in terms of domestic, agricultural, and industrial use, provokes countries in developing well-informed national policies, by extending these statistics to including data on rainwater use and volumes of water use for waste assimilation and by adding data on water use in other countries for producing imported products, as well as data on water use within the country for making export products [19]. Therefore WF can be distinguished into three components [17]: blue, green, and gray [19]. The blue WF refers to the consumption of blue water resources (surface and groundwater), whereby consumption refers to the volume of water that evaporates or is incorporated into a product. The blue WF is thus often smaller than the water withdrawal, because generally part of a water withdrawal returns to the ground or surface water. The green WF is the volume of green water (rainwater)

Fig. 13.6 Graphical representation of freshwater in rivers and lakes, with respect to the total liquid freshwater and the entire water in, on, and above the Earth [20]

consumed, which is particularly relevant to crop production. The gray WF is an indicator of the degree of freshwater pollution and is defined as the volume of freshwater that is required to assimilate the load of pollutants based on existing ambient water quality standards.

In addition to all that, there are countries that depend upon freshwater resources on other countries. Highly water-scarce countries that have a large external water dependency are in descending order, Malta, Kuwait, Jordan, Israel, the United Arab Emirates, Yemen, Mauritius, Lebanon, and Cyprus, with dependencies varying from 92–71%.

In order to eliminate the big discrepancies between nations in terms of the water use and provide an optimal water resources management, we need to incorporate all of the above valuable conclusions into an inquiry on sustainable water management, which involves (Fig. 13.6):

- Sustainable use of groundwater
- Optimal use of surface water
- Efficient use of stormwater/rainwater
- Desalinated sea water use, where available

Water is present in abundant quantities in our planet. However, as the above figure shows very brilliantly, liquid freshwater is a tiny proportion of the total available water on Earth. In the bibliography, it is well portrayed that most of Earth's water (estimated 1.4 billion km^3) is in the oceans or frozen in polar ice caps and glaciers. Less than 3% is freshwater and less than 1% of it is liquid freshwater, since of the total freshwater, more than 68% is sheltered in ice and glaciers. Another 30% of freshwater is confined in the ground. Fresh surface water sources, such as rivers and lakes, that are the main sources of water supply globally, only constitute about 1/150th of 1% of total water (about 93,100 km^3) [21]. Freshwater, containing less than 3 g/L of salts, has to satisfy all human needs. Unfortunately, it is not always available, as it is not uniformly distributed over the Earth. In many cases good quality water is deteriorated due to urban development, industrial growth, and environmental pollution. Ocean water, on the other hand, contains about 35 g/L of dissolved minerals or salts, making it unfit not only for drinking but also for most industrial or agricultural uses. It remains to be investigated whether desalinated water can satisfy all human needs. In addition to these natural resources of the water cycle that can be withdrawn, treated, and delivered to cities, there are alternative methods that should be investigated and applied in order to relieve arid or overpopulated areas. These alternative water resources include:

- Reuse of grey water
- Reuse of treated wastewater

In an effort to illustrate graphically a sustainable water network that can utilize all accessible sources and all available technology, Fig. 13.7 is created to establish a distribution network of different quality water to different end users. This optimization, of course, requires eloquent management and a robust economy, to satisfy water uses at all levels. This interrelated series of water and energy transfer is the suggested management for a sustainable city network.

Surface and Groundwater

Water was playing a significant role in the selection of the location for the earliest settled communities, and the evolution of public water supply systems is tied directly to the urban development. In the exploitation of water resources beyond their natural condition in rivers, lakes, and springs, the digging of shallow wells was probably the earliest innovation. Wells provided at the same time the necessary storage area and protection from environmental factors, as in the ancient underground storage tanks of Athens (Fig. 13.8a). The need to channel water supplies from distant sources was an outcome of the growth of urban communities as in Athens, Pisistratio aqueduct 600bC, and the Roman aqueducts of Nikolopis, first or second century aC, or Kavala, 1530aC (Fig 13.8b, c, d, respectively).

Surface water and groundwater are both valuable sources for urban water supply needs. Groundwater is less susceptible to bacterial pollution than surface water, because the soil and rocks, through which groundwater flows, filter most of the bacteria. The chemical and biological nature of groundwater is acceptable for most uses. Groundwater may contain dissolved minerals, as it dissolves them from the rocks with which it comes in contact, and gases, that give it the tangy

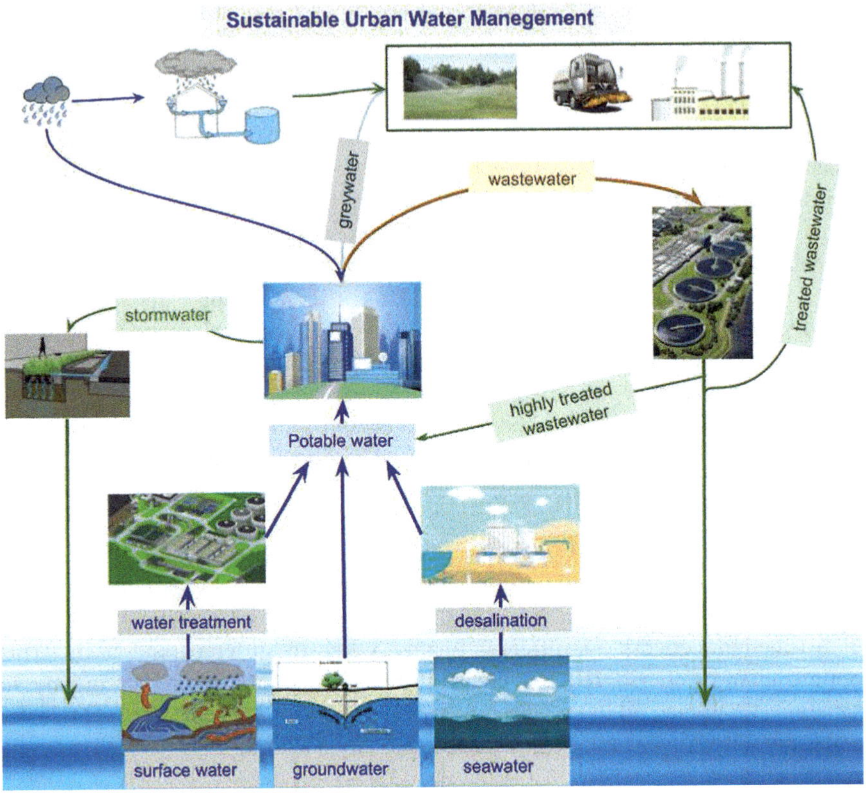

Fig. 13.7 Proposed water management for sustainable city networks

taste enjoyed by many people. Without these minerals and gases, the water would taste flat. Many unseen dissolved mineral and organic constituents are present in various concentrations, such as sodium, calcium, magnesium, potassium, chloride, bicarbonate, and sulfate. Most are harmless or even beneficial; though occurring infrequently, others are harmful, and a few may be highly toxic. The quality of groundwater, in particularly shallow grounds, is changing as a result of human activities. Bacteria occasionally find their way into groundwater, sometimes in dangerously high concentrations.

The value of an aquifer as a source of groundwater is a function of the porosity of the geologic stratum, or layer, of which it is formed. Water is withdrawn from an aquifer by pumping it out of a well or infiltration gallery. An infiltration gallery typically includes several horizontal perforated pipes radiating outward from the bottom of a large-diameter vertical shaft. Wells are constructed in several ways, depending on the depth and nature of the aquifer (Fig. 13.9). Wells used for public water supplies, usually more than 30 m deep and 10–30 cm in diameter, must penetrate large aquifers that can provide dependable yields of good quality water.

Fig. 13.8 Hydraulic works from ancient Greece: (**a**) Athens (www.tovima.gr/culture/article/), (**b**) Athens (personal archives), (**c**) Preveza (apeirosgaia.wordpress.com), (**d**) Kavala (personal archives)

They are drilled using impact or rotary techniques and are usually lined with a metal pipe or casing to prevent contamination. The annular space around the outside of the upper portion of the casing is filled with cement grout, and a special sanitary seal is installed at the top to provide further protection. At the bottom of the casing, a slotted screen is attached to strain silt and sand out of the groundwater. A submersible pump driven by an electric motor can be used to raise the water to the surface. Sometimes a deep well may penetrate a confined artesian aquifer, in which case natural hydrostatic pressure can raise the water to the surface.

Rainwater and Stormwater Management

The volume of water available for municipal supply depends mostly on the amount of rainfall. It also depends on the size of the watershed, the slope of the ground, the type of soil and vegetation, and the type of land use. Figure 13.10 shows global average annual precipitation, in millimeters and inches, where dark blue represents high and light green low precipitation, in areas that can be considered "deserts" [22].

Centuries ago, people in arid and semiarid regions used to collect rainwater for their daily needs or for irrigating their crops during extended dry periods.

Fig. 13.9 Public water fountains and wells in Greece

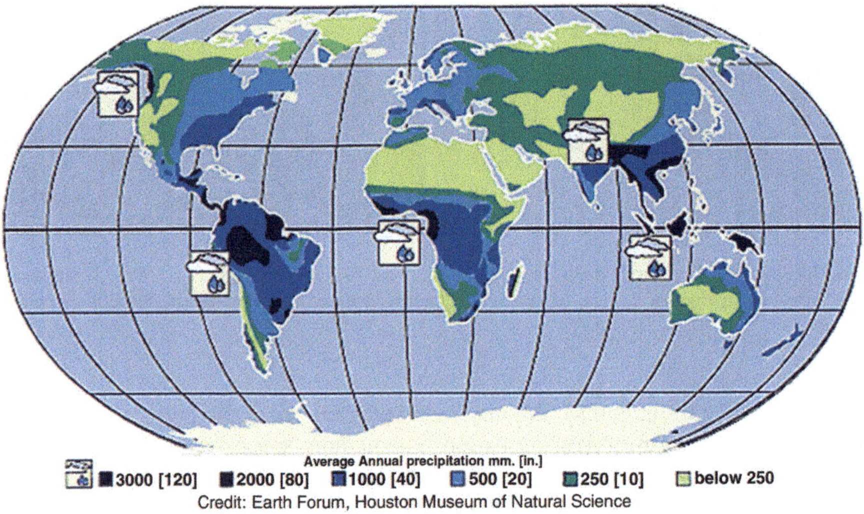

Fig. 13.10 World average annual precipitation, in millimeters and inches [22]

An excellent example is the Persian qanats [23]. Throughout the arid regions of Iran, agricultural and permanent settlements are supported by the ancient qanat system of tapping alluvial aquifers at the heads of valleys and conducting the water along underground tunnels by gravity, often over many kilometers. The traditional communal management system still in place (Fig. 13.11) allows equitable and sustainable water sharing and distribution. The qanats provide exceptional testimony to cultural traditions and civilizations in desert areas with an arid climate. Therefore, rainwater harvesting is not a new technique. It can serve however as an alternative source of water, for urban use, in modern times.

Another exquisite example of rainwater harvesting from the ninth century is Chand Baori, in India [24]. It has 3500 narrow steps arranged in perfect symmetry, which descend 20 m to the bottom of the well (Fig. 13.12a). There are similar

Fig. 13.11 Persian qanats [23]

Fig. 13.12 (**a**) Chand Baori, India. (**b**) Rainwater harvesting tank in ancient city of Orraon, Epirus, Greece (https://ellinondiktyo.blogspot.gr/2016/)

constructions in Greece from the fourth century BC (Fig. 13.12b). Many well-known examples in the Aegean islands exist, where people traditionally harvested rain runoff from building roofs and paved areas and stored it in underground cisterns for domestic use or irrigation (Fig. 13.13).

Although in most cases this practice is not in use anymore, artificial lagoons have been constructed recently for rainwater harvesting in many Greek islands [21], such as the artificial lagoon in Astypalea (Fig. 13.14a), constructed in 1998 with a volume of 875,000 m^3, providing water to the island's water supply system and irrigation demands, or the rainwater harvesting tanks in Cyclades islands (Fig. 13.14b). More and more these islands are investing on rainwater harvesting, as it is the most independent and economic way of collecting and providing good quality water to their inhabitants and visitors. Sifnos has always been using this method.

Urban stormwater runoff is generated from rain and snowmelt events that flow over land or impervious surfaces, such as paved streets, parking lots, and building rooftops, and does not infiltrate the ground. Precipitation quality, runoff surfaces, and deposits determine stormwater quality, which in turn depends on special regional characteristics, such us human activity (e.g., industrial activity, residential

Fig. 13.13 Cisterns and释 sternes in the Greek islands of Paxoi, Santorini, and Folegandros and from continental Souli (Epirus)

Fig. 13.14 (**a**) Artificial lagoon in Astypalea, Greece. (Source: www.kathimerini.gr). (**b**) Rainwater harvesting tank in Cyclades Island, Greece (http://gr.coca-colahellenic.com/en/sustainability/csr-programmes/mission-water/)

density, traffic loads), meteorological characteristics (e.g., prevailing winds), and geomorphological characteristics (e.g., materials of geological formation, proximity to the sea) [25, 26]. Furthermore, constructed materials of runoff surface and deposits from birds or other animals, atmospheric deposits and characteristics of harvesting network affect the quality of collected stormwater. Common pollutants in stormwater are heavy metals like zinc (Zn), copper (Cu), and lead (Pb), nitrates and nitrites, phosphates, and suspended solids (SS), originated from construction materials, vehicle, and industrial emissions [26]. Polluted stormwater runoff is commonly transported through municipal separate storm sewer systems and combined sewer overflows, and then often discharged, untreated, into local water bodies.

Based on all the above, stormwater is adequate for non-potable uses, such as toilet flushing, vehicle washing, irrigation of gardens, or industrial use. Research results [25, 27] show that physicochemical parameters (such as temperature, pH, total nitrogen, total phosphorus, heavy metals) are within the limits that are set by international regulations, for potable water, whereas microbiological parameters (such as total coliforms, *E. coli*) are found to be of inferior quality than required for potable use. Conclusively, harvested stormwater is suitable for non-potable use and definitely not appropriate for direct potable use without a treatment.

Nowadays, modern cities face a challenge in the field of stormwater management, as the old outdated urban stormwater drainage networks lack the potential to respond to new emerged conditions. Rapid population growth and diminution of permeable surfaces as a result of urbanization, in combination with climate change, which causes extreme rainfall events and extended dry periods, impose an extreme pressure to current urban stormwater infrastructures. At the same time, urban communities have to deal with depletion and quality degradation of conventional sources of water, and the threat that extreme weather events and continuous urbanization set to city infrastructure and especially to water supply system.

It is obvious that antiquated sewerage installations and water supply systems are required to be modernized to meet future needs and uncertainties. Under these circumstances, opportunities have arisen for effective stormwater management in combination with utilization of alternative water resources, such as stormwater and rainwater reuse. An effective strategy to implement this is the decentralized stormwater collection, treatment, and storage for reusing purposes, in small-scale infrastructures [28]. Example of these practices are detention ponds, rain barrels and cisterns, permeable pavement, and stormwater management, using vegetation like stormwater planters and rain gardens or grassed swales and filter strips. The above techniques usually are referred in bibliography as best management practices (BMPs), which according to US EPA are structural, vegetative, or managerial practices used to mitigate runoff volumes and associated nonpoint source pollution [29]. Some examples of BMPs are shown in the following figures (Figs. 13.15, 13.16, 13.17, 13.18 and 13.19) and analyzed thereafter.

Stormwater planters and rain gardens are vegetated areas that collect and treat runoff from the impermeable surrounding area such as building roofs, streets, pavements, and parking lots. These vegetated structures slow down runoff during storms, prevent flooding, and protect receiving water bodies from pollution. Grassed swales and filter strips are linear rain gardens. Depending on the available space and the indented stormwater management plan, the proper vegetated structure is chosen. Rain gardens using plants and soil are capable of capturing water and pollutants including bacteria, phosphorus, nitrogen, heavy metal, oil, and grease. They should be drained after 48 h in order to manage larger quantities of rainwater; for this reason, a soil with adequate permeability should be chosen. The infiltrated water could replenish groundwater aquifer or be collected with a drain pipe and directed to sewerage pipes or tanks for reusing. Finally, rain gardens always have an overflow pipe connected with the sewerage or storm drain. Those vegetated areas are effective

Fig. 13.15 Best management practices [28]

Fig. 13.16 BMPs: Parkside neighborhood of Camden, NJ. (Photo courtesy of Caitrin Higgins, Rutgers University) Source: nemo.uconn.edu/raingardens/

only in sites where the groundwater aquifers are deep, and they must be constructed at safe distances from buildings, sewerage networks, and septic tanks [28, 30].

Permeable pavements and streets allow groundwater recharging, by permitting rainwater to infiltrate the ground and groundwater. They contribute in reducing flooding and ponding on pavements and streets during heavy rain events and the ice formation during cold winter days. Furthermore, as water infiltrates through permeable materials to the soil, pollutants are removed by natural filtration, and

Fig. 13.17 BMPs: Grid system filled with grass (Source: http://www.plantmoreplants.com/press-photo/category/) & permeable pavements

Fig. 13.18 BMPs: dry and wet detention ponds https://www.highpointnc.gov/731/Best-Management-Practices-BMP-Devices)

stormwater loads are reduced. Permeable pavements that can be used are soft paving such as grass and mulch, permeable concrete, concrete, or other material grid system filled with soil and grass or gravel. These constructions need regular maintenance to avoid clogging, especially in regions where runoff has high suspended solids concentration [28, 30].

Detention ponds are constructions which collect stormwater for a period of time, before being released into receiving water bodies. Detention of water in a tank results in sedimentation of pollutants and therefore in mitigation of receiving water quality degradation. There are two types of detention ponds, those that stay empty after slowly releasing water to a river, lake, or sea, until the next rain event, and those that are always full of water and during each rain event the new stormwater replaces the pond water.

A common practice in the area of rainwater harvesting and reusing are the buildings' roof harvesting systems. Rainwater flows to horizontal gutters and then to vertical ones in order to be collected to tanks. Rain barrels and cisterns serve that purpose. Rainwater collection tanks can be implemented at home or community

Fig. 13.19 BMPs: Cistern at the Chicago Center for Green Technology (Source: Abby Hall, U.S. EPA; https://www.werf.org/liveablecommunities/toolbox/rainbarrel.htm)

base. Moreover in an integrated urban water management plan, a "pipeline grid" connecting regional reservoirs can be installed [31]. Determination of required tank size is achieved based on various local characteristics, for instance, rainfall height, available collection area, runoff coefficient of collection area, dry period length, and water demand. Optimal design of tank size is required, as large tanks may trigger construction difficulties and high cost; also long retention time of stormwater in a tank may cause quality degradation. On the contrary, a small tank may not be adequate during a long dry period. In [32] daily water balance method was used to dimensioning stormwater harvesting tanks. The appropriate volume of a tank to satisfy a predetermined percentage of daily residential water demand of up to 240 L/day is found to be 50 m^3 and a roof collection area up to 300 m^2. In the construction of such a system, a device for restraining large objects is required, in order to protect the system from clogging and to maintain better water quality in the tanks. Pumps send the stored water, through separate distribution network for designated uses. Usually, the system in favor of better quality of harvested stormwater is equipped with first-flush diversion procedure, to remove the first-flush, which is the first and most polluted part of the runoff, also a valve to empty the collected first-flush after each rain is needed [25].

Conclusively, rainwater and stormwater harvesting systems benefit the environment, the end users, and the community as a whole. If collected stormwater undergoes treatment, it can also contribute to the water supply system of unprivileged urban areas. Many applications of these systems can be found in international bibliography. In Beijing, China a covered rainwater storage pond has been constructed, which provides suburban farmers with good quality irrigation water for their greenhouses, resulting in improvement of crops' productivity and in lowering the cost of water supply and consequently decrease of their expenses. Also in Belo Horizonte, Brazil pilot projects were developed for harvesting rainwater for irrigation and cleaning of a school surrounding area, which also provide educational opportunities for students on water issues and for urban agriculture [33]. Singapore has managed to reduce the quantity of imported water from Malaysia, among other actions, through a comprehensive network of drains, canals, and rivers that collect rainwater and stormwater, taking advantage of its humid climatic conditions [34].

New technology trends are moving toward the collection of humidity in the air, in order to produce water in remote areas. Turbines working 24 h/7 days can create even potable water, by condensing pure water from the air, without using any power or chemicals. It can produce 27 L/d with or without air [35]. Less technological advanced methods are also used in poor countries, such as Peru, where nearly one million people in its capital lack access to good quality water; BBC World Service introduced Abel, the "fog catcher," who captures humidity, in the form of fog droplets, in nets. Water may not be drinkable, but can be used to irrigate their crops or boil water and cook soups.

Seawater and Brackish Water Management

Although Earth is a planet where the water is the dominant element, the biggest part of it is saline: approximately 97% of global water is contained in the oceans, and a small portion of groundwater and surface water is also saline or brackish. As Fig. 13.20 shows, rising sea level, due to climate change, increases groundwater's salinity. Furthermore, part of groundwater runs in depths that are not technically feasible or economically efficient to be pumped. The remaining available water has to cover all the human needs, with or without treatment. But since the quality of conventional sources of water is degraded and their quantity is depleted, and since the cost for water treatment is substantial, desalination becomes an emerging option. In addition to all that, climate change and urbanization enhance urban water demand, leading to urban water deficiency and making desalination an even more attractive option [36].

Since saline water is not appropriate for direct use for human population needs, its purification was imperative. Desalination is the eminent procedure, which produces fresh water from saline water (seawater, brackish or wastewater) by removing dissolved salts from water. Water salinity based on dissolved salts, for saline water is in the order of 3–5% (30–50 g/L), whereas for fresh water less than 0.05%. Brackish water has more salinity than fresh water, but not as much as seawater. It may result from mixing of seawater with fresh water, as in estuaries, or it may occur in brackish fossil aquifers. Brackish water contains 0.5–30 grams of

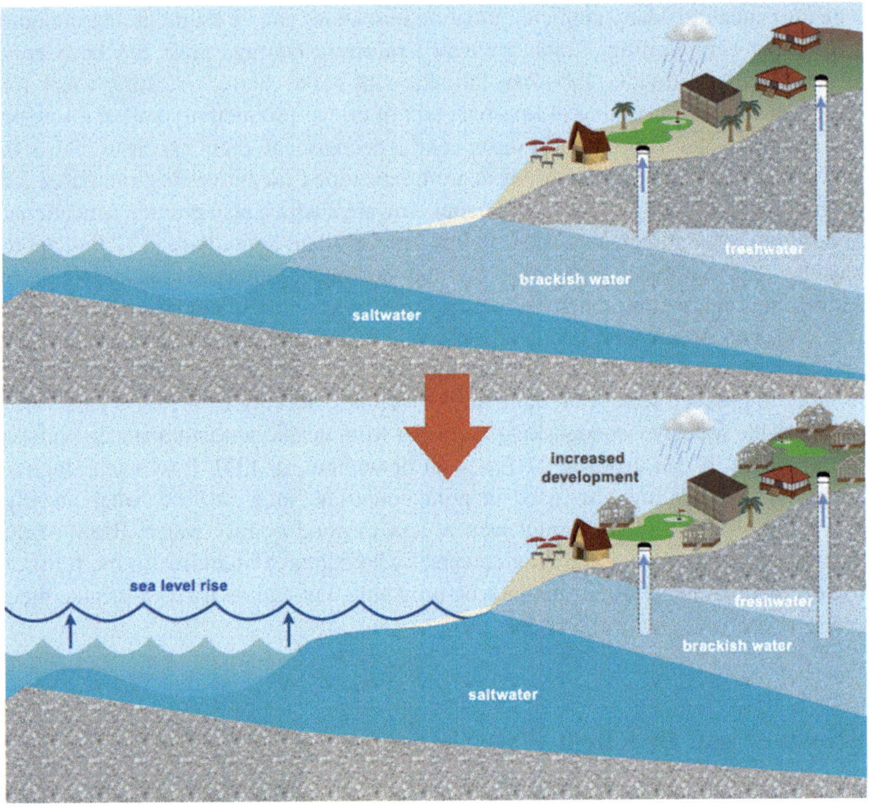

Fig. 13.20 Saline and brackish water layers (source: http://www.teachoceanscience.net/)

salt per liter (more often expressed as 0.5–30 parts per thousand (‰) or 0.05–3%), which is a specific gravity of between 1.005–1.010.

The design of a desalination plant is based on several parameters: the purpose and usage of the product water (potable or non potable use), and the available intake water and product water quality characteristics. Moreover, the available technology and financial resources in conjunction with the location of the plant contribute to the planning. Desalination units are usually placed at coasts, which are selected for water pumping, rather than leisure and tourism, therefore the surrounding land is devalued (Fig. 13.21). These parameters define the efficiency, the energy consumption, and the total cost for installation, operation, and maintenance of a desalination infrastructure. Other parameters that affect the performance are the salinity and temperature of feedwater, the required freshwater recovery percentage, and the desired quality of product water, whereas the cost and energy consumption depend on the distance between the plant and the end users as well as the engaged technology [37].

Fig. 13.21 Desalination unit in Israel (http://knowledge.wharton.upenn.edu/article/what-other-nations-can-learn-from-israels-solutions-to-the-scarce-water-challenge)

The quality of product water as well as the process result is related to the quality of input seawater or brackish water. It is worth mentioning that seawater is the receiver of domestic and industrial sewage, ship wastewater or runoff water [38]. Therefore, input water contains high percentage of total dissolved solids (chloride, sodium, sulfate, magnesium, and other ions). Furthermore, feedwater regularly also contains various microbial contaminants like pathogen bacteria, viruses, parasites, organic carbon, and chemicals such as boron, iodide, sodium, and potassium. Although saline water does not provide a friendly environment to pathogens, there are some kinds which can survive therein, like *Vibrio cholerae*. Also, some marine algae species produce toxins, potentially hazardous for humans [39].

The two major categories of desalination available technologies are membrane-based procedures and thermal-based procedures. A broadly used membrane-based desalination process is reverse osmosis (RO). With RO a large amount of total dissolved solids are removed from water by a pressurized flow of saline water through a RO membrane [40]. Some of the novel membrane technologies are thin-film nanotechnology (TFN) membranes and graphene-based membranes. Nanocomposite membranes boost water permeability while maintaining the same efficiency in salt removal and product water quality, as RO membrane. Frequent are the cases where RO membrane is coupled with another membrane technology for ameliorating installation efficiency in terms of energy consumption, recovery percentage, and quality of desalinated water. The most renowned thermal-based procedures are multistage flash (MSF), multi-effect distillation (MED), and vapor compression distillation (VCD). All these methods use evaporation and phase change in order to remove salt from water and usually need enough energy to achieve this [37].

Monitoring and pretreatment of feedwater is recommended in many cases for both protecting membranes and obtaining superior quality of product water.

Common pretreatment procedures are sedimentation, filtration (e.g., microfiltration and nanofiltration) for removing suspended solids, and some kind of disinfection, usually chlorination of saline water, for removing microbial contaminants before entering the desalination processes [39]. Chemicals that are used during pretreatment and desalination procedures, as well as for membrane cleaning, may probably be present at product water. Moreover, materials of plant structures and transportation pipes can release harmful chemicals in water which pass through or stay in touch with those materials [39]. Hence, selection of the right materials, the wise use of chemicals, the discharge of wastewater, and the flushing and discharge of wastewater after membrane cleaning is of great importance. An appropriate treatment, if needed, before the product water reaches the end user, can be applied along with the utilization of appropriate protection measures in order the product water to be of the desired quality for potable or other uses.

Posttreatment may engage blending desalinated water with partially treated seawater or untreated groundwater, adding minerals (e.g., treatment with limestone) or disinfection procedures (chlorinate, ultraviolet light, etc.) and/or adding more membranes to reduce risk of microbial infection of product water. Furthermore, during storage and distribution of product water before reaching the end user, there is high risk of growing pathogens like *Legionella*, especially when hydraulic conditions, temperature, and used materials are favorable, and when a significant amount of biodegradable organic matter and nutrients exists. To avoid these phenomena, materials that prevent biofilm development should be chosen, and a necessary quantity of residual disinfectant should be sustained in the output water [39]. Saline and desalinated water have heavy corrosion influence on metal and concrete infrastructure material. As a result treatment processes for balancing pH of water and addition of chemicals for protection construction materials are also necessary [38]. Parameters that affect the aesthetic quality of desalinated water are turbidity, taste, odor, and total dissolved solids. These parameters normally don't pose the consumers' health at risk, but they can cause nuisance and negativity against desalination [38]. Thus, taking measures, like posttreatment, is crucial to gain social acceptance for desalination.

The flowchart (Fig. 13.22) describes clearly the simple desalination procedure with RO, including pretreatment and posttreatment, as well as the connection to Sydney's water supply system. Desalination is becoming more and more a popular alternative solution for producing water for several domestic, agricultural, and industrial uses, even for potable use. More than 300 million people, in more than 150 countries depend on desalinated water to satisfy a portion of their daily needs or, in some extremely arid areas, their total everyday needs. There are 18,426 desalination installations around the globe, and they produce about 86.8 million cubic meters freshwater per day, according to International Desalination Association data for 2015. Leader countries in desalination are Spain, Australia, the United Arab Emirates, and the USA (Fig. 13.23). According to MIT's Journal Technology Review [41], Spain built Europe's first desalination plant nearly 50 years ago and is the largest user of desalination technology in the Western world. Spanish companies lead the market, operating in regions including India, the Middle East,

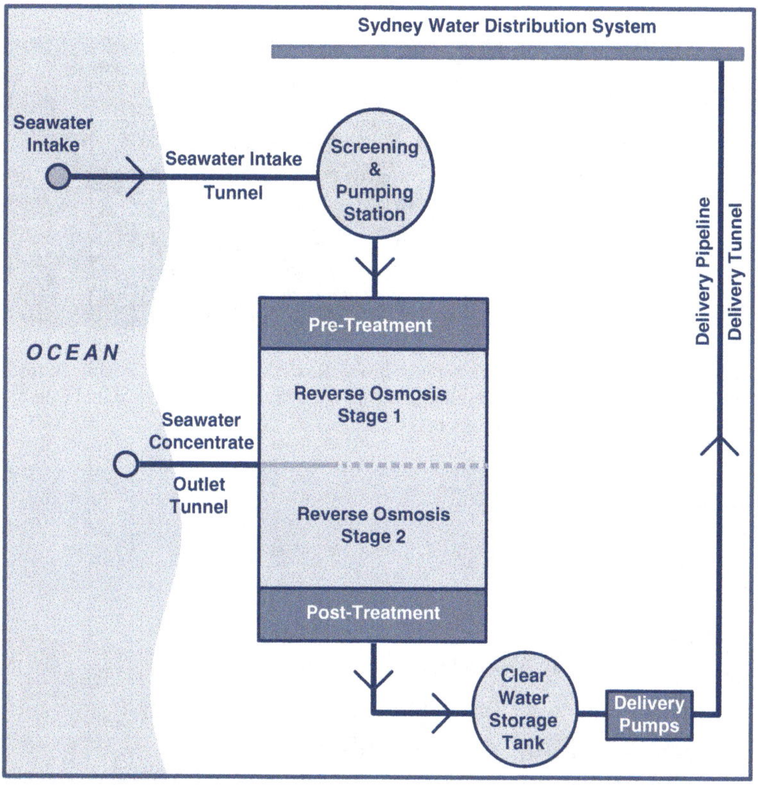

Fig. 13.22 The process of desalination through reverse osmosis at Sydney Water (http://www.awa.asn.au/AWA_MBRR/Publications/Fact_Sheets/Desalination_Fact_Sheet.aspx)

and North America. Spanish innovation contributes to advancing desalination to bring sustainable clean water to millions. On the other end, in Perth, Australia, 47% of total regional water supply comes from desalination, 46% from groundwater, and only 7% from surface water, mainly dams. One of Australia's seawater desalination plants is providing critically needed drinking water supplies for cities during floods (Fig. 13.23b) [42].

Traditionally, desalination is an expensive procedure mainly due to the engaged technology, the operational cost, and the energy consumption [37]. The major energy consumption results from achieving the appropriate pressure and for the needed pretreatment and posttreatment of the water. However, in the last years energy demand is reduced due to lower pressure requirements [40]. In the direction of reducing energy consumption, recent desalination technologies are directed toward the combination of thermal-based procedures with membrane technologies, such as membrane distillation. When brackish water is used, which is characterized by lower salinity, less energy is needed. Also, the larger the installation, the lower the cost per cubic meter of produced freshwater is [27]. Finally, when the facilities

Fig. 13.23 World's largest desalination units in (**a**) Ras Al-Khair, Saudi Arabia (https://www.betterworldsolutions.eu/the-largest-desalination-plants-in-the-world/) and (**b**) Gold Coast, Tugun, Queensland, Australia (water-technology.net)

Fig. 13.24 Hydriada, a floating desalination unit in Irakleia, Cyclades, Greece, powered by wind and solar energy

incorporate renewable or alternative sources of energy, such as solar and geothermal energy or waste heat, the cost and environmental impacts are reduced [40]. Such an example is "Hydriada," a small renewable energy operated desalination unit in Cyclades islands of Greece, which was introduced in 2007 to relieve a small island from water inadequacy (Fig. 13.24).

Even though desalination is a broadly accepted alternative for urban water supply, it has some negative environmental impacts. Desalination facilities contribute to the increase of greenhouse gas emissions, due to high energy consumption, and saline density, due to the rejection of residual salt back to the sea, but also to high level noise pollution, due to the continuously working pumps. To reduce noise volume desalination plants should be equipped with the appropriate technology and be placed at a location where the surrounding environment impact is as less as possible, such as airports [36]. On the other hand, discharges of desalination plants are very saline (approximately 50% more salt for the same water volume) in higher than normal temperatures, contain chemicals and metals, which, when discharged into the sea without any treatment, can have negative effect on the marine environment. They cause contamination in seawater, and thus the new intake water for desalination processes is polluted with entailed consequences, such as increased need for energy and chemicals additives. Also marine fauna and flora are affected because of contaminants, rising of normal seawater temperature, changes in oxygen level, and turbidity. The aforementioned changes in the marine environment can cause algae blooms or marine organism deaths. Moreover, degradation of marine environment may disturb recreation activities and cause loss of aesthetic value of urban seafront areas, because of odors and turbidity, even pose health risks to humans after contact with polluted seawater. Another concern is the possibility of leakages of pipes or storage tanks of desalination by-products or saline water, which can severely pollute soil, groundwater, and even freshwater distribution networks if they exist in proximity, with toxic constituents and solids, or induce aquifers salinization. According to an effect analysis of the operation of the desalination units in Cyprus [36], the energy use, the CO_2 emission, and the economic cost score the highest, in a scale 1–5, whereas the land use, the brine and marine environment, the aquifers, and the noise play a less significant role.

In the recent years, important advancements in desalination technology are made in the energy demand, such as the use of wind and solar energy, where available, solar thermal energy for the desalination process, and at the same time generation of electricity (STEP-EW), which is applied in Cyprus, multiple effect distillation method (MED), which requires low electrical energy, self-covered through heat source (sun), and operates without any use of chemicals. With respect to the diminution of the saline production, the latter is not environmentally sound, but reprocess of the saline water by vaporization, crystallization, and drying, which is applied in Tinos island by the National Technical University of Athens, results to an average percentage of 65% pure water and 35% pure salt [36].

In conclusion, new technological applications of desalination that reduce energy consumption in combination with implementation of environmental risk assessment and appropriate mitigation measures, make desalination a technically feasible, environmentally and economically efficient alternative source of water, in the light of upcoming water scarcity, especially in vulnerable urban areas.

Wastewater and Greywater Management

Urban wastewater in developed countries, where sanitation systems exist, is directed in wastewater treatment plants (WWTPs) and then after being treated, it is released in the environment. From an innovative point of view, this water can comprise a precious source of water for several uses and contribute with satisfaction to intensified human needs, due to population growth, urbanization, and climate change, without overexploiting water resources. More importantly, wastewater is a reliable alternative water resource, since it is not seasonal, like stormwater and rainwater.

Treated wastewater can be reused for residential, agricultural, industrial, even potable uses, and for environmental purposes such as wetlands' restoration and groundwater recharging. Domestic consumers may use recycled water for toilet flushing, indoor and outdoor washing, garden irrigation, and in extreme cases for human consumption. In commercial and industrial domain, reclaimed water is used for cooling tower, boiler feeding, and washing. Public urban uses of treated wastewater are fire protection, irrigation of parks, fields and school yards, roads, and parking lot washing.

Leading countries in the domain of wastewater reuse are Singapore where 30% of treated wastewater is reclaimed, followed by Saudi Arabia where 16% is reclaimed, Australia where 8% is reused, and the USA where 7–8% of treated wastewater is reused [43].

Urban wastewater can be of diverse origins, with different composition and as a result requiring different treatment in order to be disposed or be recycled and reused for various uses. Those types are feces, urine, or yellow water, flush water, blackwater, greywater, and stormwater. As blackwater is considered the sum of feces, urine, and flush water and as greywater the water produced from kitchens, dishwashers, sinks, bathtubs, showers, laundry tubes, and washing machines [33]. Greywater constitutes 50–80% of the total wastewater of a household [44].

Reclaimed wastewater can be used for indirect potable reuse (IPR) and direct potable reuse (DPR). In case of using treated wastewater for surface or groundwater enrichment, planned IPR is implemented. For example, in Orange County, California, USA, reclaimed water is returned to the aquifer which supplies drinkable water, and in Singapore, where reclaimed water is blended in the city's reservoir. On the other hand, DPR refers to the introduction of purified water, derived from municipal wastewater, after extensive treatment and monitoring to assure that strict water quality requirements are met at all times, directly into a municipal water supply system. Such an application comes from Big Spring, Texas, USA, where reclaimed water is provided in a raw surface water conveyance for potable reuse [43].

In developing countries, untreated wastewater is regularly used, posing at serious risk human health. Raw wastewater use is intentionally and unofficially practiced, from low-income farmers for agricultural irrigation. For these farmers, wastewater is the only source to water their crops and earn their living. In Middle East and North Africa, wastewater, which may sometimes be untreated, is used for agricultural purposes [43].

In developed countries wastewater treatment is a common practice and, in most cases, enforced by law. However, as Fig. 13.25 clearly shows, not all European countries have established a connection of households with biological treatment plants, according to the newest (2013) recorded data [45]. Wastewater treatment is performed in wastewater treatment plants (WWTPs) and is the procedure

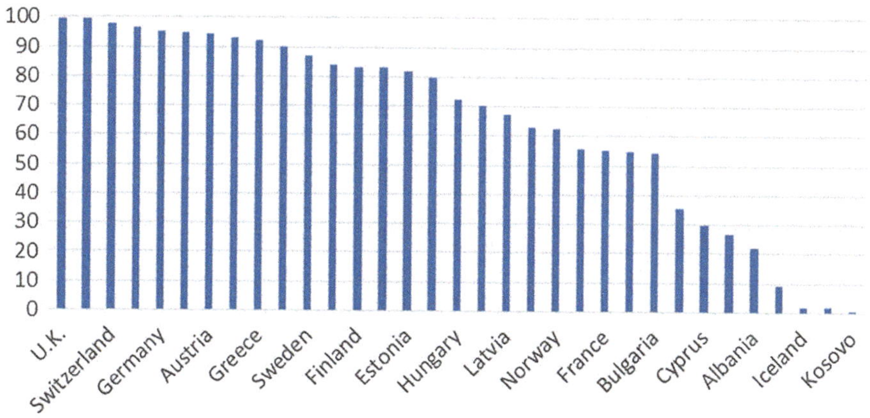

Fig. 13.25 Share of the population connected to secondary urban wastewater treatment in 2013

that mainly removes suspended solids, organic matter, and nutrients. For reusing wastewater, additional treatment is required to remove pathogens and chemical contaminants, which is performed in water reclamation plants (WRPs) and therefore is called reclamation treatment. There is a large number of available reclamation technologies, which can be categorized as intensive or conventional, and extensive or non-conventional technologies [44, 46]. Intensive technologies are physical-chemical systems (coagulation—flocculation, sand filters), membrane technologies (ultrafiltration, reverse osmosis), rotating biological contactors, and disinfection technologies (ultraviolet radiation, chlorine dioxide, ozone). These technologies engage artificial processes and require large amounts of energy. On the other hand, extensive technologies rely on natural processes and therefore require less energy, and low but considerable operation and maintenance, and more space than the conventional methods. Extensive technologies utilize waste stabilization ponds (maturation ponds, stabilization reservoir), constructed wetlands, and infiltration-percolation systems. Usually, a combination of reclamation technologies is used in order to attain the desired water quality. The criteria, based on which the reclamation technology is selected, are wastewater's quantity and quality, the water quality standards for its final use, as well as economical and environmental aspects of the project [44].

In the direction of protecting human health, environment, soil, and plants, many countries such as Australia, Canada, China, Japan; many US states such as California and Texas; European countries like Spain, Cyprus, and France; and organizations like World Health Organization (WHO) and US Environmental Protection Agency (USEPA) have adopted guidelines for the reused water quality standards. These guidelines refer to reused water designated uses, required treatment directed to specific end-uses, water quality limits for certain parameters, and guidelines for which parameters need to be monitored and in what frequency, measures for the onsite construction and maintenance in order to avoid cross connection with other water supply systems and unwanted human contact, environmental monitoring of the effect of reclamation project and communications strategies to achieve stakeholders, and public acceptance [44]. Parameters that are commonly used for reclaimed water quality evaluation are pH, BOD_5, total suspended solids (TSS), turbidity, *E. coli*, thermotolerant coliforms, and chlorine residual [47]. Thresholds for these parameters are strict, and a higher level of treatment is required when reuse involves unrestricted public exposure, with higher probability of human contact or inhalation of aerosols, such as toilet flushing, laundry, and garden irrigation. On the contrary, restricted reuse of reclaimed water means that public exposure to the reclaimed water is controlled and lower treatment may be sufficient [43]. Furthermore, guidelines comprise directions for the installation, planning, and maintenance of the wastewater recycling systems, such as keeping safe distances from other pipes, using different colors, and marking for reclaimed water apparatus like pipes, valve boxes, pumps, and outlets to distinguish potable pipes network from wastewater pipes network [47]. The American Public Work Association (APWA) in 2003 proposed purple color for reclaimed water pipes [43].

As mentioned above, greywater, which comes from kitchens, dishwashers, sinks, bathtubs, showers, laundry tubes, and washing machines [33] has lower nutrient and pathogenic loads than wastewater and as a result needs less if any treatment for being reused. Simple treatment like sand/gravel filters or constructed wetlands are sufficient for achieving high water quality. Greywater from kitchen sinks, dishwashers, and washing machines are usually more contaminated than those from showers and bathtubs, because they usually contain solids and organic substances. Therefore, regularly, the reuse of less polluted part of greywater is chosen, excluding kitchen greywater which contain fats, grease, oil, food particles and has higher amount of organic pollutants and laundry greywater that may contain fecal substances [48].

However, greywater is characterized by variability in quality and no constant flow [47], depending on living standards, the chemicals used for cleaning and personal care, pharmaceuticals products, water supply system, and water availability. Greywater volumes range between 60 and 120 L/person/day although it can be reduced to 20–30 L/person/day for low-income countries with basic water supply system and low water availability. However, it is considered adequate to cover water demand for non-potable usages [48].

Greywater collection and treatment can be implemented in decentralized systems, located in individual or multistory dwellings (Fig. 13.26) or in a centralized system collecting and treating greywater from an entire region. Centralized systems involve lower investment and maintenance cost for the households, but require the installation of larger storage infrastructure and distribution network. A case study was studied and presented for the Port Authority of Thessaloniki, Greece, which proposed the collection and treatment of greywater from 14 multistory buildings in the vicinity of the Port, to eliminate the cost of cleaning demands of the ships and the port facilities, but would not cover the potable water needs (Fig. 13.27) [49].

The cost and the characteristics of the system that is adopted for greywater reuses depend on several parameters such as the end-uses, existing pipes, available space, and pumping requirements. For instance, for subsurface restricted irrigation, greywater can be diverted and used after coarse filtration, without additional

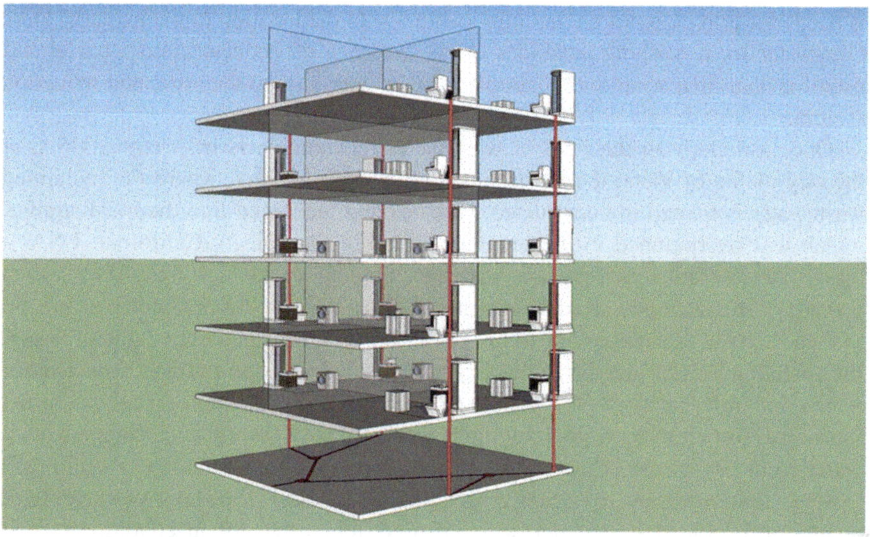

Fig. 13.26 Greywater collection in decentralized system, in multistory dwellings [49]

Fig. 13.27 Proposed greywater collection and treatment for the Port of Thessaloniki [49]

treatment, with only a few hours storage requirement, resulting in lower cost and simpler infrastructures. On the other hand, for garden irrigation and indoor uses, like toilet flushing and laundry, collection and storage is required as well as advanced treatment such as fine filtration, biological treatment, and disinfection, and thereby the cost is risen and more complex infrastructure is needed. Furthermore, the cost is higher as, in most cases, a diversion network is required and when a pumping system is needed if gravity pipes are not applicable due to the location of the irrigated area. Greywater from washing machines or showers can be collected into a barrel and used for plant irrigation with a simple hose or with a more sophisticated irrigation system.

One case study in the field of wastewater reclamation is in Arizona, USA, at the city of Sierra Vista, the Environmental Operations Park, where the reclaimed wastewater is treated in constructed wetlands and recharged into the local aquifer, which was overpumped in the previous years. In San Diego, California, USA, a wastewater reclamation program has been implemented since mid-1980, which nowadays includes advanced treatment facilities and produces water that is sent to San Vicente Reservoir and then is treated and distributed as potable water. In Australia, wastewater of a 29-story office tower is captured and recycled for reusing, in order to cover non-potable demand for cooling tower makeup and toilet flushing. Furthermore, since building's wastewater is insufficient, supplemented water from city sewer is used. This practice achieves the reduction of building's freshwater demands by 90%. Singapore utilizes NEWater project that produces high quality of reclaimed wastewater, which is used in industry, for air-conditioning and cooling in commercial and institutional complexes. Furthermore treated wastewater in Singapore is used for indirect potable reuse (IPR) by blending NEWater with raw reservoir water, following the conventional treatment that is required for raw reservoir water to produce potable water for the city. Finally, at Costa Brava, Spain, in Tossa de Mar, a small town that receives large number of tourists in the summer, reclaimed wastewater is used for non-potable uses, such as street cleaning and public irrigation [43].

However, wastewater reclamation may have certain drawbacks, such as altering land uses by changing ecosystem characteristics and water balance of an area, or by allowing development of residential, commercial, industrial, or recreational uses where it was not possible before due to water scarcity. It is also possible that it can shift the prevalent hydrologic regime of an area, by modifying wet and dry weather stream flows. Finally, surface and groundwater quality may be affected by discharging reclaimed wastewater to them [43].

Water is a finite and irreplaceable resource that is fundamental to human well-being. The water footprint is a measure of humanity's appropriation of freshwater in volumes of water consumed. Building a sustainable urban water system, in view of climate change and water pollution and scarcity, is a challenge. It depends upon the available water resources, i.e., precipitation (rainwater and snowmelt), humidity, enriched groundwater aquifers, and sensitive surface water (rivers and lakes). When these resources are not available or abundant, alternatives must be found and followed, such as water reclamation by means of treated greywater or wastewater, or desalination in coastal areas. Water is renewable if only well managed. It can pose a serious challenge to sustainable development, but, managed efficiently and equitably, can play a key enabling role in strengthening the resilience of social, economic, and environmental systems in the light of rapid and unpredictable changes. Water is a critical asset for socioeconomic development and the viability of ecosystems. Governments and leading businesses need to take first local action to improve their water use and have implicitly a global impact. Our goal is to ensure that freshwater is shared fairly to sustain thriving communities and nature's diversity.

Bibliography

1. The Guardian.: www.theguardian.com (2015)
2. World Health Organization and UNICEF Joint Monitoring Programme: Progress on sanitation and drinking-water. http://www.who.int/water_sanitation_health/publications/9789241563956/en/ (2010)
3. Water Policy in Canada: Introduction: Drinking Water Management. In: Dore, M.H. (ed.) Part of the Series Springer Water, pp. 1–10. Springer, Cham (2015)
4. World Health Organization and UNICEF Joint Monitoring Programme: Progress on drinking water and sanitation, 2015 ppdate and MDG assessment. http://www.who.int/water_sanitation_health/monitoring/jmp-2015-update/en/ (2015)
5. water.org - http://water.org/water-crisis/
6. World Economic Forum: Insight report, global risks, 10th Edition. http://www3.weforum.org/docs/WEF_Global_Risks_2015_Report15.pdf (2015)
7. United Nations: A/RES/66/288, the Future We Want – Outcome Document of the Rio+20 Conference (2012)
8. UNDESA: United Nations department of economic and social affairs. http://www.un.org/waterforlifedecade/water_and_sustainable_development.shtml (2015)
9. United States Census Bureau Estimates: US and world population clock. https://www.census.gov/ (2015)
10. Millennium Development Goals (MDGs): A/RES/64/292. (2000)
11. Chapagain, A.K., Hoekstra, A.Y.: Water footprints of nations, UNESCO-IHE, Delft, The Netherlands, value of water research report series no. 16 (2004)
12. Chapagain, A.K., Hoekstra, A.Y., Savenije, H.H.G.: Water saving through international trade of agricultural products. Hydrol. Earth Syst. Sci. **10**, 455–468 (2006)
13. Chapagain, A.K., Hoekstra, A.Y.: The global component of freshwater demand and supply: an assessment of virtual water flows between nations as a result of trade in agricultural and industrial products. Water Int. **33**, 19–32 (2008)
14. Hoekstra, A.Y., Chapagain, A.K.: Water footprints of nations: water use by people as a function of their consumption pattern. Water Resour. Manag. **21**, 35–48 (2007)
15. Water Footprint Network.: http://temp.waterfootprint.org/
16. Mekonnen, M.M., Hoekstra, A,Y.: National water footprint accounts: the green, blue and grey water footprint of production and consumption, volume 1: main report, research report series no. 50, value of water, UNESCO-IHE, Institute for Water Education (2011)
17. Hoekstra, A.Y., Mekonnen, M.M.: The water footprint of humanity. Proc Natl Acad Sci of the USA. **109**(9), 3232–3237 (2012)
18. Human Development Report.: http://www.data360.org/ (2006)
19. Hoekstra, A.Y., Chapagain, A.K., Aldaya, M.M., Mekonnen, M.M.: The Water Footprint Assessment Manual: Setting the Global Standard. Earthscan, London (2011)
20. US Geological Survey.: www.usgs.gov
21. Gikas, P., Angelakis, N.A.: Water resources management in Crete and in the Aegean Islands, with emphasis on the utilization of non-conventional water sources. Desalination. **248**, 1049–1064 (2009)
22. US Geological Survey.: https://water.usgs.gov/edu/watercycleprecipitation.html
23. UNESCO: World heritage list, The Persian Qanat. http://whc.unesco.org/en/list/1506/
24. Chand Baori.: http://www.chandbaori.org/

25. Gikas, D.G., Tsihrintzis, A.V.: Assessment of water quality of first-flush roof runoff and harvested rainwater. J. Hydrol. **466-467**, 115–126 (2012)
26. Han, J., Gao, X., Liu, Y., Wang, H., Chen, Y.: Distribution and transport of typical contaminants in different urban stormwater runoff under the effect of drainage systems. Desalin. Water Treat. **52**, 1455–1461 (2014)
27. Melidis, P., Akratos, S.C., Tsihrintzis, A.V., Trikilidou, E.: Characterization of rain and roof drainage water quality in Xanthi, Greece. Environ. Monit. Assess. **127**, 15–27 (2007)
28. Boston Complete Streets, Design Guidelines: Mayor Thomas M. Menino, city of Boston, commissioner Thomas J. Tinlin, Boston transportation department. www.bostoncompletestreets.org (2013)
29. EPA: BMP modeling concepts and simulation (2006)
30. INTEGRATED GREEN CITIES/InGreenCi: Area of intervention 1.1 – protection, management and promotion of the environmental resources, subsidy contract: no B2.11.04/20–03-20
31. Wong, F.H.T., Brown, R.R.: The water sensitive city: principles and practice. Water Sci. Technol. **60**(3), 673–682 (2009)
32. Londra, A.P., Teocharis, T.A., Baltas, E., Tsihrintzis, A.V.: Optimal sizing of rainwater harvesting tanks for domestic use in Greece. Water Resour. Manag. **29**, 4357–4377 (2015)
33. Howe, C.A., Butterworth, J., Smout, I. K. (2011). Findings from the SWITCH project 2006–2011. Sustainable water management in the city of the future
34. Mandiliotis, A: Sustainable water management in Singapore. Potential application in the Aegean islands. Undergraduate Thesis, Aristotle University of Thessaloniki (2016)
35. Collective Evolution.: waterseer.org (2017)
36. Thrasivoulou, P., Patsalosavvis, K., Zafirakou, A.: Water supply management in Cyprus under climate uncertainty. Desalin. Water Treat. **57**(5), 2279–2289 (2015)
37. Khawaji, D.A., Kutubkhanah, K.I., Wie, J.M.: Advances in seawater desalination technologies. Desalination. **221**, 47–69 (2008)
38. Cotruvo, J., Voutchcov, N., Fawell, J., Payment, P., Cunliffe, D., Lattemann, S.: Desalination technology: health and environmental impacts. ISBN 978–1–4398-2890-8 (2010)
39. WHO: Safe drinking-water from desalination, WHO/HSE/WSH/11.3 (2011)
40. Subramani, A., Jacangelo, G.J.: Emerging desalination technologies for water treatment: a critical review. Water Res. **75**, 164–187 (2015)
41. MIT Technology Review.: http://icex.technologyreview.com/articles/2009/01/desalination-in-spain/ (2009)
42. National Center of Excellence in Desalination, Australia. http://desalination.edu.au/2013/
43. EPA: Guidelines for water reuse, EPA/600/R-12/618 (2012)
44. Sanz, L.A., Gawlik, B.M.: Water Reuse in Europe, Relevant Guidelines, Needs for and Barriers to Innovation. A Synoptic Review. ISBN 978-92-79-44399-2 (PDF), ISSN 1831-9424 (online)., Luxembourg, Publications Office of the European Union (2014). doi:10.2788/29234
45. Eurostat. http://ec.europa.eu/eurostat/statistics-explained/index.php/
46. Metcalf, Eddy., Tchobanoglous, G., Burton, F.L., Stensel, H.D. (eds.): 4th edn. McGraw-Hill (2003)
47. Toifl, M., Bulteau, G., Diaper, C., O'Halloran, R.: Ch. 11: Greywater recycling: Guidelines for safe adoption. In: Memon, A.F., Ward, S. (eds.) Alternative Water Supply Systems, pp. 217–236. IWA Publishing, London, UK (2005)
48. Kemi Adeyeye Water Efficiency in Buildings, Theory in Practice, 2014
49. Tassios, F. Vantsiotis, A.: Greywater management. Case study for Thessaloniki's port authority. Undergraduate Thesis, Aristotle University of Thessaloniki (AUTH), Greece, Greek (2016)

Chapter 14
Introduction to Architectural Design Optimization

Thomas Wortmann and Giacomo Nannicini

Abstract This chapter presents black-box (or derivative-free) optimization from the perspective of architectural design optimization. We introduce and compare single- and multi-objective optimization, discuss applications from architectural design and related fields, and survey the three main classes of black-box optimization algorithms: metaheuristics, direct search, and model-based methods. We also give an overview over optimization tools available to architectural designers and discuss criteria for choosing between different optimization algorithms. Finally, we survey recent benchmark results from both mathematical test problems and simulation-based problems from structural, building energy, and daylighting design. Based on these empirical results, we recommend the use of global direct search and model-based methods over metaheuristics such as genetic algorithms, especially when the budget of function evaluations is limited, for example, in the case of time-intensive simulations. When it is more important to understand the trade-off between performance criteria than to find good solutions and the budget of function evaluations is sufficient to approximate the Pareto front accurately, we recommend multi-objective, Pareto-based optimization algorithms.

Keywords Black-box optimization • Multi-objective optimization • Architectural design • Direct Search • Model-based optimization • Metaheuristics

Introduction

Mathematical optimization is concerned with identifying the best element from a set of alternatives in terms of a specified criterion. In this sense, optimization is a fundamental aspect of design processes, although it is not always carried out in

T. Wortmann (✉)
Singapore University of Technology and Design, 138682, Singapore, Singapore
e-mail: thomas_wortmann@mymail.sutd.edu.sg

G. Nannicini
IBM T.J. Watson Research Center, 10598, Yorktown Heights, NY, USA
e-mail: nannicini@us.ibm.com

A. Karakitsiou et al. (eds.), *City Networks*, Springer Optimization and Its Applications 128, DOI 10.1007/978-3-319-65338-9_14

a formalized way. In recent years, the study of architectural design optimization (ADO) has gained increased relevance due to the integration of computational tools such as parametric design [80] and numerical simulations [54] into design processes. With parametric design, designers generate large numbers of design variants based on explicit rules and parameters. With simulations, designers evaluate design variants in terms of quantitative performance criteria such as material or energy use. Finding the best design variants according to such performance criteria is an optimization problem, which explains the increasing study of ADO [22, 60] and its introduction into architectural education (e.g., [62]). In architectural practice, mathematical optimization is currently applied only in rare cases (some of which we document), but its application can be expected to grow based on the increasing availability of relevant know-how and tools. This work contributes to this growth by providing an accessible introduction to black-box optimization, a branch of mathematical optimization that is especially suited to solve simulation-based problems in architectural design. We review black-box optimization methods, applications, and tools and present benchmark results from mathematical test problems and practical, simulation-based problems from structural, building energy, and daylighting design.

Black-Box Optimization

When applying optimization methods to problems in engineering and architecture, formulating the mathematical optimization problem is often difficult. This difficulty stems from two sources: on the one hand, some of the elements of the optimization problem (decision variables, objectives, constraints) may be hard to identify or articulate; on the other hand, even when these elements can be identified, it may not be possible to express them as mathematical formulas. The first difficulty is hard to address systematically. The second difficulty can in many cases be addressed by requiring only that the functions describing the mathematical optimization problem need to be *computable*, which is a much weaker requirement than their expression in analytical form. This computability is easily clarified with an example. Suppose a designer wants to minimize the total weight of a structure while meeting stress and displacement constraints under given loads. There is no conceptual impediment to formulate this problem, but writing mathematical formulas that represent the stress and displacement of the structure is difficult. Furthermore, even if a mathematical description can be given, it may be in the form of the solution of a system of partial differential equations, and existing mathematical optimization software typically does not allow such systems of equations as part of the problem formulation. But finite element software packages can simulate stress and displacement, i.e., provide numerical values without the corresponding formulas: the computation is easier than the formal mathematical description.

The field of black-box (or derivative-free) optimization is concerned with the solution of optimization problems in which some of the functions describing the objective and constraints of the problem can be computed but are not available in analytical form, i.e., their mathematical expression is unknown. This is a natural setting for ADO, considering the recent trend of using numerical simulations to analyze the performance of design candidates: a numerical simulation linked to a parametric model is a function that takes as input the design parameters (i.e., the values of the *decision variables*) and outputs a measure of performance of the corresponding design candidate according to a given metric (the *objective function* value). From this perspective, a single simulation run is equivalent to a single-*objective function evaluation*. In general, the output of a numerical simulation is not captured by a known mathematical expression, but in the context of black-box optimization, one can indicate that the goal of the optimization process is to minimize (or maximize) the output of a numerical simulation, e.g., the total stress and displacement of a structure or the energy use of a building.

There are several types of mathematical optimization problems. In this chapter, we are mainly interested in problems with a single-objective function and simple lower and upper bounding constraints on the decision variables (as opposed to more general constraints). A problem of this form, called *OPT*, can be expressed as follows:

$$\min\left\{f(x) : x \in \left[x^L, x^U\right] \subseteq R^n, x_i \in Z \forall i \in I\right\}, \quad (OPT)$$

where the vectors x^L, x^U represent lower and upper bounds for the decision variables, and $I \subseteq \{1, \ldots, n\}$ is the set of indices of the decision variables that are constrained to take on integer values (the remaining variables can take any fractional value). Here, a point in the space $[x^L, x^U]$, $x_i \in Z \; \forall \; i \in I$, also called a solution, represents values of the design parameters and therefore defines a design candidate.

There are several reasons why we consider only such a seemingly simple form of optimization problem. First, the optimal solution to such a problem is mathematically well-defined (which is not the case for multi-objective optimization): we want to determine one point in the solution space $[x^L, x^U]$ that achieves the lowest value of the function $f(x)$. Second, potential constraints for the problem can in principle be moved to the objective function by penalizing their violation, so that $f(x)$ becomes in fact a combination between the original objective function and penalties for violating the constraints of the original problem. (For an example of penalties applied to ADO, which penalizes the weight of a structure for exceeding stress and displacement constraints, see [83].) Third, one should keep in mind that, in practice, (*OPT*) can be very challenging to solve and that one should therefore avoid additional complications. A large part of the literature on black-box optimization deals with (*OPT*), or an even simpler, continuous problem in which $I = \varnothing$, i.e., all variables can take fractional values. Conn et al. [14] provide a comprehensive introduction to black-box optimization.

When solving (*OPT*), the main goal is to obtain a *global minimum* of the function f, that is, a point that attains the minimum over the whole *design space* (also known as the *feasible region*) defined as $\{x \in [x^L, x^U] \subseteq R^n, x_i \in Z \; \forall \; i \in I\}$. But this minimization may be difficult to achieve in theory and practice and in the worst case may require an infinite number of steps. One may therefore settle for the less ambitious goal of determining a *local minimum*, that is, a point that attains the minimum over some area of the feasible region centered on the point itself. Problems with a single local minimum that is also the global minimum are known as unimodal, while problems with multiple local minima are known as multimodal. While unimodal problems are easier to optimize, in practice most problems are multimodal. Known convergence results guarantee that one can determine a local minimum with high accuracy in a small number of steps if the function is sufficiently well-behaved. For example, if f is smooth with a Lipschitz continuous gradient, one can find an optimum point with a tolerance of at most ϵ in order of $1/\epsilon^2$ steps [75], and this bound improves to $1/\epsilon$ if f is convex [18]: i.e., some structural properties of f lead to faster convergence.

In practice, existing methodologies often employ a combination of *global search*, which aims to determine an area of the feasible region that contains the global minimum (without being able to guarantee that such an area will be found in finite time), and *local search*, which focuses on identifying a local minimum starting from a given point. This combination of (global) exploration and (local) exploitation has proven very successful: Despite the lack of strong theoretical guarantees, state-of-the-art optimization software can consistently and rapidly find near-optimal solutions to some classes of problems in the form (*OPT*) [66].

Multi-Objective Optimization

While single-objective algorithms aim to optimize only a single-objective function f—which can include a weighted sum of several performance criteria and penalty terms—multi-objective algorithms consider multiple, potentially conflicting objective functions simultaneously. One can formulate a multi-objective optimization problem as the optimization of a vector of single-objective functions:

$$\min \{F(x) = [f_1(x), f_2(x), \ldots, f_k(x)]\}$$

A multi-objective problem may not have a well-defined solution, because the set of all interesting solutions (the set of all non-dominated solutions, also known as the *Pareto front*) may have infinite size and be very difficult to represent. For a non-dominated solution, it is impossible to improve an objective value without losses in other objective values. Accordingly, a Pareto front—which for two objectives is often drawn as a curve—represents the trade-offs between conflicting objectives. Finding the true Pareto front is usually more difficult than finding the global optimum for a single-objective problem [13]. This chapter considers only Pareto-based, multi-objective algorithms that aim to approximate the true Pareto front, which is the most common type. (Marler and Arora [55] provide a more comprehensive survey of multi-objective optimization for engineering design.)

One measures the quality of Pareto-based optimization algorithms in terms of how well they approximate the "true" Pareto front and not in terms of how much they improve a single-objective value. In other words, one can think of Pareto-based optimization algorithms as pursuing a single "meta" objective: the approximation of the Pareto front. To achieve a good approximation, an algorithm needs to find not only a single, well-performing solution, but a set of solutions that is *diverse* with respect to how much the individual objectives are satisfied. This diversity ensures that the approximated Pareto front will be as wide as possible. In a second step, the designer selects solutions from the Pareto front based on the trade-offs represented by the Pareto front and/or additional considerations and preferences.

An alternative approach to multi-objective optimization is to employ a weighted sum of objectives, which reduces the problem to be single-objective. The weighted sum avoids the need for a human decision-maker by defining the relative importance of different performance criteria a priori, while choosing a solution from the Pareto front assigns a relative importance a posteriori. One can thus approximate a Pareto front by running a single-objective algorithm multiple times with different weights. However, this approach finds only solutions that are in certain parts of the Pareto front, more specifically the convex parts, unless one takes special precautions in the formulation of the weighted sum [72].

In the ADO community, many see the openness of Pareto-based optimization to human decision-making as an advantage (e.g., [23]). The shape of Pareto curves can provide valuable information about the trade-offs inherent in design problems. This information potentially allows designers to make more informed choices when faced with the conflicting objectives characteristic of architectural design problems.

However, single-objective black-box optimization problems can already be difficult to solve, and multiple objectives increase this difficulty. With the same number of function evaluations, an efficient single-objective algorithm is likely to find non-dominated solutions not found by a multi-objective algorithm. Hamdy et al. [27] find that for the tested multi-objective optimization algorithms and problem, Pareto fronts for the two criteria of building energy and cost stabilized only after 1400–1800 function evaluations. For problems in which a single function evaluation takes minutes or hours, such a large number is often impractical. Accordingly, Chiandussi et al. [11] conclude in their benchmark of five multi-objective optimization methods that "the large computational effort makes [the Pareto-based] method generally not acceptable in usual engineering problems where, e.g., the Finite Element Method is used and models with a large number of degrees of freedom are implemented."

In addition, the affinity between architectural design and the trade-offs reflected by a Pareto front is imperfect. An architectural design problem tends to have many more (qualitative and quantitative) design criteria than can be formulated—let alone solved—reasonably as a multi-objective optimization problem. On the other hand, performance criteria relevant to architectural design can sometimes be more appropriately expressed as constraints than as objectives. For example, usually structural stress and deflection should not exceed safety thresholds, rather than being as low as possible. A similar point can be made for environmental objectives such as glare and thermal comfort.

Architectural Optimization Problems

Evins [22] and Nguyen et al. [60] survey studies of black-box optimization from the perspective of sustainable building design, while Hare et al. review studies from structural design [30]. In sustainable building design, most studies are concerned with optimizing building energy consumption and in structural design with reducing structural weight. Other criteria include geometry, daylight, and cost. Relative to the substantial amount of research on the simulation-based optimization of building designs, there are only a small number of published examples from architectural design practice, which are the focus of this section.

Luebkeman and Shea [52] describe practical and experimental applications of black-box optimization at the Foresight Innovation and Incubation group of multidisciplinary design consultancy ARUP. They describe the minimization of the number of bracing elements for the Bishopsgate tower in London and the number of members in a stadium roof. Other examples consider the panelization and rationalization of curved surfaces, with the objective of using only flat and ideally repeating panels, and the Pareto-based optimization of a building envelope in terms of energy and daylight. Also at ARUP, Hladik and Lewis [34] document the optimization of the angles of the large louvers of Singapore's National Stadium in terms of shading and view. Binkley et al. [8] provide a similar example: the application of a genetic algorithm (GA) to the design of the roof of a multipurpose sports hall and athletics stadium in Saudi Arabia. The algorithm optimized the clear height below the roof, as well as overall steel tonnage.

Rüdenauer and Dohmen [68] describe their use of a GA to optimize the weight of the timber structure of a mountain shelter on Switzerland's highest mountain, which reduced cost, waste, transport, and assembly time in a difficult to reach location. Scheurer [70] describes a "proof of concept" developed in collaboration with structural engineering firm Bollinger + Grohmann that revolved around using a GA to optimize the shape of a large roof structure. In addition to the efficiency of the optimized solutions, Scheurer emphasizes their novelty: "Not one of the engineers on the project, with an impressive amount of experience between them, would have come up with the same engineering concepts that evolved from the [genetic] algorithm." Currently, Bollinger + Grohmann regularly employ Pareto-based GAs to generate efficient structures "with an aesthetic logic between order and disorder" [32]. Besserud et al. [6] document the use of gradient-based and black-box optimization algorithms for integrated structural and architectural design at architecture and engineering firm Skidmore, Owings & Merrill LLP. Similarly, Imbert et al. [40] mention the "novel iterative approach to structural optimization" developed for the design of the Louvre Abu Dhabi that achieved "a fine balance of ... structure self-weight, aesthetics, cost and buildability."

These examples illustrate the popularity of GAs for ADO and clarify that performance criteria in architectural design practice are often multiple and cover several disciplines, including a building's geometry, structure, and environmental design. This multidisciplinarity motivates studies such as Yang et al. [85] and Gerber and Lin [24], which consider multidisciplinary, Pareto-based optimization in the context of ADO. Nevertheless, most of the practical examples presented in this section employ single-objective, black-box optimization methods, but often with multiple performance criteria. The following section surveys such methods.

Types of Optimization Methods

Since the 1950s, many black-box optimization algorithms have been proposed, and the understanding of the algorithms' performance from a theoretical and practical point of view has increased considerably. This section discusses some of the main algorithm categories and outlines their characteristics and limitations. Even if these categories are presented as dichotomies, it is important to remark that many existing algorithms draw on ideas from both sides of the dichotomy; hence, the contrast is often less sharp than it seems. The first, broad distinction is between *iterative* methods and *metaheuristics*. Among iterative methods, we distinguish between *direct search* methods and *model-based* methods.

Metaheuristics

A metaheuristic is a high-level procedure designed to construct a heuristic for an optimization problem. Metaheuristics consist of general principles that can be specialized to the problem at hand. There is a vast literature on metaheuristics: some of the most well-known are simulated annealing [46], GAs [35] and particle swarm optimization (PSO) [45]. Covariance matrix adaptation evolution strategy (CMA-ES) [28] and variable neighborhood search [29] are more recent examples. Talbi provides a systematic introduction [73]. Among metaheuristics, GAs are the most well-known class of algorithms and have been applied to many problems in architecture and engineering [56]. GAs improve a set of candidate points (the population) through mechanisms that mimic biological evolution, such as mutation, crossover, and survival.

Many metaheuristics differ from iterative methods in that they generate and keep, at every iteration, a population of candidate points, whereas iterative methods generate a single point in every iteration. (But the generation of this point may involve additional function evaluations, e.g., to estimate the gradient.) Another difference is that iterative methods often are deterministic, while metaheuristics tend to be stochastic, i.e., they employ or even embrace randomness.

The main advantages of metaheuristics are their conceptual simplicity, ease of implementation and wide applicability. Their main disadvantages are the lack of sound theoretical performance guarantees besides those that derive from their inherent randomness, the need to tune many optimization parameters for every algorithm and problem [73], and the fact that metaheuristics often require thousands of evaluations of the objective function f to be effective. The last disadvantage is a source of concern when it comes to the application of metaheuristics in ADO: in many situations, the evaluation of the performance f of a design candidate requires a time-intensive numerical simulation, which makes the evaluation of thousands of design candidates impractical.

Pareto-Based Metaheuristics

Since diversity of solutions is an explicit goal of Pareto-based optimization, population-based metaheuristics appear well-suited to this task. Indeed, Pareto-based optimization algorithms typically employ a GA—for example, NSGA-II [17], SPEA-2 [86], and HypE [5]—or PSO. Pareto-based algorithms differ not only in their optimization method but also in their method for comparing solutions, because in multi-objective optimization there is no straightforward method to decide which of two solutions is better. NSGA-II and SPEA-2 rank solutions with a non-domination sorting and filter them with a crowding measure to assure a good spread of solutions on the Pareto front. HypE estimates the solutions' hypervolumes in the objective space to more directly assess their contributions to the Pareto front.

Direct Search

Direct search algorithms iteratively determine a sequence of individual points to be explored based on the search history alone, i.e., the list of previously evaluated points and without trying to estimate a model of f or its partial derivatives.

Local Direct Search

Local direct search algorithms are among the earliest black-box optimization algorithms. The basic structure of such algorithms involves a *polling step*, in which several directions (i.e., variable changes) starting from the current point are tested for improvement of the objective function, and a *search step*, in which the next point is determined based on the results of the polling step. If the directions explored in the polling step and the length of the steps taken toward those directions satisfy certain conditions, one can show that a direct search converges to a stationary point (i.e., a point which is a candidate to be a local or global optimum) of the objective function.

A classic example for this structure is pattern search, also known as Hooke-Jeeves [37]. In a seminal benchmark of six building energy problems [78], pattern search performed well in general, but failed on one the problems. A well-known variant is the Nelder-Mead algorithm [59] that samples points on a simplex (i.e., a generalization of a triangle to any dimension). At every iteration, the algorithm replaces one corner of the simplex with a better point. More recent examples are generating set search [49] and mesh-adaptive direct search (MADS) [3]. While direct search has been designed for single-objective optimization problems, extensions to multi-objective problems exist, e.g [4].

Global Direct Search

Literature on global direct search algorithms is scarce, possibly because guaranteeing global optimality for black-box optimization problems is difficult. The most well-known algorithm of this class is DIRECT [42], a variant of the Lipschitzian approach that does not require the Lipschitz constant to be specified. For a function f, the Lipschitz constant is a measure of the function's maximum change for a given interval. The algorithm proceeds by dividing the design space into boxes and further subdividing those boxes that—based on an approximation of the Lipschitz constant, which is unknown but bounded—potentially contain the optimum. If a bound on the Lipschitz constant is known, it is guaranteed that the algorithm's recursive subdivision eventually finds the global optimum (with a given tolerance). In practice, this bound typically is unknown. In that case, one runs DIRECT for a specified number of iterations, which often is sufficient to obtain a solution with a close-to-optimal objective function value. Another global direct search algorithm is multilevel coordinate search [38], which also relies on recursively subdividing the design space. It performs well especially on problems with small numbers of variables (ibid.).

Model-Based Methods

Model-based methods construct a surrogate model (i.e., a mathematical approximation) of the objective function f or its partial derivatives and use this model to inform the optimization process. Model types and usages vary significantly, yielding several classes of algorithms. Because it is in analytical form (i.e., the algorithm knows the model's mathematical formulation), computations on the model are much more efficient than on the original, unknown objective function f. But the model is typically an only imperfect approximation of f, and blindly trusting the model can lead to very poor decisions during the optimization process.

A relatively straightforward method to employ a surrogate model for optimization is to first train the model—by evaluating a sample of points—and then replace the original time-intensive simulation with the model. This replacement improves the speed of any optimization algorithm by providing much faster function evaluations and allows the application of statistical techniques that require a large number of points such as sensitivity analysis. Eisenhower et al. [20] describe such an approach for building energy problems, and Yang et al. [85] apply it to the multidisciplinary design of a sports building. A major disadvantage of this method is that the optimization result is limited by the initial accuracy of the surrogate model. To improve this accuracy, one must increase the initial sample size, which can negate the original speed advantage. Model-based optimization methods, on the other hand, employ different strategies to iteratively improve the accuracy of the surrogate model by evaluating points during the optimization process.

Local Model-Based Methods

A prominent example of a local model-based method is the *trust-region* framework for nonlinear optimization. This framework has been used for classical nonlinear programming—problems for which gradient information is available—as well as black-box problems, which are the focus of this chapter. A comprehensive monograph is the work of [15]. In a trust-region algorithm, a local model of the objective function is built around the current point. The next search point is determined inside the *trust-region radius*, and this radius is adjusted based on the difference between the actual improvement and the improvement estimated by the model. The algorithm expands the radius when the estimate is accurate; otherwise, it shrinks it. The trust region can be modeled with different types of functions, such as linear [63] or radial basis functions (RBF) [79]. The most common choice is a quadratic model (e.g., [64]), which, in effect, represents the gradient and Hessian matrix (i.e., the local curvature) of the objective function. Under suitable conditions, trust-region methods enjoy strong convergence guarantees.

Global Model-Based Methods

The need in engineering and architecture to optimize problems that require time-intensive numerical simulations, such as complex aerodynamic, building energy, or daylighting models, has led to considerable interest in global model-based methods. For example, Wortmann et al. [82] present the application of a model-based algorithm to daylighting problems. Global model-based algorithms approximate the unknown objective function f with machine learning-related techniques, such as neural networks, support vector machines, and RBF [50]. To determine a promising point to simulate next, the algorithm searches the surrogate model (deterministically, randomly, or with a metaheuristic). The algorithm then refines the model with information gained from the simulation. The most delicate question for global model-based algorithms is how to balance the need for improving the accuracy of the model—which entails exploring a problem's objective function in its entirety—with the goal of improving the value of the objective function, which entails exploiting a promising area of limited size.

One of the earliest global model-based algorithms is efficient global optimization (EGO) [43], which fits a linear combination of Gaussian functions to points with known objective function values and selects the next point by maximizing the expected improvement of the objective function. With this type of model, the variance of the Gaussian functions at a point is a measure for the model's accuracy. In practice, experimental studies indicate that other types of models typically yield better results [36], for example, RBF [26] or ensemble models [58]. Algorithms using RBF models typically choose the next point by assessing its quality in terms of two criteria: the potential improvement of the objective function value and the likelihood of improving the model's accuracy. For example, Regis and Shoemaker [65] assess the former as a candidate point's predicted objective function value and the latter as the candidate point's distance from previously evaluated points. Alternatively, algorithms following Gutmann [26] measure a model's bumpiness (i.e., a measure of how much the model oscillates) to assess the likelihood that a point will improve both the objective function value and the model's accuracy.

Pareto-Based Model-Based Methods

Increasingly, model-based optimization is employed to also approximate Pareto fronts. Brownlee and Wright [10] extend NSGA-II with explicit constraints and surrogate RBF models (one for each objective) that are updated for every generation of the GA. As an example, they optimize a building design in terms of energy and cost. ParEGO [47]—a multi-objective extension of EGO—employs only a single surrogate model. The objective values for the surrogate model are calculated as a weighted sum, and the model is recalculated with different sets of weights at every iteration. The algorithm approximates the Pareto front by finding good solutions for the different sets of weights. (ParEGO uses an augmented Tchebycheff function [72] to allow solutions from non-convex parts of the front to be selected.) Knowles and Nakayama [48] survey Pareto-based optimization with surrogate models in more detail.

Optimization Tools for Architectural Designers

Most practitioners in ADO rely on external programs for optimization, such as MATLAB or GenOpt® [77], whose connection to geometry creation and simulation packages requires special expertise or custom tools [60]. Palonen et al. [61] survey such external optimization programs in the context of building energy optimization. Turrin et al. [74] and Gerber and Lin [24] present examples of workflows that integrate architectural design, performance simulation, and black-box optimization based on custom tools. Exceptionally, modeFRONTIER® [ESTECO (2017)] allows the linkage of its optimization algorithms to external geometry creation and simulation packages via a graphical user interface (GUI). This lack of publicly available, user-friendly tools is one of the reasons why most architectural designers do not use optimization during the design process [9]. To the knowledge of the authors, there are only three publicly available architectural design tools that integrate performance simulation and black-box optimization within a single GUI and without programming skills: Grasshopper® [25], Dynamo Studio [19], and DesignBuilder [71]. In contrast to the emphasis on single-objective optimization in the mathematical optimization literature, the last two offer only Pareto-based optimization.

Grasshopper is a parametric modeling tool that is constantly extended by an active community due to its popularity and the openness of its SDK. For example, structural analysis, energy analysis, and daylight analysis are available from third parties, either as plug-ins or as easy-to-use interfaces to stand-alone software. Grasshopper includes Galapagos, a plug-in implementing a GA and simulated annealing. There are four third-party, global black-box optimization plug-ins for Grasshopper, all of which are available for free and share a similar GUI: Goat, Silvereye, Octopus, and Opossum. This breadth of available simulation and optimization tools makes Grasshopper an attractive platform for comparing optimization algorithms on ADO problems. A recent alternative is Dynamo Studio, a similar parametric design tool that has a smaller community and for now offers only one black-box optimization tool: Optimo. DesignBuilder is less modular than Grasshopper and Dynamo and offers geometry creation, simulation, and optimization in one integrated package.

Optimization Tools in Grasshopper

- Galapagos [69] is included in Grasshopper and offers two metaheuristics: simulated annealing and a GA. ARUP has employed Galapagos to optimize structural designs [8].
- Silvereye [12] offers PSO.
- Octopus [76] is the only Pareto-based optimization tool in Grasshopper and implements the SPEA-2 and HypE algorithms. Bollinger + Grohmann often use Octopus for structural design [32].
- Goat offers an interface with NLopt [41], a free library containing various gradient-based and direct search optimization methods. Goat includes the following algorithms: DIRECT, the linear trust-region method COYBLA [63], the cubic trust-region method BOBYQA [64], the Nelder-Mead variant SUBPLEX [67], and CRS2 [44], a stochastic direct search method.
- Opossum [81] is the only global model-based optimization tool available to the ADO community for free. It provides an interface to the open-source RBFOpt library [16], which offers several RBF models (linear, multi-quadratic, cubic, and thin-plate spline) and model-based algorithms.

Other Tools

- DesignBuilder [71] is a commercial software aimed at architects and engineers that integrates geometry creation, simulation, and optimization. It simulates energy consumption, daylight, carbon emissions, and thermal comfort and optimizes these criteria with the Pareto-based NSGA-II.

- modeFRONTIER® [ESTECO (2017)] is a commercial optimization software aimed at engineers that provides connections to external geometry creation and simulation packages, including Grasshopper. It implements several well-known metaheuristic and direct search, as well as proprietary (model-based) algorithms.
- Optimo [1], a free plug-in for Dynamo, also implements NSGA-II.

Comparing Optimization Algorithms

The wide range of available optimization methods and tools raises the question which ones a designer should use. Currently, the architecture community overwhelmingly applies metaheuristic methods and especially GAs [22, 60]. Given the skepticism of the mathematical optimization community about metaheuristics—Conn et al. [14] refer to them as "methods of last resort," and Hendrix and G.-Tóth [33] characterize the proliferation of stochastic methods without convergence guarantees as "evolutionary drift"—and their often poor performance in mathematical benchmarks (see below), this popularity is surprising.

There are several explanations for the popularity of metaheuristics in the architecture community: (1) Compared to other optimization methods, metaheuristics are easy to implement. (2) They are also more readily available in software packages used by architects. (3) Metaheuristics are comparatively easy to understand and use, although they can require extensive tuning of the optimization parameters to achieve good results [73]. (4) There is a consensus in the ADO literature that metaheuristics are especially applicable to non-smooth, non-convex optimization problems, like the simulation-based optimization problems often encountered in ADO (e.g., [2, 23, 27, 53, 60]). This consensus is largely based on a small number of early experimental results and has seen little recent validation. One often-cited result is the benchmark of six building energy problems by Wetter and Wright [78], who conclude that "if a user is willing to accept a slight decrease in accuracy at the benefit of fewer simulations, then the simple GA is a good choice." Before summarizing more current benchmark results, this section discusses criteria for comparing black-box optimization methods.

Criteria

Criteria for single-objective algorithms include (1) speed of convergence (effectiveness), (2) stability (reliability), and (3) algorithmic overhead (computational cost). When measuring these criteria, results depend not only on the algorithm but also on the characteristics of the tested problem. One should therefore hesitate to generalize from small numbers of examples. Next to these quantitative criteria, which are briefly discussed next, designers might want to consider criteria such as understandability and interactivity. However, such qualitative criteria are beyond the scope of this chapter.

Most often, speed of convergence is the most critical criterion. It measures how fast an optimization algorithm improves the objective value in terms of the number of function evaluations. To benchmark the speed of convergence, one graphs the values of the objective function, or an aggregate thereof, over the number of function evaluations, see [57]. Many optimization algorithms exploit a degree of randomness, which is why benchmarks often employ aggregate measures, such as the median objective function value from several (five or more) runs of the same algorithm and problem.

Optimization algorithms that exploit randomness achieve different results when applied repeatedly to the same problem. In such cases, more stable methods result in a smaller range of outcomes. Stability is an important criterion since an unstable method can sometimes yield unsatisfactory results even if it displays good performance on average. Stability is crucial especially for metaheuristics, due to their heavy reliance on randomness. A typical stability measure is the standard deviation of the objective function values attained by repeated runs of the same algorithm and problem, for a fixed number of function evaluations. Direct search methods that are fully deterministic are the most stable and always result in identical outcomes. Many optimization algorithms, including most model-based methods that employ RBFs, combine random and deterministic elements.

Measuring speed of convergence in terms of function evaluations makes it independent of (1) an optimization algorithm's complexity, (2) the speed of particular implementations, and (3) computers. Total time per iteration thus is an additional consideration. Many metaheuristics employ relatively simple calculations that require little time per iteration. Model-based methods, by contrast, typically require many more calculations to construct and search the surrogate model. Direct search methods display a broad range of workloads. To examine these differences, one can benchmark the convergence of different methods in terms of elapsed time instead of in terms of functions evaluations. When the objective function requires more than a few seconds to evaluate (e.g., long simulation runs), time per iteration quickly becomes negligible, which is why in many practical cases the number of function evaluations is the more relevant performance measure.

Benchmark Results

To further inform algorithm choice, this section summarizes benchmark results of single-objective black-box methods for both mathematical and simulation-based test problems. There is a much smaller amount of benchmark results for Pareto-based algorithms (e.g., [27]), which are not considered here.

Mathematical Test Problems

Rios and Sahinidis [66] benchmark 22 state-of-the-art black-box software implementations on an extensive set of 502 convex and non-convex, continuous and discontinuous mathematical test problems with up to 300 variables. Although no single algorithm outperforms all the others, they identify four algorithms that, in most cases, are sufficient to find the best result within 2.500 function evaluations: The commercial TOMLAB/MULTI-MIN and TOMLAB/GLCCLUSTER, and the free MCS and SNOBFIT. TOMLAB/MULTI-MIN is a multi-start algorithm that conducts repeated local searches from different starting points, while TOMLAB/GLCCLUSTER is an implementation of DIRECT hybridized with clustering techniques. The open-source SNOBFIT [39] implements a global model-based algorithm that does not employ a single global model but multiple local ones. The tested metaheuristics, which include GAs and a variant of simulated annealing, generally perform poorly, with the exception of the non-convex, discontinuous problems, on which CMA-ES and PSO performed reasonably well compared to most other implementations.

Costa and Nannicini [16] compare selected software on problems with and without integer variables, using a small budget of objective function evaluations. In this setting, which is relevant for ADO whenever simulations take a long time, efficiency is crucial. Costa and Nannicini conclude that on problems without integer variables, RBFOpt and KNITRO (a commercial implementation of a multi-start local algorithm) are more efficient than SNOBFIT and NOMAD (an open-source implementation of the local direct search algorithm MADS) and consistently find better solutions with the allowed small budget. Among the software tested in Costa and Nannicini [16], only RBFOpt and NOMAD handle integer variables, and RBFOpt outperforms NOMAD on a benchmark set.

KNITRO, NOMAD, and RBFOpt also participated in the 2015 and 2016 editions of the black-box optimization competition [51]. These extensive benchmarks show that KNITRO performs better than RBFOpt, which in turn is more efficient than NOMAD. Costa and Nannicini [16] analyze the 2015 benchmark problems and conclude that KNITRO works best because the relatively large budget of allowed function evaluations permits the estimation of gradients by finite differences, whereas with smaller budgets, the model-based RBFOpt is more efficient. Recent literature thus suggests that model-based and direct search methods are the best available methodologies, but the specific choice depends on the characteristics of the problem. Empirical evaluations, discussed below, shed additional light on the strengths and weaknesses of each algorithm.

Simulation-Based Test Problems

In a benchmark of seven metaheuristics on four discrete optimization problems from structural design with 10.000–100.000 function evaluations [31], simulated annealing and evolution strategies [7] were the two best algorithms, and the simple

GA was the worst. (However, the authors caution that "it is highly possible that recent and improved variants of GAs would produce better results.") Similarly, Hare et al. [30] observe in a survey of benchmarks from structural design that "in all of these papers, evolutionary algorithms are shown to perform comparably or worse with respect to efficiency and solution quality," which sparks "the suggestion that evolutionary algorithms may be overused, specifically for continuous problems." They recommend (local) direct search methods and highlight that the convergence theory behind them "guarantees the quality of the final solution." The benchmark by Emmerich et al. [21], which considers a four-variable building energy problem for 100 function evaluations, confirms this suggestion: The two local direct search methods converge very quickly and find better results than the evolutionary strategies, which are tested for three sets of parameters.

Wortmann and Nannicini [83] compare the performance of the Grasshopper implementations of GA, DIRECT, and RBFOpt on two structural (200 function evaluations) and two daylighting problems (100 function evaluations). They conclude that DIRECT and RBFOpt exhibit comparable performance overall—with DIRECT performing better on the structural and RBFOpt performing better on the daylighting problems—and that "the genetic algorithm seems a viable choice only when thousands of design candidates can be evaluated." Wortmann [81] presents another benchmark of Grasshopper implementations for a 40-variable problem with the competing performance criteria of daylight and glare, also testing implementations of PSO and the Pareto-based HypE. After 200 function evaluations, RBFOpt exhibits the best performance for a single, weighted sum objective and DIRECT the worst (most likely due to the relatively large number of variables), with the metaheuristics exhibiting middling performance. HypE indeed yields the overall best approximation of the Pareto front, but RBFOpt finds non-dominated solutions with a larger amount of daylight than the ones found by HypE. In other words, for high-daylight solutions, the single-objective algorithm finds a better approximation of the Pareto front.

In a benchmark of all Grasshopper implementations of single-objective algorithms, Wortmann et al. [84] revisit the seminal building energy problems from Wetter and Wright [78]. Wortmann et al. conclude that the implementation of the GA is a poor choice of default, with DIRECT and simulated annealing performing well overall. RBFOpt performed less well for 500 function evaluations, but exhibited the fastest initial improvement and therefore worked best for smaller numbers of function evaluations ($\sim$200).

Conclusion

Black-box optimization algorithms increasingly find applications in architectural design and related fields. However, most often these applications are limited to only a small selection from the wide range of available algorithms. Based on results presented in the literature as well as personal experience, optimization problems in ADO typically yield multimodal, irregular objective functions. For this reason, global methods such as RBFOpt and DIRECT, and some metaheuristics such as simulated annealing, are likely to yield the best results. When function evaluations require time-intensive simulations, model-based methods such as RBFOpt (available in Grasshopper through Opossum) typically yield the largest improvement in a small number of evaluations, but can be outperformed by other methods in the long run. When the objective function is more regular and "almost convex," local methods (direct search and model-based) should be considered. When function evaluations take a very short time, so that we can afford a large number of them, efficiency of the method is less important, and metaheuristic approaches become more viable.

Similarly, one should apply Pareto-based, multi-objective optimization when a large budget of function evaluations is available, and the efficiency of finding good solutions is less important. One should not necessarily apply a Pareto-based algorithm when one wants to optimize a problem with more than one performance criterion: single-objective algorithms focus on finding good values for the objective function, and Pareto-based algorithms focus on understanding the trade-offs between conflicting performance criteria. Single- and multi-objective algorithms thus have different applications both in terms of the goal of the optimization process and the required budget of function evaluations. The specific choice should depend on the available time and the intended goal of the optimization process.

References

1. Asl, M.R., Stoupine, A., Zarrinmehr, S., Yan, W.: Optimo: a BIM-based multi-objective optimization tool. In: Martens, B., Wurzer, G., Grasl, T., et al. (eds) Proceedings of the 33rd eCAADe Conference. eCAADe and TU Wien, Vienna, AUT, pp. 673–682 (2015)
2. Attia, S., Hamdy, M., O'Brien, W., Carlucci, S.: Assessing gaps and needs for integrating building performance optimization tools in net zero energy buildings design. Energ. Buildings. **60**, 110–124 (2013)
3. Audet, C., Dennis, J.E.: Mesh adaptive direct search algorithms for constrained optimization. SIAM J. Optim. **17**, 188–217 (2006)
4. Audet, C., Savard, G., Zghal, W.: A mesh adaptive direct search algorithm for multiobjective optimization. Eur. J. Oper. Res. **204**, 545–556 (2010)
5. Bader, J., Zitzler, E.: HypE: an Algorithm for Fast Hypervolume-Based Many-Objective Optimization. Computer Engineering and Networks Laboratory (TIK). Department of Electrical Engineering Swiss Federal Institute of Technology (ETH), Zurich (2008)
6. Besserud, K., Katz, N., Beghini, A.: Structural emergence: architectural and structural design collaboration at SOM. Archit. Des. **83**, 48–55 (2013)
7. Beyer, H.-G., Schwefel, H.-P.: Evolution strategies – a comprehensive introduction. Nat. Comput. **1**, 3–52 (2002)
8. Binkley, C., Jeffries, P., Vola, M.: Design computation at Arup. In: Andia, A., Spiegelhalter, T. (eds.) Post-Parametric Automation in Design and Construction, pp. 113–119. Artech House, Boston/London (2014)
9. Bradner, E., Iorio, F., Davis, M.: Parameters tell the design story: ideation and abstraction in design optimization. In: Proceedings of SimAUD. SCS, p. 26 (2014)

10. Brownlee, A.E.I., Wright, J.A.: Constrained, mixed-integer and multi-objective optimisation of building designs by NSGA-II with fitness approximation. Appl. Soft Comput. **33**, 114–126 (2015)
11. Chiandussi, G., Codegone, M., Ferrero, S., Varesio, F.E.: Comparison of multi-objective optimization methodologies for engineering applications. Comput. Math. Appl. **63**, 912–942 (2012)
12. Cichocka, J., Browne, W., Rodriguez, E.: Evolutionary optimization processes as design tools: implementation of a revolutionary swarm approach. In: Proceedings of 31th International PLEA Conference. Bologna, IT, (2015)
13. Coello, C.A.C.: A comprehensive survey of evolutionary-based multiobjective optimization techniques. Knowl. Inf. Syst. **1**, 269–308 (2013)
14. Conn, A., Scheinberg, K., Vicente, L.: Introduction to Derivative-Free Optimization. Society for Industrial and Applied Mathematics, Philadelphia (2009)
15. Conn, A.R., Gould, N.I.M., Toint, P.L.: Trust Region Methods. Society for Industrial and Applied Mathematics, (2000)
16. Costa, A., Nannicini, G.: RBFOpt: an Open-Source Library for Black-box Optimization with Costly Function Evaluations. Singapore University of Technology and Design, Singapore (2014)
17. Deb, K., Pratap, A., Agarwal, S., Meyarivan, T.: A fast and elitist multiobjective genetic algorithm: NSGA-II. IEEE Trans. Evol. Comput. **6**, 182–197 (2002)
18. Dodangeh, M., Vicente, L.N.: Worst case complexity of direct search under convexity. Math. Program. **155**, 307–332 (2016)
19. Dynamo Studio: Autodesk. (2017)
20. Eisenhower, B., O'Neill, Z., Narayanan, S., et al.: A methodology for meta-model based optimization in building energy models. Energ. Buildings. **47**, 292–301 (2012)
21. Emmerich, M.T.M., Hopfe, C.J., Marijt, R., et al: Evaluating optimization methodologies for future integration in building performance tools. In: Proceedings of the 8th Int. Conf. on Adaptive Computing in Design and Manufacture (ACDM). p. 7 (2008)
22. Evins, R.: A review of computational optimisation methods applied to sustainable building design. Renew. Sust. Energ. Rev. **22**, 230–245 (2013)
23. Evins, R., Williams, C., Pointer, P., et al.: Multi-objective design optimisation: getting more for less. Proc. ICE – Civ. Eng. **165**, 5–10 (2012)
24. Gerber, D.J., Lin, S.-H.E.: Designing in complexity: simulation, integration, and multidisciplinary design optimization for architecture. Simulation. **90**, 936–959 (2014)
25. Grasshopper®: Robert McNeel & Associates (2014)
26. Gutmann, H.-M.: A radial basis function method for global optimization. J. Glob. Optim. **19**, 201–227 (2001)
27. Hamdy, M., Nguyen, A.-T., Hensen, J.L.M.: A performance comparison of multi-objective optimization algorithms for solving nearly-zero-energy-building design problems. Energ. Buildings. **121**, 57–71 (2016)
28. Hansen, N., Ostermeier, A.: Completely derandomized self-adaptation in evolution strategies. Evol. Comput. **9**, 159–195 (2001)
29. Hansen, P., Mladenović, N.: Variable neighborhood search: principles and applications. Eur. J. Oper. Res. **130**, 449–467 (2001)
30. Hare, W., Nutini, J., Tesfamariam, S.: A survey of non-gradient optimization methods in structural engineering. Adv. Eng. Softw. **59**, 19–28 (2013)
31. Hasançebi, O., Çarbaş, S., Doğan, E., et al.: Performance evaluation of metaheuristic search techniques in the optimum design of real size pin jointed structures. Comput. Struct. **87**, 284–302 (2009)
32. Heimrath, M.: ADO at Bollinger+Grohmann (2017)
33. Hendrix, E.M.T., Tóth, B.G.: Introduction to Nonlinear and Global Optimization. Springer, New York (2010)
34. Hladik, P., Lewis, C.J.: Singapore national stadium roof. Int. J. Archit. Comput. **8**, 257–277 (2010)
35. Holland, J.H.: Genetic algorithms. Sci. Am. **267**, 66–72 (1992)

36. Holmström, K.: An adaptive radial basis algorithm (ARBF) for expensive black-box global optimization. J. Glob. Optim. **41**, 447–464 (2008)
37. Hooke, R., Jeeves, T.A.: "Direct search" solution of numerical and statistical problems. J. ACM. **8**, 212–229 (1961)
38. Huyer, W., Neumaier, A.: Global optimization by multilevel coordinate search. J. Glob. Optim. **14**, 331–355 (1999)
39. Huyer, W., Neumaier, A.: SNOBFIT – stable noisy optimization by branch and fit. ACM Trans. Math. Softw. **35**, 9:1–9:25 (2008). https://doi.org/10.1145/1377612.1377613
40. Imbert, F., Frost, K.S., Fisher, A., et al.: Concurrent geometric, structural and environmental design: louvre Abu Dhabi. In: Hesselgren, L., Sharma, S., Wallner, J., et al. (eds.) Advances in Architectural Geometry 2012, pp. 77–90. Springer, Vienna (2013)
41. Johnson, S.G.: The NLopt nonlinear-optimization package. (2010)
42. Jones, D.R., Perttunen, C.D., Stuckman, B.E.: Lipschitzian optimization without the Lipschitz constant. J. Optim. Theory Appl. **79**, 157–181 (1993)
43. Jones, D.R., Schonlau, M., Welch, W.J.: Efficient global optimization of expensive black-box functions. J. Glob. Optim. **13**, 455–492 (1998)
44. Kaelo, P., Ali, M.M.: Some variants of the controlled random search algorithm for global optimization. J. Optim. Theory Appl. **130**, 253–264 (2006)
45. Kennedy, J., Eberhart, R.: Particle swarm optimization. In: IEEE International Conference on Neural Networks, 1995. Proceedings. pp. 1942–1948 vol. 4 (1995)
46. Kirkpatrick, S., Gelatt, C.D., Vecchi, M.P.: Optimization by simulated annealing. Science. **220**, 671–680 (1983)
47. Knowles, J.: ParEGO: a hybrid algorithm with on-line landscape approximation for expensive multiobjective optimization problems. IEEE Trans. Evol. Comput. **10**, 50–66 (2006)
48. Knowles, J., Nakayama, H.: Meta-modeling in multiobjective optimization. In: Branke, J., Deb, K., Miettinen, K., Słowiński, R. (eds.) Multiobjective Optimization, pp. 245–284. Springer, Berlin/Heidelberg (2008)
49. Kolda, T., Lewis, R., Torczon, V.: Optimization by direct search: new perspectives on some classical and modern methods. SIAM Rev. **45**, 385–482 (2003)
50. Koziel, S., Ciaurri, D.E., Leifsson, L.: Surrogate-based methods. In: Koziel, S., Yang, X.-S. (eds.) Computational Optimization, Methods and Algorithms, pp. 33–59. Springer, Heidelberg (2011)
51. Loshchilov, I., Glasmacher, T.: Black-Box Optimization Competition. bbcomp.ini.rub.de. (2017.) Accessed 14 Feb 2017
52. Luebkeman, C., Shea, K.: CDO: computational design + optimization in building practice. Arup J. **3**, 17–21 (2005)
53. Machairas, V., Tsangrassoulis, A., Axarli, K.: Algorithms for optimization of building design: a review. Renew. Sust. Energ. Rev. **31**, 101–112 (2014)
54. Malkawi, A.: Performance simulation. In: Kolarevic, B., Malkawi, A. (eds.) Performative Architecture beyond Instrumentality, pp. 85–95. Taylor and Francis, Hoboken (2005)
55. Marler, R.T., Arora, J.S.: Survey of multi-objective optimization methods for engineering. Struct. Multidiscip. Optim. **26**, 369–395 (2004)
56. Miles, J.: Genetic algorithms for design. In: Waszczyszyn, Z. (ed.) Advances of Soft Computing in Engineering, pp. 1–56. Springer, Vienna (2010)
57. Moré, J. J. and Wild, S. M.: 2009. Benchmarking derivative-free optimization algorithms. SIAM Journal on Optimization, **20** (1), 172–191
58. Müller, J., Shoemaker, C.A.: Influence of ensemble surrogate models and sampling strategy on the solution quality of algorithms for computationally expensive black-box global optimization problems. J. Glob. Optim. **60**, 123–144 (2014)
59. Nelder, J.A., Mead, R.: A simplex method for function minimization. Comput. J. **7**, 308–313 (1965)
60. Nguyen, A.-T., Reiter, S., Rigo, P.: A review on simulation-based optimization methods applied to building performance analysis. Appl. Energy. **113**, 1043–1058 (2014)
61. Palonen, M., Hamdy, M., Hasan, A.: MOBO a new software for multi-objective building performance optimization. In: Proceedings of the 13th International Conference of the IBPSA. pp. 2567–2574 (2013)

62. Pasternak, A., Kwiecinski, K.: High-rise Building Optimization: A Design Studio Curriculum. In: Proceedings of the 33rd eCAADe Conference. eCAADe and Faculty of Architecture and Urban Planning, TU Wien, Vienna, AUT, pp. 305–314 (2015)
63. Powell, M.J.: A direct search optimization method that models the objective and constraint functions by linear interpolation. In: Gomez, S., Hennart, J.-P. (eds.) Advances in Optimization and Numerical Analysis, pp. 51–67. Springer, Dordrecht (1994)
64. Powell, M.J.: The BOBYQA Algorithm for Bound Constrained Optimization without Derivatives. University of Cambridge, Cambridge, UK (2009)
65. Regis, R.G., Shoemaker, C.A.: A stochastic radial basis function method for the global optimization of expensive functions. Inf. J. Comput. **19**, 497–509 (2007)
66. Rios, L.M., Sahinidis, N.V.: Derivative-free optimization: a review of algorithms and comparison of software implementations. J. Glob. Optim. **56**, 1247–1293 (2013)
67. Rowan, T.H.: Functional stability analysis of numerical algorithms. Ph.D. Dissertation, The University of Texas (1990)
68. Rüdenauer, K., Dohmen, P.: Heuristic methods in architectural design optimization. In: 25th eCAADe Conference Proceedings, Frankfurt am Main, DE, pp. 507–614 (2007)
69. Rutten, D.: Evolutionary Principles Applied to Problem Solving. www.grasshopper3d.com/profiles/blogs/evolutionary-principles (2010) Accessed 14 Feb 2017
70. Scheurer, F.: Getting complexity organised. Autom. Constr. **16**, 78–85 (2007)
71. Singh, S., Kensek, K.: Early design analysis using optimization techniques in design/practice. In: ASHRAE Conference Proceedings. ASHRAE, Atlanta, GA, (2014)
72. Steuer, R.E., Choo, E.-U.: An interactive weighted Tchebycheff procedure for multiple objective programming. Math. Program. **26**, 326–344 (1983)
73. Talbi, E.-G.: Metaheuristics: from Design to Implementation. Wiley, Hoboken (2009)
74. Turrin, M., von Buelow, P., Stouffs, R.: Design explorations of performance driven geometry in architectural design using parametric modeling and genetic algorithms. Adv. Eng. Inform. **25**, 656–675 (2011)
75. Vicente, L.N.: Worst case complexity of direct search. EURO. J. Comput. Optim. **1**, 143–153 (2013)
76. Vierlinger, R.: Multi objective design interface. Msc Thesis, Technische Universität Wien (2013)
77. Wetter, M.: GenOpt–a generic optimization program. In: Proceedings of the Seventh International IBPSA Conference. Rio de Janeiro, BR, pp. 601–608 (2001)
78. Wetter, M., Wright, J.: A comparison of deterministic and probabilistic optimization algorithms for nonsmooth simulation-based optimization. Build. Environ. **39**, 989–999 (2004)
79. Wild, S.M., Regis, R.G., Shoemaker, C.A.: ORBIT: optimization by radial basis function interpolation in trust-regions. SIAM J. Sci. Comput. **30**, 3197–3219 (2008)
80. Woodbury, R.F.: Elements of Parametric Design. Routledge, London/New York (2010)
81. Wortmann, T.: Opossum—a model-based optimization tool in grasshopper. In: Proceedings of the 22nd CAADRIA Conference, Hong Kong, CN, pp. 283–292 (2017)
82. Wortmann, T., Costa, A., Nannicini, G., Schroepfer, T.: Advantages of surrogate models for architectural design optimization. AIEDAM. **29**, 471–481 (2015)
83. Wortmann, T., Nannicini, G.: Black-box optimization for architectural design: An overview and quantitative comparison of metaheuristic, direct search, and model-based optimization methods. In: Chien, S.-F., Seungyeon, C., Schnabel, M.A., et al. (eds) Proceedings of the 21th CAADRIA Conference. CAADRIA, Hong Kong, pp. 177–186 (2016)
84. Wortmann, T., Waibel, C., Nannicini, G., et al: Are genetic algorithms really the best choice in building energy optimization? In: Proceedings of the Symposium on Simulation for Architecture & Urban Design. Toronto, CA (2017)
85. Yang, D., Sun, Y., Turrin, M., et al: Multi-objective and multidisciplinary design optimization of large sports building envelopes: a case study. In: Proceedings of the International Association for Shell and Spatial Structures (IASS) (2015)
86. Zitzler, E., Laumanns, M., Thiele, L.: SPEA2: Improving the Strength Pareto Evolutionary Algorithm. Computer Engineering and Networks Laboratory (TIK). Department of Electrical Engineering Swiss Federal Institute of Technology (ETH), Zurich (2001)

MIX
Papier aus verantwortungsvollen Quellen
Paper from responsible sources
FSC® C105338

Printed by Books on Demand, Germany